RESOURCES FOR THE FUTURE LIBRARY COLLECTION
AGRICULTURE AND FISHERIES

Volume 6

Governing Soil Conservation
Thirty Years of the New Decentralization

Full list of titles in the set
AGRICULTURE AND FISHERIES

Governing Soil Conservation
Thirty Years of the New Decentralization

Robert J. Morgan

Washington, DC • London

Publisher's note

The publisher has made every effort to ensure the quality of this reprint, but points out that some imperfections in the original copies may be apparent.

At Earthscan we strive to minimize our environmental impacts and carbon footprint through reducing waste, recycling and offsetting our CO_2 emissions, including those created through publication of this book. For more details of our environmental policy, see www.earthscan.co.uk.

GOVERNING SOIL CONSERVATION:

Thirty Years of the New Decentralization

GOVERNING
SOIL CONSERVATION:
Thirty Years of
The New Decentralization

ROBERT J. MORGAN

Published for RESOURCES FOR THE FUTURE, Inc.

By THE JOHNS HOPKINS PRESS

RESOURCES FOR THE FUTURE, INC.,
1755 Massachusetts Avenue, N.W. Washington, D.C. 20036

Resources for the Future is a nonprofit corporation for research and education in the development, conservation, and use of natural resources. It was established in 1952 with the co-operation of The Ford Foundation and its activities since then have been financed by grants from that Foundation. Part of the work of Resources for the Future is carried out by its resident staff, part supported by grants to universities and other nonprofit organizations. Unless otherwise stated, interpretations and conclusions in RFF publications are those of the authors; the organization takes responsibility for the selection of significant subjects for study, the competence of the researchers, and their freedom of inquiry.

This book is one of RFF's studies in land use and management, which are directed by Marion Clawson. The author, Robert J. Morgan, is professor of political science at the University of Virginia. From 1960 to 1962 he was research associate with RFF while on leave from the University. The manuscript was edited by Virginia D. Parker. The illustrations were drawn by Clare O'Gorman Ford.

Director of RFF publications, Henry Jarrett; *editor,* Vera W. Dodds; *associate editor,* Nora E. Roots.

Preface

IN THE MIDST of a grave national emergency during the 1930's the U.S. Department of Agriculture initiated so-called action programs of soil conservation and land use adjustment. These programs provided mainly technical assistance and cash subsidies, in contrast to information or education which the land-grant agricultural colleges had provided to farmers since 1914. Starting in 1936, the Department also encouraged farmers to organize soil conservation districts authorized by state law to carry out certain soil conserving activities. When the Department created such new agencies as the Soil Conservation Service and the Agricultural Adjustment Administration to administer these new programs, it profoundly altered existing federal-state-local relations in agriculture.

In the 1960's, it is possible to look back thirty years and see great changes in agriculture, but many competent observers expect even more radical ones in the future. Certainly, the number of farms and farmers will continue to decline drastically. In this event, should sparsely settled rural areas continue to be served by the present units of government? Or should the number of government units be reduced for efficiency and economy, without loss of responsiveness and responsibility? The shift of population to urban areas can be expected to continue. New problems and opportunities for land use are appearing. Pressure on the existing water supply is mounting at a geometric rate. Political power is shifting rapidly from rural to urban areas in state legislatures and the U.S. Congress. What do these changes portend for the ability of the Department of Agriculture to concentrate its funds and manpower on national conservation goals, when the supply of land and manpower is already in excess of need? Has the experience of soil conservation districts in planning and administering conservation programs during the past thirty years fitted them to assume new responsibilities in meeting the future needs of an urbanized and industrialized society?

In this book, I provide data and some interpretation which may help to answer these and other questions. I concentrate chiefly, but not exclusively, on the Soil Conservation Service and soil conservation districts because it is widely believed that they were created to have unique responsibility for carrying out national soil conservation objectives.

The first seven chapters trace the origins and evolution of several complex issues reflected in the efforts of successive Secretaries of Agriculture to organize conservation administration in accord with the Department's objectives. Some of the issues discussed are suggested by such questions as: What were the origins and aims of the Department's national program of soil conservation? What prompted the Secretary of Agriculture to urge the states to organize soil conservation districts, thereby changing the Department's traditional ties with the state agricultural colleges and their extension services? How did this change affect political power in agriculture? What functions were the districts expected to perform? Why, at present, is the Department's technical conservation assistance given to farmers by three different agencies and subsidies given by yet another agency? Do these agencies overlap and duplicate each other, as critics sometimes charge, or has the Department adequately coordinated their activities? What have been the respective roles of Congress and the executive branch in determining these matters? How have the Department's conservation objectives changed over the past thirty years?

In the remaining chapters of the book, I look at the way in which districts have actually functioned to help or hinder realization of the Department's conservation objectives. I show why some districts are organized to coincide with counties and others are not; how district governing boards are selected and how they administer district operations. I examine the type and extent of state and local financial assistance to, and administrative supervision of, districts—although this kind of information is impressive more for its elusiveness than its availability. I show that districts and the state committees which supervise their administration do not represent the broad spectrum of interests which ought to characterize multiple-purpose resource districts. I describe in some detail the functions of district boards as policy-making agencies; consider their relations with other state and federal agencies; and ask whether districts perform distinctive functions which justify the view that their mission is unique in fact as well as theory. I raise such questions as these: What public responsibilities do districts perform? Do they coordinate the specialized conservation agencies; or do they tend to serve the special purposes of some, but not others? Do they serve as centrifugal or centripetal forces acting upon national conservation agencies? Are districts multiple-purpose conservation units or, if not, are there real prospects that they may be in the future?

I have been consciously selective in stressing and analyzing the political processes through which soil conservation policy has been made and executed. Moreover, I have preferred analysis to prescription with rare exceptions. I have ventured beyond a conventional analysis of structure and function to focus on the interaction between governmental administration (structure, missions, specialization of skills, procedures, and so on) and political power. It is essential that this relationship be understood as long as resources conservation is a governmental objective. Political ecology is as important to the life of a unit of government as physical ecology is to a living organism. It takes more than technical and administrative skills to achieve the purposes of government. The U.S. experience with a broad range of domestic programs and aid to underdeveloped nations points to this fact of life. Consequently, in the last chapter, I offer both an analysis and a thesis which justify the subtitle of this book: "Thirty Years of the New Decentralization."

Despite the risk of oversimplifying some complex matters, a few of my conclusions are summarized here:

1. The Soil Conservation Service and soil conservation districts never have had exclusive, or even unique, responsibility for carrying out the Department's soil and water conservation programs.

2. Farmers participate voluntarily, if at all, in these programs, although districts originally were organized under state law to compel all farmers eventually to follow sound conservation practices.

3. Intense and widespread opposition from the land-grant agricultural colleges and their extension services forced the Department of Agriculture progressively to abandon its intention to make the district boards truly responsible for planning and enforcing sound conservation programs adapted to local needs.

4. Much of the hostility to districts reflected a complex struggle for political power among members of Congress, the executive branch, the agricultural colleges, and major farm organizations. The colleges proved not to be strong enough to enforce their demands for decentralization of the conservation programs because influential members of Congress shared with the Secretary of Agriculture an understanding of the effects of federal line administration upon the distribution of political power.

5. Since 1951, the Department of Agriculture generally has reduced the

intensity of interagency competition within the agricultural community by coordinating technical planning and vesting program responsibility in its own line agencies and not districts—except as they act in an advisory capacity.

6. Patterns of district organization, financing, state supervision, and the level and character of district board activities are extremely variable. Only a small percentage of all districts engage in genuine program planning to adapt national plans to local needs.

7. Districts generally are so intimately connected with the Soil Conservation Service that their activities tend to be indistinguishable, except in a very formal way. This fact minimizes their effectiveness as multiple-purpose units capable of coordinating the work of all relevant agencies.

8. The Department of Agriculture is actively promoting conservation planned around small watersheds which will be attractive to urban interests as well as farmers. The Department's practical need for the support of urban interests and the legal requirement that units of state or local government sponsor these projects prevents the Department from concentrating these projects in some areas of greatest agricultural need.

Two of my remaining conclusions are, in effect, my thesis. The first is that the centrifugal forces in American government and politics invariably restrict the power of federal agencies to concentrate national resources of funds and trained manpower upon those areas and people in most critical need of assistance. There is a tendency for all federal "welfare" or "benefit" programs to become "national" only in the sense that they are available in suitable form in a maximum number of congressional constituencies.

My final conclusion is that the problems of administering agricultural conservation programs is inseparable from broader constitutional and political issues. It is usually believed that there has been an enormous shift of power to Washington during the past thirty years that has greatly enhanced the power of the federal bureaucracy and the President. In some respects this view is correct, but it overlooks the fact that the bureaucracy shares power with Congress. Members of Congress have developed many ways of sharing intimately in the processes and power of program formulation, execution, and review with the executive agencies. Congressmen insist on sharing these powers because they understand how administrative action affects the distribution of political power in their constituencies. Each member of Congress acts to obtain the maximum distribution of federal benefits in his constituency. At the same time, he acts to minimize administrative controls under the direction of the President as a party leader—or in the hands of state and local party rivals—to the end that he not be driven from congressional office involuntarily. The political neutrality of the federal service is essential to congressmen; it cannot be used to help his rivals openly, but he can, in considerable measure, make it responsive to his local needs.

Before I express my gratitude to the many people who have helped to make this book a reality, I want to mention one intended limitation. It is not my purpose to assess or evaluate the Department's physical accomplishments on the land. This job has been done by R. Burnell Held and Marion Clawson in *Soil Conservation in Perspective,* also a Resources for the Future book.

Joseph Fisher, Irving Fox, and Marion Clawson of Resources for the Future have been both helpful and kind in many ways. They encouraged me to undertake this study, and gave invaluable advice, when it was needed the most, to improve the substance and style of this work. RFF generously supported my research for more than two years. Several national, regional, and local officers of the National Association of Soil and Water Conservation Districts granted extensive interviews, permitted me to attend meetings of their organization, and introduced me to fellow workers sharing their concern for resources conservation. I received equally generous help of the same kind from officials in the Soil Conservation Service, the Agricultural Stabilization and Conservation Service, and other agencies of the Department of Agriculture, as well as from administrators and other specialists in the many colleges of agriculture and state agencies which I visited to conduct field interviews in a widely scattered sample of states. The personnel of soil conservation districts gave invaluable information which was not available anywhere else, when they responded to a special questionnaire which Resources for the Future sent to them to provide data for this study. The same thing must be said for the officers and employees of the state soil conservation committees who responded to another questionnaire.

The following persons read the review draft of the study: Fred A. Clarenbach, Aubrey D. Gates, John M. Gaus, Philip M. Glick, Russell G. Hill, Frank Miller, Dillon S. Myer, and Donald C. Swain. I thank them for their comments, but they are to be absolved of any responsibility for the work in its present form. Officers of the National

Association of Soil and Water Conservation Districts and the Soil Conservation Service did not comment formally upon the review drafts submitted to them.

I am most emphatic in taking all of the responsibility for the substance of this book. It reflects my interpretations and no other's. I hope that my data generate reflection, even where they invite dissent.

Readers may often wonder whether authors who are, first of all, university teachers are not too extravagant in their expressions of gratitude to their families who suffer neglect by fathers and husbands during long periods of scholarly gestation. For my part, I think not. The praise and the thanks are hard-earned, especially where one's spouse is concerned.

ROBERT J. MORGAN

Charlottesville, Va.
August 1965

Contents

List of Tables

List of Figures

Figure

GOVERNING SOIL CONSERVATION:
Thirty Years of the New Decentralization

CHAPTER I

A Crusade for Erosion Control

ON MAY 11, 1934, a dense cloud of reddish-brown dust boiled up out of the parched and abused wheatlands of Texas, Oklahoma, Colorado, and Nebraska. After darkening the main streets of prairie towns at midday so that lights had to be turned on, the cloud swept eastward across the continent and out over the Atlantic Ocean, where it was still visible hundreds of miles at sea. It left behind land scarred and scorched with the remaining topsoil drifted over fence posts, livestock dead, and the people of the Plains choking with "dust pneumonia." Again, in April of 1935 and repeatedly during 1936, the topsoil of the Plains was lofted skyward into dust clouds by the harsh and unremitting winds that ravaged a defenseless land. More than words could, these awesome storms dramatized the disastrous consequences of unchecked human exploitation of the land. Pictures of the aftermath of shrivelling drought so shocked the American people, demands were heard on all sides for relief of the victims and repair to the stricken land.[1]

It was in this atmosphere of natural disaster and national emergency that the Soil Conservation Service (SCS) was created in the U.S. Department of Agriculture in April 1935, to provide a permanent, national program of direct technical assistance to farmers to control erosion. It is not at all clear, however, that this new organization or its program would have been established either in 1935, or later, were it not for circumstances peculiar to President Franklin Delano Roosevelt's early years in office. It is widely believed that the dust storms of 1934 and 1935 provided the impetus for initiating the Soil Conservation Service; in fact, erosion control was started as an emergency federal public works project to relieve unemployment under the direction of

[1] See Arthur M. Schlesinger, Jr., *The Age of Roosevelt: The Coming of the New Deal* (Boston: Houghton Mifflin Co., 1958), pp. 319–53, esp. 335–41; Murray R. Benedict, *Farm Policies of the United States, 1790–1950: A Study of Their Origins and Development* (New York: The Twentieth Century Fund, 1953), esp. chaps. 7 and 13; John M. Gaus and Leon O. Wolcott, *Public Administration and the United States Department of Agriculture* (Chicago: Public Administration Service, 1940), chap. 8.

1

Harold L. Ickes who was both Secretary of the Interior and administrator of the federal works program.

In August 1933, Ickes approved a program to demonstrate erosion control to farmers as the mixed result of what Rexford G. Tugwell, later called his "empire-building compulsion" and Hugh Hammond Bennett's messianic campaign to awaken farmers to the danger of erosion.[2] Ickes' decision to put Bennett in charge of this work in the Department of the Interior still affects the organization and content of soil conservation programs. Indeed, from 1933 to 1953, rival state and federal agencies competed vigorously for control of the various programs created during the emergencies of the thirties. This struggle has left its mark on the professional agricultural workers to the point where many firmly hold that only their own respective agencies have scientifically sound programs of conservation. So sensitive are many of these partisans that they cannot even agree which agency first recognized the true dimensions of the erosion problem and initiated remedial action on a scale commensurate with it. This issue would scarcely seem to be of great importance today but it was crucial to those scientists and administrators in the state experiment stations and agricultural extension services who opposed the organization of the Soil Conservation Service in 1935 and of soil conservation districts under state enabling legislation first passed in 1937.

EROSION CONTROL BEFORE 1933

As late as 1933, the U.S. Department of Agriculture had no program of direct technical assistance to farmers for conserving soil and water. The Department was organized and functioned principally to conduct research and to disseminate findings through extension education. It commenced research into various agronomic problems as early as 1894, when the Bureau of Chemistry and Soils was created. That year it also issued a farmers' bulletin describing means of curbing erosion. In 1908, a division of soil erosion was established within the Bureau of Chemistry and Soils. In 1917, the Department issued another bulletin dealing with terracing as a means of erosion control, and in 1931 the Bureau of Agricultural Engineering was set up.[3]

[2] Rexford G. Tugwell, *The Democratic Roosevelt* (Garden City: Doubleday and Company, Inc., 1957), p. 320. On p. 360, Tugwell called Ickes a "whiner, an egoist and an incorrigible empire-builder."

[3] T. Swann Harding, *Two Blades of Grass* (Norman: University of Oklahoma Press, 1947), pp. 193–209, 263–67.

The state agricultural experiment stations and extension services, aided with federal funds, worked with farmers by demonstrating means of erosion control, but their programs varied widely in scope and content within and among states. In 1936, C. W. Warburton, Director of the federal Extension Service, reported that between 1914 and 1933 much experimental work and many demonstrations were conducted by the extension services in Alabama, Texas, and Iowa. Warburton ventured the opinion that in these three states erosion was recognized as "more important than [in] some other states within their respective areas." It was, he said, "only within recent years that soil erosion control has been an extension project in the Northeastern and Western states, and this is true also of the more Northern States in the North Central Region." He added that "practically all" states "are now doing some work in this field." Warburton pointed with special pride to work done in the Southern Great Plains during 1936 with federal emergency funds administered by the state extension directors.[4]

In 1928, according to the Director of the federal Extension Service, terraces and "soil-saving dams" had been constructed with some assistance by the state extension services on 45,058 farms affecting 1,349,000 acres in various states. In 1932, the number of farms on which terraces were constructed had increased to 48,704, but a 1933 report indicated the total had dropped to 38,576. In his 1928 report, the Extension Service Director had pointed out that erosion control "continued to be the major agricultural engineering activity in all areas where the problem of soil erosion is serious," and he had reported that various agronomic practices such as the use of lime, rotations, and green manure were applied.[5] There is some disagreement on what prompted the 1928 comment on soil erosion. It is possible, but by no means certain, that his remark was stimulated by the increasingly aggressive efforts of a few specialists in the Department of Agriculture and some experiment stations to focus attention on the loss of agricultural land through erosion. The warmest and most partisan supporters of Hugh Bennett incline to this view, while his critics deny it, saying that the

---

[4] National Archives, Record Group 33 (Extension Service: Soil Conservation), C. W. Warburton to W. C. Lowdermilk, Oct. 5, 1936. (National Archives records cited hereinafter usually will be identified by the abbreviation NARG.)

[5] U.S. Department of Agriculture, *Annual Report of the Director of the Extension Service, 1928*, pp. 41, 61, and 80; also his *Annual Report, 1932*, p. 55; and U.S. Department of Agriculture, *Serving American Agriculture: A Report of Extension Work in Agriculture and Home Economics in 1933* (Washington: Government Printing Office), p. 12.

state agricultural colleges were giving the matter sufficient attention where the problem merited it.

Quite possibly this stale controversy can be put partially to rest by recalling that in 1928 the state extension services were only fourteen years old and not fully organized in all counties. The Department of Agriculture had not shown much more concern, if any, with erosion than the agricultural colleges. Far more important, in the long run, was the fact that American agriculture was then chronically depressed, and suffering the birth pangs of a technological revolution in which production was shifting from handicrafts to the assembly line. In the face of this changing technology it was as difficult then, as it is now, to measure the importance of the loss of agricultural lands through erosion in relation to the total acreage needed for production. Held and Clawson have made this point quite clear in their book on soil conservation.[6]

In any event, Hugh Hammond Bennett of the Bureau of Chemistry and Soils, U.S. Department of Agriculture, was convinced that existing governmental efforts to conserve soil were inadequate. He launched a personal crusade to make professional workers, farmers, and the public understand the dangers of soil erosion on agricultural land and to accelerate research and action to combat it. Eventually, his success was rather spectacular, but his methods were so abrasive in the judgment of those who opposed, or differed from, him that he had earned a singular reputation even before Ickes put him in charge of the erosion work in the Department of the Interior.[7] Even before 1933, Bennett and his supporters were deeply divided over two basic questions which still remain unanswered: what are the extent and character of the erosion damage on agricultural lands which require the investment of public resources for preventive measures; and what are the best techniques for preventing or minimizing any losses that are identified?

Both widespread interest in erosion control among Texas farmers and Bennett's friendship with A. B. Conner, Director of the Texas Experiment Station at Spur, were very important to Bennett's success over the years. Experiments with terraces had been conducted at Spur, and between 1916 and 1933 Texas county agents had assisted farmers in terracing more than 7 million acres. During a meeting at the Spur Station in 1927 it was generally agreed, however, that there was a lack

---

[6] R. Burnell Held and Marion Clawson, *Soil Conservation in Perspective* (Baltimore: The Johns Hopkins Press for Resources for the Future, Inc., 1965), chap. 1.
[7] Wellington Brink, *Big Hugh: The Father of Soil Conservation* (New York: Macmillan, 1951), pp. 15–22, 48–60.

of research data upon which sound recommendations could be based.[8] Therefore, a two-stage campaign was decided upon; it was to be both educational and legislative.

In 1928, Bennett launched the crusade to dramatize the problem with a report, *Soil Erosion: A National Menace.*[9] In January and early February, in a lecture to co-workers in the Department of Agriculture he laid great stress on the differences between geologic erosion and "accelerated" erosion. By his definition, accelerated erosion, resulting from removing trees, grass, and shrubs from the soil, is the excess which occurs when measured by "normal erosion . . . taking place under natural conditions of vegetative cover and ground equilibrium." There were spectacular evidences of accelerated erosion in upwards of 10 million acres of cultivated upland, according to Bennett, but even more dangerous in his judgment was sheet erosion which "is doing the greatest damage." He warned that "we really stand upon the threshold of vast erosional losses to our agricultural lands, unless we proceed to do very considerably more about restraining the wastage than we have done in the past."[10]

Soon, Senator Morris Sheppard of Texas introduced a bill authorizing the Secretary of Agriculture to undertake studies to find out how to control erosion, conserve soil fertility, and minimize the deposition of silt in reservoirs. It passed the Senate without debate or a record vote, but died in the House Committee on Agriculture where the members, most of whom represented constituencies of the Corn Belt and Northeast, apparently saw no need for this legislation—or, at least, no need for the U.S. Department of Agriculture to administer it.[11]

[8] "Thirty Years of Pioneering in Soil Conservation Work by the Texas Extension Service, 1903–1933," mimeographed, College Station, Texas, March 17, 1948. M. L. Wilson kindly gave me a copy of this report from his personal files.

[9] Hugh H. Bennett and W. R. Chapline, *Soil Erosion: A National Menace*, U.S. Department of Agriculture Circular No. 33, Washington, 1928. See also Hugh H. Bennett and William C. Pryor in *This Land We Defend* (New York: Longmans, Green and Co., 1942), pp. 35–36. Bennett claimed that the "estimated annual bill which America pays in one way or another for erosion is $3,844,000,000." Of this amount three billion dollars was the "charge" for the loss of three billion tons of "soil material." The remainder of the cost he attributed to direct loss of income by farmers ($400 million) and damage to irrigation and drainage works, highways and railroads damaged, and the siltation of navigable streams and harbors.

[10] Hugh H. Bennett, *Lectures before the U.S.D.A. Graduate School*, SCS-TP-7, mimeographed, Soil Conservation Service, U.S. Department of Agriculture, Washington, Jan. 30, Feb. 1 and 3, 1928, pp. 1–15, *passim*.

[11] *Congressional Record*, Vol. 69, Pts. 4, 8, 9, 70 Cong. 1 sess. (1928), pp. 4064, 9044, 9583, and 10007; see also S. 3484 and Senate Report 1211, 70 Cong. 1 sess. (1928).

When this direct method of securing authorization for a research program failed, Bennett and his Texas friend, Conner, conferred with James P. Buchanan who represented the 10th Congressional District in Texas and was a ranking Democrat on the House Subcommittee on Agricultural Appropriations. Conner called Bennett to Buchanan's attention after convincing the Congressman that he ought to be willing to vote federal funds to help the farmers of his district combat soil erosion as long as he voted to build battleships in other states.[12] Early in 1929, therefore, during hearings on the Department of Agriculture appropriation bill, Buchanan interrupted the testimony of the Department's spokesman so that he could "develop some facts in the record on soil erosion." He described his conversations with Conner and said that his intention was "to lay the foundation to procure an adequate appropriation for the department in cooperation with the States where possible to conduct experiments." A. G. McCall testified for the Department that there was a need for four or five experiment stations costing $22,000 each, but Buchanan interrupted this testimony and gave Bennett his floor. Bennett then immediately asked for at least $150,000 for the coming fiscal year. He again sounded the alarm, venturing the guess that 25 per cent of the runoff could be prevented in 500,000 square miles of the Upper Mississippi Basin, if proper techniques were used. He was convinced, however, that soil erosion would never be successfully stopped unless farmers were shown the proper methods and "county agents, leading farmers, business men, bankers, chambers of commerce, railroads and the press" were to give their "unceasing assistance" to practical programs directed by men "determined to stop the wastage."[13] It took Bennett nearly twenty years to organize exactly this kind of support against both opposition and apathy, but he succeeded to a remarkable degree. Indeed, his testimony on this occasion was virtually a blueprint for his later crusade: shock the nation into the conviction that erosion is a menace requiring immediate national action, accelerate research into proper methods of control, and gain support of the influential leaders of every community to work with Bennett's "determined" men.

The funds Bennett had labored to secure, with Buchanan's valuable

[12] Brink, op. cit., pp. 74–77. Bennett dedicated his book, This Land We Defend, op. cit., to "James Paul Buchanan of Texas . . . who early recognized the great value of America's soil resources and was instrumental in launching the nation on a course of practical soil conservation."

[13] Department of Agriculture Appropriation Bill for 1930, House, 70 Cong. 2 sess. (1929), pp. 310–30, passim. The appropriation made to the Bureau of Chemistry and Soils in support of Bennett's erosion stations was $160,000; 45 Stat. 1207.

aid, were made available, but were divided among the Bureau of Chemistry and Soils, the Bureau of Agricultural Engineering (after its establishment), and the U.S. Forest Service. Bennett remarked that his failure to get the whole appropriation for his own experimental work "was a shock to me that we need not discuss at this point further than to say that out of this move a tremendous amount of grief has originated." He claimed that there were endless delays and quibbles over the line between agronomy and agricultural engineering, although none could be established in reason. Bennett also charged that some agricultural engineers waged a "deliberate" campaign of distortion of his work to the point where Bennett was constrained to correct "certain misrepresentations placed upon our experimental work."[14]

THE SOIL EROSION SERVICE ORGANIZED

Here the differences between Bennett and some of the agricultural engineers rested until shortly after the advent of the Roosevelt Administration. On June 9, 1933, Rexford G. Tugwell, then Assistant Secretary of Agriculture, sent to Harold L. Ickes, in his capacity of administrator of emergency public works, a program of erosion control by terracing prepared by S. H. McCrory, Chief of the Bureau of Agricultural Engineering, for the Department of Agriculture. This proposal to use public works funds was discussed by the Secretary of Agriculture, Henry A. Wallace, in general terms with the President in May. Some of Ickes' subordinates in Interior had talked over various possibilities with staff members in the Agriculture Department, including Bennett, who at the time had not worked out a proposal. Apparently Wallace had approved the McCrory plan, since his assistant, Paul H. Appleby, urged John Collier in Interior to get in touch with McCrory "as well as with the other people you have been contacting in our Department."[15]

McCrory argued that terracing was the most effective means of erosion control. He suggested that landowners be formed into cooperative organizations for terracing, "preferably under the guidance or with the assistance of the State extension service." Each farmer participating would sign an agreement specifying the number of acres he proposed to terrace, and also binding himself to furnish the necessary labor and

[14] NARG 16 (Office of the Secretary of Agriculture: Correspondence of the Immediate Office of the Secretary, 1933–35; Erosion File), H. H. Bennett to R. G. Tugwell, Sept. 20, 1933.

[15] NARG 16 (Erosion File), letters as cited, May 30 and June 9, 1933.

power to perform the work in accordance with specifications prepared by the government and to provide for its future maintenance. When the area covered by agreements in any one locality totaled 6,000 acres, the local organization and the federal government could agree that the latter would provide loans, equipment, and an engineer to lay out terraces and supervise their construction. When the job was finished, the equipment would be given to the local organization to maintain the improvements. The state agricultural extension services would then institute a program to restore humus to the soil. To protect the federal government and the farmers, McCrory recommended that his Bureau of Agricultural Engineering be made the administrative agent for Ickes in his capacity as administrator of the federal public works program.

Although McCrory's plan caught Bennett napping, some of his friends were supporting him. In February 1933, A. B. Conner had written to President Roosevelt to mention "H. H. Bennett, who has charge of erosion work in the Bureau of Chemistry and Soils, and who is widely recognized as the best informed man on the subject in the country."[16] Congressman Buchanan also wrote to the President and strongly recommended Bennett as the most knowledgeable erosion specialist in the country. He pointedly reminded Roosevelt that he and Bennett had initiated the federal erosion experiment stations.[17]

While McCrory's proposal was pending before Ickes, Bennett prepared a counterproposal and, on June 16, 1933, sent it to his superior officer, Henry G. Knight, Chief of the Bureau of Chemistry and Soils of the Department of Agriculture.[18] Bennett urged that any erosion control work undertaken under the Administration's new public works program should utilize "all the practical information acquired by the Department, not merely part of it." He was especially concerned that terracing should not be so emphasized as to overlook the losses of soil productivity caused by sheet erosion "now affecting something over 200,000,000 acres of the 350,000,000 now in cultivation." Discussion should not be foreclosed on the assumption that control of erosion and floods could be achieved only with "trees and terraces," Bennett stressed. Vegetative cover could be secured quickly and cheaply on almost any degree of slope and could be integrated into the over-all

[16] NARG 16 (Erosion File), A. B. Conner to Franklin D. Roosevelt, Feb. 28, 1933.

[17] NARG 16 (Erosion File), J. P. Buchanan to Franklin D. Roosevelt, Feb. 28, 1933; A. B. Conner to Henry A. Wallace, May 24, 1933, and Wallace to Conner, June 5, 1933.

[18] NARG 114 (Soil Conservation Service, General Correspondence; 1933–35), H. H. Bennett to H. G. Knight, June 16, 1933.

management of a farm. It would aid, moreover, the development of a better balanced agriculture by decreasing acreages of the major surplus-producing crops. Bennett argued that "mechanical means of erosion control can never be as effective as vegetative methods, but they can be very helpfully employed, provided due attention is given the matter of their applicability." Terraces could produce "disastrous" results, however, if used without consideration of their limitations. Proof was to be found, he asserted, in a study by the Georgia College of Agriculture, showing that more than 100,000 acres in a single county had been "ruined" by improper terracing. Bennett warned that, if the Department pushed erosion control without an "orderly, discriminative procedure," such efforts "will eventually do far more harm than good."

Bennett recommended also that a reasonable acreage of badly eroded steep lands be selected by soil type and seeded permanently to grass and tree cover crops as far-reaching demonstrations of good land use practices. This, he suggested, should be carried out cooperatively by the Division of Soil Erosion in the Bureau of Chemistry and Soils and the Division of Forage Crops in the Bureau of Plant Industry. He shrewdly urged that areas selected be distributed among a variety of regions to demonstrate the methods appropriate to local conditions.

Six weeks later Bennett's proposals were forwarded to Secretary Wallace. Knight explained to Wallace that Bennett's program would be administered jointly by the Bureaus of Chemistry and Soils and of Plant Industry "and also with the Extension Service wherever practicable."[19]

Bennett's plan apparently was not officially submitted by the Department of Agriculture. In fact, on July 17, a week before Knight sent Bennett's plan to Secretary Wallace for consideration, the McCrory plan had been approved by the Special Board for Public Works under Ickes' chairmanship. One can only speculate on Knight's motives for this delay. But, in any case, Bennett was not foiled, because Tugwell wanted Bennett's program set up in Interior. Tugwell—who had become the first Under Secretary of Agriculture on June 19—was the Department's representative on the Special Board of Public Works. Bennett's ideas were brought to the attention of Frederic A. Delano, the President's uncle who became Chairman of the National Planning Board when it was established July 30 by Ickes to plan public works, regional surveys, and the expenditure of funds by the Federal Emergency Administration of Public Works (PWA). Some of the members

[19] NARG 114 (General Correspondence, 1933–35), memorandum and letter, H. G. Knight to Henry A. Wallace, July 25, 1933.

of the Special Board made inquiries about Bennett and finally decided that he should be permitted to go ahead with his program of erosion control demonstrations, utilizing what Bennett always called "all practicable methods"—meaning not merely terracing or tree planting.

On August 25, 1933, Ickes in his capacity as Administrator of PWA wrote a letter to himself as Secretary of the Interior in which he allocated $5 million of emergency funds for a program of soil erosion control in the Department of the Interior. Four days later the Special Board of Public Works "corrected" the minutes of its July 17 meeting by substituting the Department of the Interior for the Department of Agriculture as the one to administer this program.[20] Within a fortnight, Ickes had talked with Bennett, requested his transfer from the Department of Agriculture, and secured Secretary Wallace's approval on the assumption that the move was for not more than two years.[21]

The available evidence on why the Soil Erosion Service (SES), as Ickes designated it, was set up in the Department of the Interior, is spotty. It indicates that Bennett was put to work there upon Rexford Tugwell's recommendation. But there are also indications that Ickes planned to transform the Department of Interior into a department of conservation, and that he expected to use Bennett's organization for trading with the Department of Agriculture when the time was ripe.[22] So far as Tugwell's motives are concerned, they are not known directly. Both Milburn L. Wilson and Bennett later said that Tugwell approved of Bennett's crusade for erosion control, but expected some bureau chiefs in the Department of Agriculture and the "land grant college crowd" to block Bennett's plan to demonstrate techniques in an action program instead of emphasizing research.[23] President Roosevelt's direct

[20] NARG 114 (Soil Erosion Service File), Administrator of Public Works to Secretary of Interior. See also Special Board of Public Works, *Minutes,* July 17 and Aug. 29, 1933.

[21] NARG 16 (Erosion File), Henry A. Wallace to the Secretary, Department of the Interior, Sept. 12, 1933.

[22] Jane Ickes (ed.), *The Secret Diary of Harold L. Ickes* (New York: Simon and Schuster, 1953), Vol. I, pp. 37, 47, 52, 169–72, 194, 250, and 292; Brink, *op. cit.,* pp. 83–84; and NARG 16 (Erosion File), Harold L. Ickes to R. G. Tugwell, June 6, 1934.

[23] Interview, Oct. 27, 1960, with Milburn L. Wilson. Wilson was Director of Subsistence Homesteads in the Department of Interior in 1934, and from 1934 to 1953 served as Assistant Secretary of Agriculture, Under Secretary of Agriculture, and as Director of the Extension Service in the Department. Also, Dillon S. Myer, who served in the Soil Conservation Service from 1935 to 1942 as Chief of the Division of Cooperative Relations and Planning and Deputy Chief in an interview on Sept. 12, 1962, said that Bennett told him Tugwell was responsible for having the Soil Erosion Service established in the Department of the Interior. According to

role in these events is not clear from the record. However, his interest in conservation generally was evident in his reforestation program when he was Governor of New York, and by his leadership early in 1933 to secure legislation for the Civilian Conservation Corps (CCC) and the Tennessee Valley Authority (TVA). The camps established by the CCC and given technical supervision by Agriculture and Interior were enormously popular, particularly with congressmen who deluged the White House with requests for camps to be located in their districts.[24] The President needed to satisfy these demands in order to gain congressional support for other important measures.

In any event, the first federal program of erosion control was neither initiated nor approved by the Department of Agriculture. The most important result of this game of administrative chess was that the Department of Agriculture eventually had to accept a program of erosion control which it did not originate, along with an organization which it did not create, to serve its objectives.

The Soil Erosion Service immediately launched a national program of erosion control on agricultural lands with two objectives: to demonstrate that erosion on croplands largely can be controlled, and to lay the foundation for a permanent program. The first demonstration project was started at Coon Valley, Wisconsin, in November 1933; by March 1934, $10 million of emergency funds had been spent on 24 projects. Much of the manpower for the nontechnical work was supplied under agreements with the Federal Civil Works Administration (CWA) and with the Emergency Conservation Works which provided 51 CCC erosion camps under the technical supervision of SES. Also, SES entered into five-year agreements with farmers who agreed to supply certain labor and materials and to carry out recommended land use

Bennett's version of his conversations with Tugwell, Tugwell wanted Bennett to head the new agency, and he wanted the Soil Erosion Service in the Department of the Interior to avoid the endless strife among the bureaus of the Department of Agriculture. On Sept. 30, 1933, Bennett wrote to Tugwell expressing "great appreciation" for the interest he had taken "in my behalf in connection with the erosion program." NARG 16 (Erosion File). One historian has said that Tugwell delighted in irritating people of the "Farm Bureau—Extension sort." He is said to have called the American Farm Bureau Federation the "most sinister influence in America," and to have warned Secretary Henry A. Wallace against working with that organization. Tugwell, in Christiana M. Campbell, *The Farm Bureau and the New Deal* (Urbana: University of Illinois Press, 1962), pp. 173–74, is quoted as saying that, when Wallace ignored his advice by trying to work with the Farm Bureau, it "cut Wallace's throat."

[24] E. G. Nixon (ed.), *Franklin D. Roosevelt and Conservation, 1911–1945* (Hyde Park: General Services Administration, National Archives and Records Service, Franklin D. Roosevelt Library, 1957), Vol. I, pp. 161–62.

practices. In turn, SES built check dams, new fences, contour strips, and terraces; planted trees; seeded eroded areas; and performed other necessary operations.[25]

Soon, however, Bennett found his own program caught up in the rival ambitions of Ickes and Harry Hopkins.[26] President Roosevelt had placed Hopkins, rather than Ickes, on the Federal Emergency Relief Administration (FERA), which was responsible for providing relief to the millions of unemployed through the winter of 1934–35. In September 1934, only six weeks before the congressional elections, the FERA, under the direction of Hopkins, organized its own conservation program. Bennett considered work planned by FERA to be a revival in many respects of the McCrory Plan which had rivaled his own in 1933. The FERA work would be principally terracing and building ponds under the technical direction of the state extension services, and at a staff meeting on September 27 Bennett characterized this as "a body blow to our program." He thought that the work would not be coordinated with his own, and would be poorly conceived and executed by either inexperienced persons or specialists who disagreed with Bennett's prescriptions for control—especially his opposition to indiscriminate terracing. Poor results would reflect harmfully on the Roosevelt Administration and Soil Erosion Service, he feared, since farmers and the public generally might not distinguish among the "alphabet" agencies. The crusade to mobilize the nation against erosion might fall flat at the very time when the dust storms were dramatizing the issue. If the state extension services were to guide the programs, they would be in a position to argue forcefully that they were fully prepared to assume major responsibility for any future permanent programs. In addition, since FERA administered its work through the established organs of state and local government, there would be opportunities to use erosion control for partisan gain and for expanding ties between the state extension services and their supporters.

At this staff meeting, Bennett and his principal aides considered requesting a transfer to FERA under Hopkins' direction, but rejected this idea in favor of staying in Interior, at least for the time being. Some of Bennett's subordinates feared that FERA would receive the lion's share of emergency funds and leave SES in Interior with no money to

[25] *Annual Report of the Secretary of the Interior, 1934* (Washington: Government Printing Office), pp. 353–61.

[26] See Robert Sherwood, *Roosevelt and Hopkins, An Intimate History* (New York: Harper and Bros., 1948), esp. pp. 44–69. See also Arthur MacMahon, John Millet, and Gladys Ogden, *The Administration of Federal Work Relief* (Chicago: Public Administration Service, 1941), pp. 17–120 (esp. pp. 66–72).

continue its program. Bennett, however, thought that President Roosevelt personally determined the distribution of emergency funds between the agencies run by Ickes and Hopkins, and that the soundest strategy was for Ickes to argue the wisdom of making SES responsible for the technical direction of the FERA program. While a decision on this matter was pending, Bennett told his supporters to "build a fire" at the grass roots by having friendly members of the staffs of agricultural colleges write testimonials to the work of the Soil Erosion Service. He told them to get up a report to convince Ickes and the President "they have started something that has caught the fancy of the people. . . . Here is a chance for the President to put a feather in his cap, by taking advantage of this favorable reaction." Bennett also directed his staff to accelerate operations. He wanted 20 new regional directors hired at once even if they had to "set up in a hotel room." These 20 men should be sent out "to find 20 watersheds right now." One aide conjectured that SES could employ 5,000 college men "tomorrow," but Bennett warned that, if they were hired, "don't make it merely engineers." Various phases of operations were to be speeded up, to include control of gullies. Bennett charged his staff: "Plug up every damn one."[27]

TRANSFER TO THE DEPARTMENT OF AGRICULTURE

In November 1934, Bennett worked out a report for the National Resources Board, which had replaced the National Planning Board on June 30, 1934 and been given the responsibility to prepare for the President a plan for the development and use of land, water, and other resources. Bennett said his objective was to restrict sheet and gully erosion and to rehabilitate badly eroded lands wherever they adversely affected valuable agricultural lands. According to his report, national, state, and local governments should be employed and, to be successful, the efforts must be "practicable" and enjoy popular backing and the support of public leaders. Bennett insisted that erosion could not be prevented completely, nor could the conditions of the land be re-established as they were before settlement. He did consider it possible, however, to "reduce erosion on most of the declining areas of more valuable soil to such an extent that the land may continue to be used almost indefinitely." He anticipated that within ten years

[27] NARG 114 (General Correspondence, 1933–35), memorandum, "Staff Meeting, September 27, 1934," 9 pages.

control would be initiated on all land suffering seriously from erosion; reasonable controls should be established within twenty years on such land, and they should cover "practically all" the better lands within a generation. Bennett further argued that such a program should avoid "unnecessary restrictive legal regulations on the use of the land." In many situations there would be voluntary cooperation from the landowners and users, but in others the need "would undoubtedly arise" for enacting "state or local land-use zoning ordinances based on the specific adaptability of soils, as [affected] by topography, climate, type of agriculture and other factors." The enactment of such regulations "could be advantageously encouraged through the granting of Federal aid for the control of erosion to those regions in which necessary restrictions are set up, and the withholding of such aid from regions which refuse to cooperate."[28]

To effectuate these aims, Bennett proposed seven recommendations in his memorandum: (1) To coordinate action and to supply proper technical direction, the United States should maintain a single agency "charged with the responsibility for directing and coordinating erosion work generally, whether on Federal, state or private lands. Such an organization would also coordinate and give general supervision to the erosion work of other federal agencies." (2) A "fairly detailed survey" should be made to determine the needs for control. (3) Legislation should be enacted authorizing the federal agency charged with erosion work to carry out necessary work on state and private lands "on a cost-sharing federal aid basis" with states providing adequate "contributions, cooperation, regulation and agreements controlling the use of the land." The federal government's share of the cost should be in proportion to the federal benefit contributed. (4) Federal appropriations should be made for work on the Indian reservations. (5) Federal, state, and local governments should acquire badly eroded lands unsuited to continuing safe agricultural use and convert them into forest or grazing preserves. (6) Federal legislation should withhold federal loans and

[28] NARG 16 (M. L. Wilson File, 1934–40), Oscar Chapman, Assistant Secretary of Interior, to Charles W. Eliot, 2nd, covering a memorandum of four pages "from Dr. Bennett." Bennett's report, as edited, appears in *Report of the National Resources Board on National Planning and Public Works in Relation to Natural Resources and Including Land Use and Water Resources with Findings and Recommendations* (Washington: Government Printing Office, December 1, 1934), Pt. II, Sec. III, pp. 161–74. The published version does not mention giving preferential treatment to cooperating farmers or withholding federal credits from new agricultural developments on submarginal lands (see p. 173). In 1935, the National Resources Board was replaced by the National Resources Committee and it, in turn, was replaced by the National Resources Planning Board in 1939.

other types of credit from new agricultural developments on submarginal lands, including those subject to destructive erosion; farmers who cooperated with erosion control measures under the federal program should receive *"definite advantages and preferences"* [emphasis added]. (7) Through various procedures, especially "selective extension of Federal aid" for control of erosion, states should be encouraged to pass legislation authorizing: (a) cooperation with the federal government in erosion control; (b) the "organization of conservancy districts or similar legal sub-divisions with authority to carry out measures of erosion control; and (c) the establishment of state or local land-use zoning ordinances where lack of voluntary cooperation makes such ordinances necessary."

At the time Bennett prepared his report for the National Resources Board, he raised with Ickes the question whether the Soil Erosion Service ought to remain in the Department of the Interior or be shifted to Agriculture. Bennett elaborated considerations on both sides of the issue, and then decided that from a "purely theoretical point of view, there would seem to be no completely logical location for the Service." Bennett insisted, however, that "it is of the greatest importance that all erosion control work, perhaps with the exception of that on the national forests, be coordinated through one agency."[29] A few weeks later, in mid-December, a committee which had been appointed by Ickes supported Bennett's policies generally and expressly recommended that all responsibility for erosion control on private lands be vested in the Secretary of Agriculture, provided that he consolidate all research and control activities within a single agency.[30]

The need to coordinate the growing federal erosion control programs in a single agency was recognized also in both the Department of Agriculture and several of the state agricultural colleges. Pressure on this subject had been building up by the time the annual meeting of the Association of Land-Grant Colleges and Universities was held in Washington on November 19–24, 1934. The Association's Special Committee on Duplication of Land-Grant College Work by New Federal Agencies reported "ample evidence to convince your committee that duplication of land-grant college work by new Federal agencies exists now and that it will increase unless effective action is taken to remove

[29] NARG 16 (Land Policy File), memorandum, H. H. Bennett to Secretary of the Interior, Nov. 8, 1934.

[30] Committee on Soil Erosion, "Report to the Secretary of Interior on the Soil Erosion Service on a Permanent Coordinated Program of Soil Erosion Control," mimeographed, December 18, 1934.

it." The "present trend" must be reversed, the Committee avowed, or it would lead to "waste and confusion" that would impede the work of both the federal agencies concerned and of the colleges as "agencies of research and education and of coordination." The Committee reported it had not yet been able to identify all the sources of difficulty, but especially objectionable were the activities of the FERA in rural home demonstration work (but not conservation work) and the work of the Soil Erosion Service in "research, demonstration and extension work in soil erosion control." It was recommended that the Executive Committee of the Association bring this matter to the attention of President Roosevelt and other appropriate federal officers, and that the Association affirm its "willingness" to make available the "trained and experienced personnel, the extensive organization and the benefits of long experience of the land-grant colleges as the agencies through which Federal-State cooperation should function."[31] Finally, the Executive Committee was directed to formulate a statement of proper relations with the Department of Agriculture, to negotiate an improvement in existing relations, and to seek additional federal appropriations to assist the colleges.[32]

Some leading administrators in the colleges assured Secretary Wallace that the soil conservation program of the Federal Emergency Relief Administration was based on a sound conception of federal-state relations, even if the work of the Soil Erosion Service were not. The FERA relied upon the state agricultural extension services to give technical direction to its conservation projects. In a letter to Secretary Wallace, Dean C. L. Christensen of the Wisconsin College of Agriculture called the work under Robert Fechner of FERA to be "admirable," and went on to say that it "has been very disconcerting to have the Federal government set up duplicating and competing . . . agencies in the field of erosion control." He and his colleagues at the University of Wisconsin were convinced when the Soil Erosion Service was established in the Department of Interior that the move was "a grave mistake." The colleges and the Department of Agriculture, he noted, had built up a fine working arrangement over more than fifty years of experience, but no such tradition had been developed between the colleges and the Department of the Interior. Moreover, farmers in Wisconsin wanted to know why, when the Bureau of Chemistry and Soils, the Bureau of Agricultural Engineering, the Forest Service, and the

---

[31] *48th Proceedings of the Association of Land-Grant Colleges and Universities, 1934* (Washington), p. 240.
    [32] *Ibid.*, p. 279.

Wisconsin Agricultural Experiment Station coordinated their research at the Upper Mississippi Valley Experiment Station at La Crosse, the SES "comes into the picture as an entirely independent agency, and by inference, if not directly, gives the impression that it has a different program of erosion control, and a better program." Christensen warned that "embarrassing questions" would soon be asked in Congress unless it was recognized that all the "pioneer work" in erosion control had been done in the Department of Agriculture. The existing confusion would also be compounded, if supervision of the soil erosion camps of the CCC were transferred from the Forest Service to the Soil Erosion Service. He urged Wallace to "oppose as vigorously as you can any efforts . . . to make this transfer," and to push for incorporating SES into the Department of Agriculture.[33]

R. K. Bliss, Director of Extension at Iowa State College, opposed the transfer of supervision of CCC camps to SES, also, because Iowa State College had just completed agreements to furnish technical assistance to the camps "to guide the work along sound and constructive lines." He, too, argued in writing to Secretary Wallace that "we know how to work with the Department."[34]

Robert Fechner of FERA joined in the chorus of protest, writing to Wallace that he had received a number of letters from the agricultural colleges and other groups to protest reports in circulation that SES was about to assume responsibility for all erosion work by federal agencies. He claimed that every letter makes "vigorous protest" because SES "does not consult with nor cooperate with State Agricultural Colleges or any other agency" in carrying out its work which, he said, was sometimes undertaken contrary to the wishes of state authorities. In addition, SES was bearing the entire expense of its work and "does not require the land owner who is benefited to contribute in any way to the cost." Fechner wanted Wallace to know of these views so that the Secretary might take such action as "you may feel is desirable in the matter."[35]

T. C. Richardson, Acting Chairman of the Texas Soil and Water Conservation Committee (the members of which were also members of the Board of Regents of Texas Agricultural and Mechanical College), repeated all of these arguments and added that Texas farmers had

[33] NARG 16 (Land Policy File), C. L. Christensen to Secretary of Agriculture, Dec. 15, 1934.

[34] NARG 16 (Land Policy File), R. K. Bliss to H. A. Wallace, Dec. 29, 1934.

[35] NARG 16 (Land Policy File), Robert Fechner to the Secretary of Agriculture, Dec. 27, 1934.

heard of work being done by the Soil Erosion Service and were reluctant to proceed with erosion control under the guidance of the Texas Extension Service. Richardson complained that loans, grants, and contributions "directly to individuals cannot be continued without inequities and excessive costs of administration."[36]

Wallace apparently took these complaints under consideration and decided that it was time for him to ask Ickes to transfer the Soil Erosion Service to the Department of Agriculture so that federal and state erosion control activities could be coordinated in a single agency. Wallace told Ickes SES was tending increasingly to duplicate the "functions and responsibilities" of the Department of Agriculture. It was, he asserted, duplicating research in no fewer than seven bureaus within the Department and was disrupting relations with the state experiment stations and extension services. Moreover, SES was hiring away from the Department and the land-grant colleges their best scientists at a time when the "most effective use of the limited trained personnel available can best be obtained through a unified administration in the Department."[37]

Ickes refused to transfer SES to the Department of Agriculture without a fight, or at least some guarantee of a reasonable return. Wallace, therefore, asked Donald Richberg, Executive Director of the National Emergency Council, to take the issue to the President. Secretary Wallace said he wished to make it clear that, although land use is overwhelmingly an agricultural problem . . . we do not have any desire simply to acquire additional administrative units." His concern was wholly with the "coordination and unification of those things that are importantly agricultural, and I should like to see the possibilities explored and shall be interested in having your comments as to possible steps in that direction." Wallace assured him that his objective was only to further the Administration's policy of agricultural production adjustment through appropriate land use changes.[38] Privately, Ickes conceded: "I am half inclined to agree with him that it properly belongs in Agriculture, but the other day the President brought up the subject and I found him inclined to believe it belonged

[36] NARG 16 (Erosion File), T. C. Richardson to Henry A. Wallace, Feb. 5, 1935.

[37] NARG 16 (Land Policy File), Secretary of Agriculture to Secretary of Interior, Dec. 10, 1934. This letter was prepared by H. G. Knight for Wallace on Nov. 15, 1934 during the meeting of the Association of Land-Grant Colleges in Washington. Ickes demurred in a reply to Wallace, Dec. 17, 1934.

[38] NARG 16 (Land Policy File), Secretary of Agriculture to Donald Richberg, Dec. 22, 1934.

here."[39] Wallace was convinced that Ickes "has no illusions whatever as to the character of the functions of the Soil Erosion Service and where it belongs, but he is holding onto it because he thinks it is good trading stock."[40]

Officially the matter rested in Richberg's hands for a month before he sent a memorandum, on February 11, to Wallace, advising him of the President's desire to have all erosion control functions transferred to, and administered in, the Department of Agriculture.[41]

Richberg called the whole matter to the President's attention late in February, urging that it be decided because SES was entering into agreements with several of the state experiment stations and they might later cause confusion. He told the President that all of the work could not be put in Agriculture without "some legislation," but the principal work being carried out by SES "can be transferred to the Department of Agriculture by Executive Order."[42] Roosevelt briefly noted his desire to see Wallace when the President returned from a pending trip. A conference was arranged and Hugh Bennett and Secretary Wallace were included in the discussions for the first time.[43] According to Ickes, the issues were canvassed with the President on March 5, and at that time Ickes asked him if he would "let nature take its course" if "anything should happen on the Hill." Ickes understood Roosevelt to say that he would, so that "I instigated the filing of several bills in both the Senate and the House . . . setting up in this Department on a permanent basis a Division of Erosion Control."[44] He did this despite Roosevelt's remark in an earlier conversation over Ickes' ambition to head a department of conservation that it is sometimes hard to distinguish between conservation and agricultural activities.[45] Wallace then reported to the President that Congressman Marvin Jones of Texas (Chairman of the House Committee on Agriculture) had called "me up early this week saying he had a bill providing for setting up the Soil Erosion Service in a permanent way in the Department of Agriculture."

[39] Jane Ickes, op. cit., Vol. I (entry of Dec. 29, 1934), p. 258.

[40] Nixon, op. cit., Vol. I, p. 362, Wallace to President Roosevelt, March 7, 1935. Also NARG 16 (Land Policy File), Harold Ickes to Henry A. Wallace, Jan. 5, 1935.

[41] NARG 16 (Erosion File), H. A. Wallace to Donald Richberg, Feb. 19, 1935, and attached memorandum from Seth Thomas, Solicitor of the Department of Agriculture, to Paul H. Appleby, Feb. 16, 1935.

[42] NARG 16 (Erosion File), memorandum, Donald Richberg to the President, Feb. 21, 1935.

[43] Nixon, op. cit., Vol. I, pp. 357–59. The President also had a copy of a memorandum in which Bennett criticized the rival erosion program of the FERA, ibid.

[44] Jane Ickes, op. cit., Vol. I, pp. 310, 325.

[45] Ibid., Vol. I, p. 300.

Jones had said he understood that action was in prospect "from this end" and did not want to get his wires crossed if "you were perhaps contemplating handling the Soil Erosion Service a while longer on a temporary basis." Wallace said that he had counseled delay for the time being, but that Jones was eager to push his own bill in the absence of presidential action.[46]

ENACTMENT OF PUBLIC LAW 46

Several bills, introduced in Congress early in 1935, added to the confusion on the status of the Soil Erosion Service. Three senators— Carl Hayden, Democrat of Arizona, Edward P. Costigan, Democrat of Colorado, and Carl A. Hatch, Democrat of New Mexico—introduced S. 2149 to provide for SES to be in the Interior Department, and the bill was referred to the Senate Committee on Public Lands for a friendly discussion by its authors.[47] Virtually identical bills, also placing SES under Interior, were offered in the House of Representatives by Jack Nichols, Democrat of Oklahoma, John J. Dempsey, Democrat of New Mexico, and Mrs. Isabella S. Greenway, Democrat of Arizona,[48] and were referred to the House Committee on Public Lands where hearings were started on March 20.[49] But before the hearings had gone very far Representative Marvin Jones, Democrat of Texas, Chairman of the Committee on Agriculture, introduced H.R. 6872 to place SES in Agriculture; the bill was referred, of course, to his committee.[50]

These legislative maneuvers, inspired partially by Ickes, forced the President to act at once. He wrote to Ickes, saying that he wanted SES transferred to the Department of Agriculture immediately and needed a resolution of the PWA Board which had created it. Harry Slattery, an assistant, called Ickes, who was out of town. Ickes was outraged because he thought the President had not yet foreclosed the issue against him without a further hearing. He sent the President a protest, but on the next day received a report that the President was firmly insisting on the transfer, even in Ickes' absence. Ickes again asked for a

[46] Nixon, *op. cit.*, Vol. I, p. 361.
[47] *Congressional Record*, Vol. 79, Pt. 3, 74 Cong. 1 sess. (1935), p. 2821.
[48] H.R. 6319, H.R. 6432, H.R. 6439, and H.R. 6440; see also *Congressional Record*, Vol. 79, Pt. 3, *op. cit.*, pp. 2807, 2983, and 2984.
[49] *Soil Erosion Program*, Hearings on H.R. 7054 by a Subcommittee of the House Committee on Public Lands, 74 Cong. 1 sess. (1935).
[50] *Congressional Record*, Vol. 79, Pt. 4, 74 Cong. 1 sess. (1935), p. 4133.

delay of the decision until he had returned on Saturday, March 23, but he was refused. Consequently, the PWA Board met on the afternoon of Friday, March 22, and passed the necessary resolution.[51]

During these three days of executive activity the House Committee on Public Lands continued hearing the bills to establish SES in Interior on a permanent statutory basis. The hearings, however, brought out little new information. Bennett cited the Muskingum Conservancy District in Ohio as the sort of local unit which would "possibly" serve his plan for the expansion of the erosion control program. He repeated the major points of his program set forth on the previous November 8 and added the remark that in addition to conservancy districts "other erosion control associations or similar organizations" might be utilized. Most of the testimony recorded for publication consisted of resolutions, speeches, articles, and similar evidence of public support inserted by Bennett with the approval of the Committee.[52] Representative William Lemke, Republican of North Dakota, voiced the only note of concern. He feared that people might be "pushed" too far, and that the Department of Agriculture might come under heavy fire in 1936 in the Midwest because of the issue of "coercion." He asked Bennett to tell him what SES would do on the farm of an operator who would not cooperate. Lemke said he was against commands. Bennett agreed that coercion was not desirable and said that SES had succeeded in carrying through some farmers in economic distress by employing them on their own farms which were then called "demonstration farms." Lemke said he thought that there was great bitterness in the West Central states over the land acquisition policies of the Department of Agriculture, because population was being shifted out of the affected states. These factors, he thought, were likely to injure SES, if it were in the Department of Agriculture, or if it attempted to use coercion. Rufus Poole, who testified for the solicitor, said that the only device they had yet thought useful would be a restrictive covenant on land treated with the assistance of the federal government. Aside from these admonitions of caution, there was no hostile questioning of the witnesses. On the final

[51] Jane Ickes, *op. cit.*, Vol. I, pp. 325–26 and 398; Vol. II, pp. 37–41. Nixon, *op. cit.*, Vol. I, pp. 363–64. In the meantime, the Master of the National Grange, L. J. Taber, had written to Secretary Wallace and the President on March 13 and 14, calling for the transfer of SES to Agriculture, in accord with a resolution adopted at the annual Grange convention held at Hartford during November 1934; see NARG 16 (Erosion File), Taber to Roosevelt, March 13, 1935, and (Land Policy File), Taber to H. A. Wallace, March 14, 1935.

[52] *Soil Erosion Program*, Hearings on H.R. 7054, *op. cit.*, p. 87.

day of hearings the only witness was G. H. Collingwood of the American Forestry Association who favored placing SES in Agriculture.[53]

With the transfer a fact, the following congressional action was largely a formality. On March 27, Representatives Dempsey and Jones introduced new bills to create the Soil Conservation Service in the Department of Agriculture and both were referred to the Committee on Agriculture. As submitted they merely changed the word "Interior" to "Agriculture" in the earlier bills considered by the Committee on Public Lands.[54] The new measure sped through the House, but was kept two days in committee, where consideration was given to some amendments suggested by Bennett at Jones' request. According to Bennett, the declaration of policy in the act was intended to authorize control of soil erosion, preserve natural resources, control floods, protect reservoirs, maintain the navigability of streams, protect public lands, and relieve unemployment. Bennett wanted language to insure that the power of the Secretary to conduct investigations, surveys, and studies would be a continuing one.[55] Other changes of a more technical character were made to insure that the Secretary would have flexible authority over any transferred personnel and that the new agency might make use of unexpended balances available after June 16, 1935.

On March 29, the bill was reported by the Committee on Agriculture and Jones moved to suspend the rules and pass it. The sponsors offered explanations of the bill and several members, especially those from the Southwest, gave warm eulogies in support of it. With little discussion and no apparent opposition the rules were suspended and the bill passed.[56]

Passage through the Senate was also speedy. Senator Hatch explained that the purpose of the bill was to coordinate all the agencies concerned with soil erosion control in the past—FERA, Forestry, and others—and make the funds to be expended by the new agency applicable to the provisions of the $5 billion work relief bill which had just passed the Senate. Senator William H. King, Democrat of Utah, expressed his concern that 200 million acres of public land would be transferred from the Interior Department. King was apparently given

---

[53] *Ibid.*, pp. 97–99.

[54] *Congressional Record*, Vol. 79, Pt. 4, *op. cit.*, pp. 4577 and 4496.

[55] NARG 114 (Soil Conservation Service, General Correspondence, 1933–35), memorandum, H. H. Bennett for the Secretary of Agriculture, March 23, 1935.

[56] *Congressional Record*, Vol. 79, Pt. 4, *op. cit.*, p. 4577, and Pt. 5, pp. 4803–09.

assurances of some kind that the Interior Department would continue to be responsible for administering these lands under the new Taylor Act, so he withdrew his opposition. The bill was then passed on April 19 without a record vote. It became the Soil Erosion Act of 1935, Public Law 46, on April 27.[57]

Support for this law had a distinctly western flavor. Members who were active in securing its passage were familiar with the ravages of wind erosion, the shortage of water, the need to improve grazing conditions on badly eroded public lands, and the flooding and siltation of irrigated farmlands. Some contended that the bill was intended to control floods through watershed treatment. This argument was offered partly to demonstrate that the act was constitutional and would pass any test in the Supreme Court. It was probably used also because it was widely believed to be correct by a variety of people with conflicting interests. "Watershed" is a word which runs like a silver thread through the history of the Soil Conservation Service.

This new legislation recognized the waste of soil and water on agricultural lands as a "menace to the national welfare" and directed the Secretary of Agriculture to "coordinate and direct all activities" related to soil erosion. For this purpose he was given broad authority to conduct surveys, investigations, and research into control methods and to disseminate findings to the public. He was authorized to conduct demonstration projects, to carry out a wide range of preventive operations, and to give aid to private persons and public agencies subject to conditions which the Secretary might set. The act specified that he might require state and local laws imposing permanent restrictions on land use to prevent soil erosion, and covenants regulating land use and local contributions of money, services, or materials. For the purposes of this act, the Secretary was given power to acquire lands, or rights and interests in them, by purchase, gift, or condemnation.

This act directed that the Secretary establish the Soil Conservation Service to exercise these new powers and authorized him to transfer funds and personnel from other agencies of the Department to this new agency to coordinate erosion control operations.

In sum, Public Law 46 in 1935 gave the Secretary of Agriculture a very broad grant of authority to carry out the kind of program which the National Resources Board was in the process of publishing as this act was clearing Congress. As a matter of law, the Secretary was free to

[57] *Ibid.*, Pt. 5, pp. 4830, 4900, 5411, 5644–45, 5734; Pt. 6, pp. 6011–18, and 6222. The citation is Soil Erosion Act of 1935, Public Law 46, 74 Cong. 1 sess., 49 *Stat.* 163; 16 USC 590 a-f.

reorganize the Department by consolidating old agencies, or parts of them, with the Soil Conservation Service. In addition, he was free to carry out erosion control operations either through existing channels, such as the state extension services, or by creating new lines of authority to each individual farmer.

SUMMARY AND COMMENT

The mixture of executive orders and legislation which launched a permanent national program of erosion control was conceived and initiated in the Department of the Interior in an atmosphere of grave national emergency. The first and most pressing objective of this work was to provide employment through a federal public works program. When Harold L. Ickes established the Soil Erosion Service, he by-passed specialists in agriculture who differed deeply over the extent and character of erosion, the relative emphasis that should be placed on research as distinguished from preventive measures, and the best methods of control. Ickes ignored the plan which the Secretary of Agriculture approved, stressing terracing under the joint guidance of the Department and the state agricultural extension services although he should have known that this plan differed from the one which Hugh H. Bennett had urged the Secretary of Agriculture to adopt. As soon as the Soil Erosion Service demonstrated the effectiveness of massed technical assistance used to plan and reorganize whole farms in selected watersheds, administrators in the agricultural colleges demanded that this work be transferred to the Department of Agriculture and administered through them. The Soil Erosion Act of 1935, supported by the President and passed quickly by Congress, made the Soil Conservation Service and the program which SES had been carrying out permanent. Legally, the Secretary of Agriculture was given great discretion to coordinate the Department's soil conservation operations; in fact, the problem proved to be so unmanageable that solutions evaded the grasp of successive secretaries for the next two decades.

CHAPTER 2

The Search for Organization

AT THE TIME the Soil Erosion Service was transferred to the Department of Agriculture, Secretary Henry A. Wallace was aware that its operations had upset the established relationships, rituals, and interests of the old-line bureaus of the Department of Agriculture and the land-grant agricultural colleges. Deans of some of the agricultural colleges and chiefs of nine of the bureaus in the Department implored Secretary Wallace to curb Hugh Hammond Bennett and the new Soil Conservation Service (SCS).[1]

Shortly after transfer of the Soil Erosion Service (SES), Wallace constituted an interbureau committee (called the Secretary's Committee on Soil Conservation) of four members who were "thoroughly familiar with the Department's traditional lines of cooperation, history, and present policies . . . (to) be helpful to Mr. Bennett and his associates in becoming a permanent, major unit of the Department." He assured Bennett that his choice of procedure was "in no way critical of the Soil Conservation Service."[2] Although this procedure gave the established agencies an opportunity to protect their own interests, it was true then, as it is now, that the task of fixing limits on the empire-building tendencies of young, healthy, and vigorous organizations is a necessary task of responsible department heads. Decisions to allocate resources of manpower and funds and to fix program goals and agency relations are matters of too much public importance to be left to the unrestrained ambitions of bureau chiefs devoted to the successful pursuit of their own limited objectives.

DEPARTMENT OF AGRICULTURE—STATE EXTENSION SERVICES

While the interbureau Committee on Soil Conservation deliberated, several state extension services and their supporters stepped up their

[1] National Archives, Record Group 16 (Land Policy File), H. G. Knight to Paul H. Appleby, Jan. 8, 1935. (National Archives records cited hereinafter usually will be identified by the abbreviation NARG.)

[2] NARG 16 (Committees File), memorandum of the Secretary of Agriculture, May 4, 1935.

efforts to gain a major share in the administration of the new federal erosion control programs. Clyde W. Warburton, Director of Extension Work in the Department, reminded Assistant Secretary Milburn L. Wilson that the "1914 Agreement" provided for all "extension work" of the Department to be done through the state extension services and not independently. Warburton argued that the work being done by SCS on erosion control demonstration projects "is in large part extension, and it is without question extension so far as efforts are made to get farmers outside these areas to adopt the practices followed thereon." Future relations between SCS and the extension services should be conducted, therefore, through traditional channels.[3]

Dean C. L. Christensen of the University of Wisconsin wrote Wallace saying: "As you know it is very much our desire that the Soil Erosion Service be transferred . . . to the Department. . . . You can imagine our surprise, when we learned today for the first time that it was your desire . . . to have all administration and work direction of erosion control programs handled by federal authorities, and that no authority was to be left with the states." Wallace replied that the Department "has no intention of departing from its traditional policy of cooperating with state agricultural agencies in every way it can." He said that the Department would have to move "slowly on this thing" but that the committee he had appointed was thinking "entirely in terms of State cooperation and execution of soil erosion control programs."[4]

Meanwhile several state extension services worked out agreements with the Federal Emergency Relief Administration (FERA) which was operating under Harry Hopkins. Colorado and Oklahoma submitted their plans to Wilson for comment and he replied that such proposals would serve only as emergency measures suitable until a permanent national program was prepared.[5] When Oklahoma created a conservation commission with a dean from Oklahoma A. & M. as a member, Bennett commented that it was intended to have a "more or less independent program" with political implications that ought not to be ignored. He warned Under Secretary Rexford G. Tugwell that the matter "which is most likely to give us most difficulty is the one of establishing cooperative relations with the state colleges of Agricul-

[3] NARG 16 (M. L. Wilson File), April 8, 1935.

[4] NARG 16 (Reorganization File), C. L. Christensen to Secretary Wallace, April 11, 1935; and Wallace reply, April 16, 1935.

[5] NARG 16 (Erosion File), M. L. Wilson to D. P. Trent, May 24, 1935; memorandum, C. W. Warburton to Wilson, May 1, 1935, and attached file.

ture." It could be done, he said, "but it is going to take careful handling. We know pretty well the conditions . . . where the sore spots are and where the troublemakers are."[6]

Leonard J. Fletcher of the Caterpillar Tractor Company in Peoria, Illinois, reported to Wilson that two of his company's representatives in the Southeast recommended that "soil conservation work be carried thru the state extension service. . . . It is our belief that there is going to be a great demand for decentralization of Federal activities. . . . Nothing but trouble will result from a further carrying out of the so-called demonstration small watershed plan of operations. . . . There has already been a great deal of criticism of the Federal Government not treating all farmers alike." Many observers were critical, Fletcher said, of the SCS requirement that farmers sign five-year agreements to execute complete farm plans in return for its services. Fletcher contended that "75 to 90 per cent of the farmers in the southeast signed these agreements . . . to secure the terracing work which is promised them if they will sign." Because the Department had not yet settled the organizational issue, Fletcher said the Caterpillar Company was marking time "in the demonstrating of terracing equipment since we do not know thru what organization we should work." He concluded that if the Soil Conservation Service "must exist as a separate organization" its leaders should consider it as "an additional technical aid to the extension service. We believe it is quite unfair to give this Service great sums of money to expend independently of the careful advice and direction which they could secure from the extension service."[7] Wilson's reply was brief and noncommittal.

These attitudes grew out of relationships which had grown up between the Department of Agriculture and the state extension services during the two decades since passage of the Smith-Lever Act in 1914. This act and an agreement by the Secretary of Agriculture and the Executive Committee of the Association of American Agricultural Colleges and Experiment Stations in 1914 provided for basic relationships.[8]

The legislation authorized the distribution of federal funds to the states in consideration of a variety of factors, and the "1914 Agreement"

[6] NARG 16 (Erosion File), memorandum for R. G. Tugwell from H. H. Bennett, April 22, 1935.

[7] NARG 16 (Erosion File), Leonard J. Fletcher to M. L. Wilson, May 23, 1935.

[8] 38 *Stat.* 372. This association of colleges, organized in 1887 and better known as the Association of Land-Grant Colleges and Universities, has had several changes of name. With the most recent, in November 1964, it became the National Association of State Universities and Land-Grant Colleges.

stipulated that all "extension work" in the Department, whether supported by Smith-Lever funds or not, was to be carried out by the state
colleges of agriculture. Research was to be a joint activity with the state
agricultural experiment stations, but with no other state agencies. (In
actual fact, there were some exceptions to the principle of the 1914
agreement, especially insofar as the U.S. Forest Service worked
through state forestry agencies and administered the national forests
without agreement with the extension services.) Another agreement,
reached in 1923, involved the same organization, but with its new
name, the Association of Land-Grant Colleges and Universities, and
also the National Association of Commissioners of Agriculture, and the
Secretary of Agriculture. According to this agreement, programs of a
regulatory character were to be administered by the state departments
of agriculture and not by the colleges. The Office of Extension Work
under a Director was established in that year to represent the Secretary
and provide federal supervision of work done by the state services.

The Secretary of Agriculture was authorized to examine state annual
work plans before distributing funds, to require annual reports, and to
take other actions which gave him supervisory powers somewhat
broader in legal contemplation than in actual practice. In a study made
at the time of these events, Miss Gladys Baker found that the Department rarely commented on state programs until it reviewed them.
Initiative came principally from the state extension services. Even in
the case of the Agricultural Adjustment Act, administered with the
direct assistance of the extension services, the Department had only
limited control over important matters of policy.[9] Despite the fact that
the Agricultural Adjustment Administration (AAA) transferred more
than $30 million to extension in 1934–35 and enabled the states to hire
1,171 workers between June 1933 and June 1935,[10] the zeal with which
the various state services "educated" farmers varied widely. In the
South, where Miss Baker characterized it as a "moral crusade," when
the AAA proposed to sponsor cooperative group discussion projects in
nine states during 1933–34, some of the states did not appoint leaders,
some appointed a few who were not interested, and others called the
discussion material thinly disguised propaganda. In the Northeast,

[9] John M. Gaus and Leon O. Wolcott, *Public Administration and the United
States Department of Agriculture* (Chicago: Public Administration Service, 1940),
pp. 96–98, 265–66, and 373–74; Gladys L. Baker, *The County Agent* (Chicago:
University of Chicago Press, 1939), pp. 102–04.

[10] Baker, *op. cit.*, pp. 78–83.

county agents were not interested in propagating the first AAA program which was extremely unpopular with dairy and poultry farmers in the region. One state refused outright to participate in AAA county planning projects in 1935. The director of extension in Pennsylvania refused to permit county agents to do more than hand out information.[11]

Many individual county agents refused to cooperate regardless of the position of their state directors—and the control by state directors of their organizations varied from virtually dictatorial in Pennsylvania to almost nothing in other states. In some states, the extension services had developed ties with politically influential local sponsors to the point where county agents were responsible to the local sponsors and not to their state directors. All states provided for local sponsoring organizations in accord with the Department of Agriculture's recommendation. In 15 states, local units of the American Farm Bureau Federation were the sponsors by law, and by formal or informal association in 14 other states. In the other states, county governing bodies were the sponsors. In all, however, the county was the basic unit of the system.

With county agents operating in such varied environments it is not surprising that Miss Baker could find no single pattern of activities. In some states, many agents lacked precise long-term objectives and devoted most of their time to organizing local sponsoring groups or farmer cooperatives or performing services for individual farmers. To some degree, this last activity was necessary simply to "sell" farmers on "book farming." One way to convince a farmer of the value of extension work, as one old hand remarked, was to cull his chickens or innoculate his hogs. Many agents found numerous opportunities to be of service to local office holders as a means of building support for extension work. Other agents who lacked buffers from the cruder forms of local partisan pressures were caught up in political campaigns in which they either backed the right candidates and held their jobs, or backed the wrong ones and then found new employment.[12]

Miss Baker found that local sponsors of extension were interested chiefly in getting personal services from the county agents and not in planning programs or supervising the agents' work. Often the extension supervisors on intermediate levels were chiefly concerned to smooth over disruptions in the county agents' relations with politically influen-

---

[11] *Ibid.,* pp. 71, 83–85, and 96.
[12] *Ibid.,* pp. 120–39 (esp. pp. 130–31).

tial people. The supervisors did "not pretend to spend much time checking on the quality of the work done in the various counties."[13]

The Great Depression, which shattered the financial capacity of many rural counties, especially in the South and West, left its mark on the extension services. Between June 1931 and June 1933, extension employment declined by 2,222 man-years. Between 1930 and 1933, membership in state Farm Bureaus plunged disastrously from 321,195 to 163,246. In the Northeast, however, the decline in the number of county agents was not so great as in the distressed South and West, since the urban tax base in the Northeast provided a somewhat more constant level of support for extension.[14]

Aside from financing, however, the extension services were not universally popular at this time. The Iowa Extension Service was under legal attack for its tie with the Farm Bureau, and extension was under political attack in the Northeast for supporting the New Deal's AAA. In fact, the AAA was so unpopular in the Northeast that relations between state Farm Bureaus and the American Farm Bureau Federation were seriously strained until late in 1934. At that time, an understanding was apparently reached that the Federation's national organization would launch a membership drive as urged by the leaders of the New York State Farm Bureau.[15] It was also understood that the American Farm Bureau Federation would exert pressure on Congress for increased appropriations for the state extension services.[16]

When, in November 1934, the Executive Committee of the Association of Land-Grant Colleges and Universities met in Washington, it nevertheless demanded that the Secretary of Agriculture end "overlapping" in the programs of the Soil Erosion Service and others which they claimed "properly" belonged in the Department of Agriculture and the colleges. The Association's Committee on Extension Organization and Policy recommended to the Department an immediate increase of $4 million for fiscal year 1936. In January, the extension leaders learned that the Bureau of the Budget had not recommended the increase (an increase of 25 per cent over the current year) and the matter was

[13] *Ibid.*, p. 126.

[14] *Ibid.*, pp. 58–64.

[15] Christiana M. Campbell, *The Farm Bureau and the New Deal* (Urbana: University of Illinois Press, 1962), pp. 35–36, 68–74, 78, and 133. See also William J. Block, *The Separation of the Farm Bureau and the Extension Service* (Urbana: University of Illinois Press, 1960); Grant McConnell, *The Decline of Agrarian Democracy* (Berkeley: University of California Press, 1953); and Charles M. Hardin, *The Politics of Agriculture* (Glencoe, Ill.: The Free Press, 1952).

[16] Campbell, *op. cit.*, pp. 80–82.

"entirely in the hands of the President."[17] Starting in the spring of 1935, there was evidence that the Farm Bureau was stepping up pressure on Congress to be more generous with the state extension services.[18]

In the face of these circumstances, it is not surprising that extension leaders were appalled at the growth of such new agencies as the Resettlement Administration and the Soil Conservation Service (both of which were started in Interior and moved to Agriculture). The new agencies had the funds and manpower to hire away extension's best specialists and administrators, and they could woo farmers with assistance which could be massed in strategic locations on a scale which extension simply could not match. As traditional ties between the Department and the colleges were crumbling in 1937, Dean A. R. Mann of Cornell University passionately eulogized past practices as "so nearly a perfect ideal that agencies of government and agencies of light and learning which have close contact with the people shall freely cooperate in the public interest . . . that it must also become as nearly as possible an impregnable ideal, not blown about easily by the winds."[19]

REPORT OF THE SECRETARY'S COMMITTEE

On June 5, 1935, the interbureau committee which Wallace had appointed to help fit the Soil Conservation Service into the Department of Agriculture made its report.[20] The Secretary's Committee on Soil Conservation recommended detailed guides for organization, procedures, and relationships of a fundamental character which were so clearly intended as curbs on the Soil Conservation Service that furious battles over Bennett's program were waged for nearly two decades after issuance of its report.

Most of the recommendations dealt with relationships between SCS and the research agencies of the Department and its relations with the Forest Service in respect to the administration of public lands and forestry as erosion control on private lands. Only passing mention was

[17] Committee on Extension Organization and Policy, *Proceedings of the Association of Land-Grant Colleges and Universities,* mimeographed (Washington, Nov. 18, 1934), and *Proceedings,* mimeographed (Jan. 11–12, 1935).

[18] Campbell, *op. cit.,* p. 82.

[19] *51st Proceedings of the Association of Land-Grant Colleges and Universities,* 1937 (Washington), p. 63.

[20] "Report of the Secretary's Committee on Soil Conservation," mimeographed, U.S. Department of Agriculture, June 5, 1935. (A copy is on file in the Library of the U.S. Department of Agriculture.)

made of the relationship of SCS to AAA, although it was recognized that SCS would have to be fitted into the framework of general land use planning, proper crop rotations, controlled livestock grazing, and the application of other sound farm-management practices. Erosion control was to be the concern of all agencies of the Department and not of SCS alone. Although these recommendations and others were of interest when they were made, they may be passed over except for the matter of relations with the state extension services and experiment stations. For many years after this report was distributed, some scientists and administrators in the state agricultural colleges opposing the Soil Conservation Service considered the recommendations of the Secretary's Committee to be permanently binding. When Secretary Wallace and his successors later departed materially from some of them, there were college leaders who talked as if the Secretary were bound to accept the advice of a study committee appointed under his own authority. They also ignored the terms of Public Law 46 in which Congress had provided the Secretary with authority to carry out this program without any action by the state colleges, experiment stations, or extension services.

The Committee recommended that SCS be fitted into the Department's traditional organization for functioning with the state extension services. In the federal agency, there would be a division of extension in SCS responsible for developing cooperative relations between SCS regional offices and the state extension services. In the states, a soil conservation advisory committee would be established to include the state directors of extension and the experiment station, a representative of SCS, and the heads of other interested state agencies. The state advisory committees would correlate technical information from all interested agencies and examine preliminary plans for all control projects before clearing them for operations. Also, SCS would encourage state experiment stations to conduct research into the problems and methods of erosion control. Such projects would be reviewed by the state committee which would be expected to coordinate all conservation activities by negotiations among the various groups.

The Committee suggested the addition of extension specialists in soil conservation to the staffs of the colleges. Soil conservation specialists would augment the staffs of county agents. The county agents were to be appointed with the concurrence of the regional director of SCS and the state director of extension. These extension conservation specialists on the state and local levels would be expected to conduct the "educational phases" of the program worked out in each state by its soil conservation advisory committee. The county extension specialists, especially, were to encourage farmers to organize into voluntary or "legally constituted" soil conservation associations such as those which had been functioning in the South, particularly in the Tennessee Valley, since 1933. Farmers who organized such groups would be considered to have shown sufficient interest in soil conservation to justify public assistance. It was assumed by the Secretary's Committee that federal funds appropriated to the Department would be distributed to the extension services for these increases in staff. It seemed obvious to all concerned that SCS would have to share its funds with the extension services.

The Committee equivocated, however, on two crucial organizational issues. It suggested organizing either "cooperative control associations or Governmental agencies, which should be permanent in character, and legally empowered to own and dispose of real estate, to lay assessments on their members, and otherwise to obtain compliance in a complete erosion-control program on the area owned or controlled by the members of the Association." It recommended that voluntary soil conservation associations be organized in connection with all new Emergency Conservation Work projects—that is, Civilian Conservation Corps (CCC) projects—until "legally constituted" associations could be formed. The deadline would be July 1, 1937, after which "all erosion control work on private lands, including demonstration projects, [should] be undertaken by the Soil Conservation Service only through legally constituted soil conservation associations or Governmental agencies empowered to function as indicated above."[21]

The Secretary's Committee also avoided making clear-cut recommendations for relations between SCS and extension work in the Tennessee Valley, except to forbid SCS operations there without the consent of the Valley Coordinating Committee. The Tennessee Valley Authority (TVA) already had agreements with the state extension services in its area to supervise its soil conservation program—chiefly the use of new fertilizers. The committee contented itself with recommending merely that SCS and the extension services reach "cooperative agreements" in states where the extension service already had a "basis of cooperation" with the state relief agency and the FERA. In short, the Committee made no settlement, for it could not, of the question whether the Soil Conservation Service was to become the single organization through which all the federal erosion control programs would be unified.

[21] *Ibid.*, p. 42.

These organizational issues were of critical importance to Bennett for two major reasons. The relations recommended by this Committee would have required SCS to modify its recommendations for erosion control practices to suit the convictions of specialists in the state agricultural colleges. Yet, here was the crux of Bennett's disagreement with others, and it is no exaggeration to say that this is still a matter which arouses sharp differences in some places. Second, SCS would not be allowed to work with individual farmers unless local sponsors were organized under state law with the good will and active help of the extension workers. The only important exceptions to this rule were that SCS could conduct operations on public lands by direct contact with ranchers and might carry out certain responsibilities related to flood control without using associations or local governments.

The organization and procedures recommended by the Committee were intended to fit SCS into an administrative system which was almost exactly the one described by Fletcher of the Caterpillar Tractor Company. Since he was relaying ideas to Wilson from employees who were in touch with extension leaders, it is reasonable to conclude that the Secretary's Committee was responsive to dominant extension attitudes. L. N. Duncan, Director of Extension in Alabama, said in 1935 that soil conservation would progress "best" if left to "state and local" organizations. He reported that the county agents in fifty Alabama counties had organized soil conservation associations and were terracing land with assistance from CCC camps.[22]

NATIONAL LAND USE PLANNING

Agricultural land use planning was viewed in 1935 as part of a comprehensive program for extending national well-being by properly utilizing the country's natural and human resources. Agricultural land policy had gone through several phases with traces of each lingering in the early thirties when crisis had struck across the nation. During most of previous U.S. history it was assumed that all lands should be open for settlement and agricultural production. Even in areas where the lack of normal rainfall inhibited or prevented production without irrigation, federal funds were invested to make land arable. Some portions of the public domain were abused, wasted, or managed to give special privileges to operators interested in exploitation for short-run gain. This situation, together with destruction of much of the supply of

[22] NARG 114 (Soil Conservation Service, General Correspondence, 1935–36), L. N. Duncan to W. C. Lowdermilk, Assistant Chief of SCS, Oct. 18, 1935.

wildlife, led to demands for continued public ownership of some lands and the acquisition of others under the management of trained specialists.

In the 1920's, however, some careful observers had argued that the trend of agricultural land use must be toward intensive utilization of the soils best suited for production by their physical characteristics, topography, moisture conditions, and related factors. Marginal lands would have to be removed from production. If reliance were not to be placed exclusively on the market place to effect adjustments, it was assumed that these lands would have to be placed in public ownership, their occupants resettled, and the land retained under public management. In some instances, according to these views, some lands could be restored and leased for private use with restrictions administered by governmental agencies.

As the economic plight of farmers generally grew worse during the 1920's, it became evident that public acquisition and ownership of marginal agricultural lands would not solve all land use problems. The conviction grew that knowledge of good land use practices was the key. Emphasis was placed on scientific research both in the Department of Agriculture and the land-grant colleges. Scientists and extension educators believed that landowners who were aware of the wastefulness of some practices and the profitability of others would conserve their land because it was in their long-term interest to do so. It was believed that the enlightened pursuit of self-interest would simultaneously serve a public interest. By 1933, however, this idea was giving way in some quarters. It was becoming clear that a land use policy resting upon individual private choice, rather than public decision-making as the culmination of a process of careful planning, was inadequate.

Two objectives of Hugh Bennett's personal crusade contributed to the complexity of these ideas. First, he believed that "accelerated" erosion was occurring on farm lands at an alarming rate, but neither farmers nor many agricultural specialists were awake to the dangers. Consequently, his first task was to educate both groups by convincing them that they had a problem and showing them how to solve it. His second objective was to get specialists and farmers to act to control erosion; to him action meant providing technical assistance to all farmers eventually, and not just to those who would permit specialists to demonstrate practices on their farms. Both of Bennett's objectives clashed with the older tradition of research and extension education. His methods were equally challenging, and not solely because they were novel.

Bennett's aims coincided generally, although not necessarily with respect to all specifics, with those of some of President Roosevelt's principal advisers. They were persuaded of the truth of yet another idea affecting land management: to be effective a program would have to be the product of national, regional, and local plans coordinated, although not necessarily executed, through a single responsible agency. Many agencies had engaged in research and even action contributing to an understanding of, and solutions to, land use policy. But no single agency had ever been responsible for integrating and coordinating such policy. It is for this reason that the point was missed by so many specialists who later engaged in heated arguments over who was first to recognize the true dimensions of the problem or had the special competence necessary to deal with it. The first step toward coordinated planning was taken when the National Planning Board was established in the Public Works Administration in 1933. It was followed in 1934 by the National Resources Board, which in 1935 was followed by the National Resources Committee (NRC). Within this context, the Soil Erosion Service was created and transferred to the Department of Agriculture, which was made responsible by Public Law 46 for coordinating the agricultural phases of national land use policy.

When Congress provided the legislative authorization for a comprehensive national attack on the agricultural phases of land use, it also embraced another set of ideas then growing in popularity: that water and land use are ultimately inseparable. For example, the National Resources Board recommended in 1935 that plans be made for multi-purpose water developments to remove the flood menace and reduce soil lost by uncontrolled erosion. Yet, contradictory ideas also had much influence. One, for example, was that every owner of land in fee simple is free to use his land without acting to conserve it as a long-term capital asset of both private and public value. Another hardy perennial was the Malthusian assumption that population growth will eventually outstrip the capacity of available land to produce sufficient amounts of food to prevent mass starvation. Where lands were known to have been productive in the past there was a tendency to believe that conservation should consist of measures to restore the land to productivity, regardless of any consideration but the desire of people in the area to revive trade. Such people assumed that the land and labor which had been used on the land ought to be retained in agriculture.[23]

All of these ideas converged to some degree in the middle thirties.

[23] Gaus and Wolcott, *op. cit.*, pp. 115–59, *passim.* See also Gilbert F. White, "A Perspective of River Basin Development," *Law and Contemporary Problems,* Vol. 22 (Spring 1957), pp. 157–84.

For a brief time, the Roosevelt Administration tried to harmonize and redirect these conflicting conceptions of proper means and ends. It even added an ideal which was undoubtedly the central goal of the optimistic and pragmatic liberalism of the age: "The Land Report presents a complete reversal of the attitude of heedless and unplanned land exploitation. It reflects the point of view that public policy should aim at effecting such ownership and use of land as will best subserve general welfare rather than merely private advantage."[24]

Secretary of Agriculture Wallace summed up much of the nature of the change in means and ends then contemplated when he said that "research and action must fit together and come to dynamic focus on the farm and watershed."[25] Wallace, in effect, was given an agency in 1935 which he had not created and was directed to develop national, state, and local governmental relations which harmonized the traditional with the novel. It was in this atmosphere that a small group of men in the federal government turned their talents to the task of creating new instrumentalities to serve the national purpose.[26]

SOIL CONSERVATION DISTRICTS CONCEIVED

Two weeks after the executive order was issued transferring the former Soil Erosion Service to the Department of Agriculture, Bennett asked Secretary Wallace for lawyers to study existing state legislation authorizing the organization of "conservancy districts." He explained that a "highly important" future phase of his program would "probably" involve cooperation with such units of local government. Since many state legislatures were then in session and would not meet again for two years and, if the kind of cooperation he visualized were to be carried out "under any funds which may be granted for soil erosion work under the new Relief Act, it is essential that suitable State legislation be passed this year." He was convinced that such legislation "will probably have to be initiated by suggestions from the Department."[27] Under Secretary Tugwell, acting while Wallace was out of

[24] *Report of the National Resources Board on National Planning and Public Works in Relation to Natural Resources and Including Land Use and Water Resources with Findings and Recommendations* (Washington: Government Printing Office, December 1, 1934). See also supplements to the NRB *Report*, especially Supplementary Report No. 5, *Soil Erosion: A Critical Problem in American Agriculture,* report of the Land Planning Committee (1935), p. 8.

[25] Quoted in Gaus and Wolcott, *op. cit.,* p. 59.

[26] *Report of the National Resources Board, op. cit.,* p. 2.

[27] NARG 16 (Erosion File), memorandum to the Secretary of Agriculture from H. H. Bennett, April 10, 1935.

town, replied that he was forwarding the memo to the Solicitor and also giving a copy to Assistant Secretary Wilson so he could lay the matter before the Land Policy Committee. He urged Bennett to get in touch with Wilson "at the earliest opportunity."[28] Wallace was persuaded to act on this suggestion.

Bennett's suggestion for legal research into suitable state legislation was consistent with the long-range program (which he and his associates in the Soil Erosion Service had prepared in December 1934) published in April 1935 as one of several supplementary reports of the National Resources Board. He proposed: (1) to expand the number of erosion demonstration projects to "educate" farmers in all "representative" regions; (2) to complete treatment on all federally owned or controlled lands and tributary watersheds draining into them; (3) to execute erosion control plans on large watersheds in the agricultural regions of the United States in cooperation with "conservancy districts or other subdivisions of States, and . . . erosion-control associations or similar organizations." In addition, there were to be an adequate research program, a "cooperative" public education program, and, if the need for "work relief" continued for some years, soil conservation offered a "rich field of endeavor." It was explicitly anticipated that as "time goes on" and the demonstration program reached its "full expansion," the "tendency" would be for any additional work to be carried out through conservancy districts, local governments, and erosion-control associations.[29]

Bennett lacked an exact idea of the best organization to carry out the program, but he had discussed it with members of the Land Planning Committee of the National Resources Board and personnel of the Interior Department, when he was there. The matter was under discussion particularly because the Soil Erosion Service had demonstration projects that included both private and public lands, with the latter administered mainly by federal agencies. Ickes required that public works funds be allotted to units of state and local government for public improvements, and it was also thought that individuals receiving the benefits ought to be organized collectively to assume various responsibilities locally. For example, someone needed to enforce the terms of the five-year agreements which SES made with individual farmers and ranchers by which they committed themselves

[28] NARG 16 (Erosion File), memorandum to H. H. Bennett from R. G. Tugwell, "Acting Secretary," April 17, 1935; and Tugwell's memorandum to Seth Thomas, Solicitor, April 17, 1935.
[29] Soil Erosion, supplement to Report of National Resources Board, op. cit., p. 54.

to follow up their conservation plans and to maintain structures built by federal technicians. The Bureau of Reclamation had found that it had to assume some undesirable burdens to enforce contracts with individual farmers. It was for this reason that irrigation districts were authorized and organized under state enabling legislation. Special districts might also provide some cooperative local financing of projects. The idea that a conservancy district ought to be organized on a natural watershed without regard for the boundaries of existing governmental subdivisions was widely advocated among the people with whom Bennett was in touch.[30]

Assistant Secretary M. L. Wilson, who was the Department's representative on the Land Planning Committee of the National Resources Board and chairman of a departmental Land Policy Committee, devoted considerable attention to this matter, especially since he and Bennett had discussed the problem on various occasions. A month before the Secretary's Committee completed its recommendations for the SCS, Wilson discussed with Representative Marvin Jones, chairman of the House Agriculture Committee, the prospects for allotting AAA funds to some proposed wind erosion districts in Texas. Jones was particularly interested, since some of these districts would be organized in his congressional district if the Texas legislature passed enabling legislation. In relating these events many years later, Wilson said that his talk with Jones gave him the idea of soil conservation districts.[31]

In August, a draft of a model state law for creating soil conservation districts was ready for discussion. When it was submitted to the National Resources Committee in October 1935, M. L. Wilson told the NRC chairman, Charles W. Eliot, 2nd, this proposed legislation rested on three major assumptions.[32] First, genuine erosion control would not be achieved unless conservation practices such as terraces and check dams were used as parts of a comprehensive plan of farm management. That is, no program would be effective if practices were applied piecemeal according to each individual farmer's preferences. In this respect, the Department's new program was fundamentally different

[30] *Department of Agriculture Appropriation Bill for 1937*, Senate, 74 Cong. 2 sess. (1936), p. 401.

[31] Interview with M. L. Wilson, Oct. 27, 1960. See also NARG 16 (M. L. Wilson File 1934–40), for note from Marvin Jones, of May 2, 1935, enclosing a draft of the Texas wind erosion districts bill, S.B. No. 115.

[32] NARG 16 (Erosion File), M. L. Wilson to Charles W. Eliot, 2nd, Oct. 2, 1935; and "Memorandum Summarizing Attached Copy of Draft of Proposed Standard Soil Conservation Districts Law for Adoption by State Legislatures," 13 pages, dated Aug. 23, 1935, and stamped in the Secretary's File Room, Jan. 28, 1936.

from the erosion control activities recommended by the extension services. Second, Wilson said, even if a "complete erosion control program" were installed by scattered individual farmers, the effort would be ineffective unless other farmers practiced control measures. "This fact is one of the more important reasons why the present demonstration project program of the Department . . . will be inefficient by itself." Wilson expected a great many farmers to adopt practices which they saw demonstrated, but said "a great many others will not." He also assumed that farmers would practice proper land use only if they were first "educated to cooperate voluntarily." Since farmers would have to change many familiar ways of doing things on their farms, he was certain that "an attempt to impose regulation from above may make the statute as unenforceable as was the prohibition legislation." Success could be expected only if democratic procedures were used to institute the program in "regions where the farmers have been educated to the necessity and the wisdom of the steps called for."

Wilson explained further that the draft of the law for model soil conservation districts contained three "administrative procedures which are intended to supplement each other." Soil conservation districts would be created only upon the affirmative vote of "land occupiers" in an area, provided that a majority participating in the referendum on the issue approved. The district governing board would consist of some elected members and others appointed; districts would have power to conduct research, disseminate information, hire personnel, carry out erosion control demonstrations and control operations. The second key feature of this scheme was to be a comprehensive erosion control plan prepared in each district to serve as the basis upon which the governing board would "formulate tentative regulations governing land use" followed by public hearings and a referendum on proposals to prohibit some land practices and to require others. The third procedure incorporated in this draft, but not later explicitly included in the final version, was for districts to enter into contracts with landowners, which would stipulate forbidden and required practices and bind future owners and occupiers of the land affected. Districts were expected to obtain funds from state governments or other sources, since it was not recommended that they have the power to tax, borrow, assess and collect benefits, or issue obligations because these would "make the entire legislation unpopular." According to Wilson, for the "next two years, at least, it is likely that the operations of these . . . Districts must be financed chiefly with Federal funds."

Wilson cautioned Eliot to treat the draft as confidential because the "proposal to issue a standard statute with the recommendation of the Department that the several states adopt either that . . . or one very similar thereto, has not yet been finally approved by the Secretary of Agriculture."[33] In December 1935, however, Dillon S. Myer of SCS informed state advisory soil conservation committees that, after they had consulted with appropriate state agencies, the Department would be willing to assist them in drafting legislation. Myer emphasized that the districts must be "governmental sub-divisions of the state" similar to counties, cities, towns, and incorporated places, but having broader powers over soil conservation than existing special districts. They were not to be soil conservation associations, however.[34]

By the time the final version of the standard act had been virtually completed, it was delayed further, when the Supreme Court invalidated the first Agricultural Adjustment Act on January 6, 1936.[35] For the next several weeks, attention was focused on legislation to replace the AAA. The decision was made to make future payments to farmers for reducing "soil depleting" crops (most of them in surplus), calling the program soil conservation. The question then remained: would the soil conservation districts contemplated by the Department be used in any way to administer the new "conservation payments"? How, if at all, were the two new federal soil conservation programs to be coordinated?[36] The new bill, called the Soil Conservation and Domestic Allotment Act of 1936, was enacted as an amendment to Public Law 46 of 1935 which created the Soil Conservation Service. Myer of SCS and Howard R. Tolley of AAA discussed the problems, and Tolley concluded that with slight modifications of the standard law the districts could be used to administer AAA conservation payments. There was legal opinion, however, that the districts could not be required to do so. The issue seemed especially troublesome because the new law provided that after July 1, 1938, the Secretary might turn over the administration of the conservation payments program to states enacting suitable legislation. Myer therefore asked Wallace whether legislation substantially based on the standard soil conservation districts law

[33] NARG 16, Wilson-Eliot letter, Oct. 2, 1935, *op. cit.*

[34] NARG 16 (Miscellaneous; Regional Meetings File), memorandum, D. S. Myer to All Regional and State Conservators and State Advisory Committees, Dec. 19, 1935, with memorandum attached. See also NARG 16 (Soil Conservation Districts and New Agricultural Act File), memorandum, H. H. Bennett to Mastin White, Solicitor, Feb. 24, 1936.

[35] *United States* v. *Butler*, 297 U.S. 1.

[36] NARG 16 (Soil Conservation Districts and New Agricultural Act File), memorandum, Mastin White, Solicitor, to the Secretary of Agriculture, Feb. 25, 1936.

would satisfy this requirement. If it did, would separate agencies be established to administer it on the local level—or one?[37]

The first of these two questions was not answered unequivocally in 1936 and, although it agitated the leaders of the land-grant colleges for the next few years, it eventually was forgotten. The answer to the second question was that there were to be two separate programs and organizations. Technical assistance would be given by SCS and subsidies by AAA. There were several obvious reasons for this choice. The Roosevelt Administration's immediate need was to distribute checks to farmers over the entire country and there was not enough time available to wait until state legislatures could enact enabling legislation. Most of them would not meet in regular session during 1936 and no one knew how they would act in an election year. The AAA already had its own organization, procedures and experience for administering the crop control programs and, since the new payments system was intended to control production, and only incidentally to inspire improved land use, it made sense to let AAA handle the new measure. At this time, SCS was carrying on 141 demonstration projects and directing 450 CCC camps.

The Department of Agriculture would have had to undertake a major, and probably painful, struggle to fuse SCS and AAA at that time. Such a step probably would have stung the leadership in agricultural colleges into furies of opposition, since it would have cut them off from a share of the AAA conservation program. The AAA depended largely on them for its recommendations on practices, such as liming, fertilizer, rotations, and grasses. The county administrator for the AAA committees was the county agent, and in many states the director of extension was chairman of his state AAA conservation committee. Of course, there were no soil conservation districts then, and no one knew whether there ever would be any. Wallace's decision was sound under the circumstances, but eventually it led to trouble for the Department's later secretaries.

The final version of the standard districts law, published May 13, 1936, retained many of the main features of the 1935 draft submitted to the National Resources Committee, but there were also some important differences. Comparison of these two documents shows that Secretary Wallace and his advisers were concerned about relations between districts and the state extension services. The final version of the

[37] NARG 16 (Miscellaneous; Regional Meetings File), memorandum, D. S. Myer to Paul H. Appleby, March 4, 1936.

standard act contained two key changes touching upon this prickly issue.

The standard act recommended that a state soil conservation committee be created to consist of the directors of extension and the experiment station, the state commissioner of agriculture, an a representative of the state planning board or conservation department. The fifth member was to be appointed by the Secretary of Agriculture upon invitation from the state committee; and it was assumed that this member would be the chief officer of the Soil Conservation Service in each state. This differed from the draft version which had called for a single commissioner to be appointed by the governor. The commissioner would have no regulatory powers, according to M. L. Wilson, so that farmers would view districts programs as their own and not as programs dictated from the state capital. It is evident that this provision in the draft would have created an office through which districts might be organized in areas where the extension service was not actively advocating them. And, it was quite as obvious that the system of selection would tie the question of organizing districts to partisan politics in each state.

The final version of the model act provided that the primary function of the state soil conservation committee would be to encourage farmers to organize districts and to conduct the proceedings required to establish them. The state committee was to help the districts carry out their operations and "coordinate" their programs, but only insofar as it was possible through the "interchange of advice and experience" and by "advice and consultation."

The final version changed the draft by altering the method of selecting the five members of the district governing body. In the recommended act, each board was to consist of two supervisors fitted by "training and experience" to be appointed by the state committee and three elected members. The elected members could be nominated by petitions requiring the signatures of only twenty-five "land occupiers," and would be elected by "land occupiers" in the district, all of whom were to be eligible to vote in their district. In the October draft, the district governing board was also to consist of five members but, of the five, two were to be popularly elected and three were to be appointed by the governor, if a district included lands in more than one county, or by the presiding officer of the county governing board, if the district were wholly within a single county. All of the appointed members were to be chosen from a list supplied by the dean of the state college of agriculture.

From the available evidence, it appears that Secretary Wallace feared that the method of selecting district boards suggested in the draft version might result in majorities consisting of warm friends of the extension services. In some situations, this arrangement could restrict the influence of the Department over the members. The decision to recommend a state committee including the directors of the state experiment stations and extension was probably more or less necessary, since Wallace had already established an advisory conservation committee in each state including these two officers and the SCS state coordinator.

The major powers to be granted to districts under state law if the standard act were adopted indicate the activities which the Department expected them to undertake. Aside from the power to conduct research, demonstrations, control operations, and to enact and enforce land use regulations and enter into contracts, the districts would have the authority to enter into agreements with land occupiers to carry out operations subject to such conditions as the governing board might make. They were to have the power to purchase or receive lands as gifts, manage them by receiving and spending income or acting as the agent of the federal government in doing so. They could also lease or sell such lands.[38] Many of these features were included not only to carry out recognized control operations, but also to place under public ownership and administration lands which might be withdrawn from crop production. In this fashion, submarginal lands could be withdrawn from private ownership, restored to productive capacity, and leased under public management.

REASONS FOR RECOMMENDING DISTRICTS

The decision to recommend that states permit interested parties to organize soil conservation districts was a signal that the Department of Agriculture was prepared to change its traditional relations with the extension services. The county had served as the basic unit in extension work, whether local sponsors were merely quasi-public entities or county governing bodies. The standard districts law, however, was not drawn to discard or replace extension entirely, but to enable various other agencies to contribute to the new soil conservation program.

[38] U.S. Department of Agriculture, *A Standard Soil Conservation Districts Law* (Washington: Government Printing Office, 1936).

Districts were to provide a mechanism through which both SCS and the extension services could directly contact and assist farmers.

Many features of the model law were determined by legal considerations necessary in drafting statutes carefully to withstand possible attack in the courts. Most of these are too numerous and detailed to be considered here because they are not germane to the central problem. Before they decided to recommend districts organized under state law, however, Wilson, Philip M. Glick, Chief of the Land Policy Division of the Office of the Solicitor, and others considered the legality of organizing seventy-six major drainage basins as federal soil conservation districts by authority of Public Law 46 of 1935. Glick later said they discarded this idea because they were afraid that Congress did not have the power to authorize such districts to make and enforce land use regulations. They considered the spending and commerce powers, but they decided that the Supreme Court would not construe either of these broadly enough at that time.[39] In any event, it is certain—whether arising out of distaste for coercive controls or faulty memories—that the Department's central objective at that time was to administer its erosion control program through a unit of government possessing the power to make and enforce land use regulations. Wallace and his associates apparently became convinced also that controls could not be imposed by a federal agency, so that "democratic," and not "bureaucratic," procedures would be essential to success.

Wilson explained to extension leaders why the districts were necessary. The demonstration projects then administered by SCS, he said, were "impractical" means for extending proper treatment to all the land. The Department might enter into contracts with individual farmers, but this choice was not workable because the federal government "does not now have power . . . to regulate land use . . . so as to bring within the program the lands of a recalcitrant minority." State governments, he conceded, had the power to regulate land use, but no program could succeed, he argued, if it were "bureaucratic in form or substance." Counties were inadequate because they lacked "the necessary traditions of scientific administration to carry on the work expertly" and their boundaries were drawn to have "political significance, but often very little if any significance in land use management." Existing special districts had neither the legal powers nor the experience to serve adequately, he pointed out, and cooperative

[39] Philip M. Glick, "The Soil and the Law," *Soils and Men—The Yearbook of Agriculture, 1938* (Washington: Government Printing Office), pp. 296–318 (esp. pp. 299–300); see also, pp. 241–45 for a discussion of rural zoning.

terracing associations were not governmental agencies. They could neither receive state appropriations nor enact land use regulations. The Department's position, therefore, was that the most suitable instrumentality would be the soil conservation district, organized under state law "over a naturally bounded area like a watershed," with a majority of its governing body elected by local farmers and possessed of the power to receive state appropriations and to enact land use regulations.[40]

Spokesmen for the Department insisted that it would be necessary to educate farmers to an awareness of erosion and to show them how to cope with it while using democratic procedures. The provisions for referendums to establish districts and to enact land use regulations were necessary, they emphasized, not only to meet requirements of state law, but also to gain acceptance of the program through education. Secretary Wallace said that the "informed and voluntary participation of the farmers is the chief goal at which the Standard Law has aimed." The referendums, he felt, would "prevent the growth of a feeling that farmers are being regimented." For this reason to "exclude renters and tenants and lessees from participation is to deprive them of the educational opportunities which come from such participation."[41]

Since counties could have been authorized by state legislatures to enact land use regulations, why didn't the Department recommend this? Counties are usually eulogized as units of "grass roots democracy," certainly as "close to the people" as special districts organized over the same areas. Wilson and several of his key associates, including Bennett, were convinced that most of the extension administrators and specialists were not sufficiently concerned with erosion to make its control a major program in keeping with the scope of the problem.[42]

[40] *50th Proceedings of the Association of Land-Grant Colleges and Universities, 1936* (Washington), pp. 187–91.

[41] NARG 114 (SCS, General Correspondence, 1937), Henry A. Wallace to R. W. Brown, President, Missouri Farm Bureau, May 3, 1937.

[42] Interview with M. L. Wilson, Dec. 15, 1960. Wilson said that at the time of these events he was much influenced by John Dewey's "instrumentalism." See David G. Smith, "Pragmatism and the Group Theory of Politics," *American Political Science Review*, Vol. 58 (September 1964), pp. 600–10. Charles E. Merriam, who also influenced Wilson and others associated with him on the National Resources Board and its committees, was a vigorous critic of state government, but inclined to think that national and city governments could be invigorated. In 1928, Merriam concluded: "Most states do not now correspond to economic or social unities, and their validity as units of organization and representation may be and has been seriously challenged. The nation and the cities are vigorous organs, but the state is not." Quoted in W. Brooke Graves, "The Future of the American

Paul H. Appleby, Assistant to the Secretary, especially, and others in the Department were convinced that the rural county was the most poorly governed unit in the United States and that it needed either to be rejuvenated or by-passed in resources administration. Appleby, Walter Lowdermilk, Bryce Browning, and many others around Wallace were convinced that the watershed was the only natural unit of organization for resource programs.[43] The Department did not want to work through local soil conservation associations, partly because they were inadequate legally and partly because it could not control them.[44] Any one of these reasons alone might have been sufficient, but there were others.

The Department was afraid that it could not hold extension accountable for federal funds to insure program goals. M. L. Wilson and others were concerned that if the extension services administered the program through county governments it would become enmeshed in courthouse politics, subject to favoritism, unvarnished partisanship, and other abuses of an obvious sort. The likelihood was that farmers who supported the county agent, the "right" county commissioners, or officers of the local farm bureaus would be the first to enjoy the benefits

---

States," *American Political Science Review,* Vol. 30 (February 1936), pp. 24–50. Graves cited several other critics who spoke in the same tenor. The Department's proposal to work directly with soil conservation districts and not state agencies reflected this general pattern of criticism. See also Karl Barry, *Executive Reorganization and Reform in the New Deal* (Cambridge: Harvard University Press, 1963), pp. 37–79.

[43] See Arthur MacMahon, John Millett, and Gladys Ogden, *The Administration of Federal Work Relief* (Chicago: Public Administration Service, 1941), p. 201. They point out that soil conservation districts were not the only governmental units urged by the Roosevelt Administration to be organized without primary regard to county boundaries because of their effects on the distribution of political power. When the Works Progress Administration organized its field services in 1935, it created state and "district" offices—as its predecessor, the Federal Emergency Relief Administration, had previously done. Usually these districts embraced several counties. The explanation offered was that officials in Washington appreciated the "political influences always present when local areas of operation must be established. The headquarters intention that there be only a limited number of districts within a state might well have been compromised by local pressures playing upon state administrators."

[44] In 1936, there were about 430 soil conservation associations, with membership varying from 30 to 250. T. L. Gaston, assistant to Dillon S. Myer in SCS, said that SCS was having difficulties with associations "doing terracing and guiding their operations in a satisfactory manner, and my guess is we may have more difficulties in the future. Some disagree in this matter." See NARG 114 (SCS General Correspondence, 1936), memorandum, Gaston to George Barnes, Director of Information for SCS, June 20, 1936.

of CCC labor, materials supplied with federal funds, and technicians whose pay came as much from Washington as elsewhere.[45]

Moreover, the full dimensions of a major political challenge to the Department were just beginning to emerge. On September 21, 1935, Edward A. O'Neal, President of the American Farm Bureau Federation, sent a telegram to the chairman of the Committee on Extension Organization and Policy of the Association of Land-Grant Colleges and Universities, then meeting in Washington. O'Neal called the attention of the assembled directors to the "vital necessity of closer cooperation between the Extension Service and the Farm Bureau in getting farmers fully organized as soon as possible for the mutual welfare of both." The Board of Directors of the Farm Bureau "respectfully request more aggressive cooperation of the Land Grant Colleges through their extension divisions in a definite campaign to more thoroughly organize the men and women of the farm." Three days later, the chairman of the Committee on Extension Organization and Policy replied that the Committee recognized the need for more thorough organization of farmers and is "directing its support to this end." He elaborated the Committee's position by saying that although it was "not the function of county extension agents to conduct membership campaigns of farmers' organizations, they rightly can be expected to make available to such organizations such information as will be helpful to them and contribute to the success of their work." Within these limits, the Committee "encourages county extension agents to give their fullest cooperation in assisting farm organization leaders to build and maintain strong membership."[46]

Shocked by the trend of events, the American Farm Bureau Federation and the Committee on Extension Organization and Policy made common cause against the New Deal's Department of Agriculture. From this time forward, the Farm Bureau made the extension services of the land-grant colleges its official "chosen instrument" for administering some agricultural programs. Frightened by drastic losses of

[45] NARG 114 (SCS, General Correspondence, 1936), D. S. Myer to State Coordinator, SCS, Mississippi, Feb. 25, 1936. It had been proposed that in Mississippi the county boards be allowed to purchase equipment and appoint an advisory soil conservation committee to guide its use. See also NARG 114 (SCS, General Correspondence, 1937), D. S. Myer to K. G. Harmon, State Coordinator, SCS, Missouri, March 31, 1937, in which Myer explained that if county agents were made district supervisors they would be in an "anomalous position . . . and the result in some instances is more or less obvious."

[46] Committee on Extension Organization and Policy, *Proceedings of the Association of Land-Grant Colleges and Universities,* mimeographed (Washington, Sept. 24, 1935).

membership, the American Farm Bureau Federation turned to the colleges and demanded that they use their county agents to organize a Farm Bureau in every state to the point where the Federation would have no serious rival as the voice of American agriculture. The college extension directors, frightened by the rise of the AAA, the Resettlement Administration, and the Soil Conservation Service as rivals with superior fiscal resources and popular programs, committed the extension apparatus to the task of increasing the membership of the American Farm Bureau Federation. The alliance, which had existed legally or in fact in more than half the states, by this exchange of telegrams, was made a national alliance. In the years which followed, this alliance waged a determined, although often muted, battle to "decentralize" the Department's soil conservation program.

This move on the part of the college and Farm Bureau leadership signaled a lack of confidence in the Department at the very time when Wallace and his associates were beginning to understand the difficulties of administering "action" programs through extension. The dominance of a single party in rural county governments, especially in the Republican Northeast and Midwest, made extension unreliable. If the Farm Bureau and extension alliance could control the conservation program with the understanding that the county agents were to encourage farmers to join the Farm Bureau, the Department would soon be faced with a single axis of power in agriculture capable of exercising a paralyzing influence over national policy. Gaus and Wolcott observed in 1939 that the overriding task of the Department was to redirect the drives of agricultural pressure groups so as to serve not only their limited interests but also to contribute to improved rural life and national economic welfare.[47]

SUMMARY AND COMMENT

When the Soil Conservation Service became part of the Department of Agriculture, the Secretary appointed a committee of bureau chiefs to advise him on fitting SCS into the Department. The Secretary's Committee on Soil Conservation recommended that SCS execute its erosion control operations on private lands by sharing its funds and responsibility for program planning and technical assistance with the

[47] Gaus and Wolcott, op. cit., p. 379.

state extension services. These arrangements would be in accord with the traditional patterns established by the "1914 Agreement." Extension would provide all conservation "education" and technical assistance through assistant county agents. Assistance would be given only in areas where farmers had organized into soil conservation associations or established an appropriate unit of government.

The Secretary decided to exercise his new authority under Public Law 46 of 1935, through units of government organized under state law. The Department published a model state law which would authorize farmers to organize soil conservation districts. They were to have power to carry out a wide range of erosion control operations and to enforce proper land use practices based upon plans made for each entire district and each farm within it. Thus, national, state, and local governments were to be organized so that each would contribute a share to a national plan of erosion control on agricultural lands.

The Secretary chose to administer erosion control through a new unit of local government, instead of exclusively through the state extension services, for several reasons, but three stand out. First, under the "1914 Agreement" state extension services were not effectively responsive to the Secretary for program planning and execution. Second, the President and Congress had directed the Secretary of Agriculture to plan and execute a new program which would consist not merely of demonstrations of techniques which farmers might adopt or ignore as they chose. Rather, it should be based on national, district, and individual farm conservation plans to be enforced, ultimately, by the police power of the states. And, third, the state extension services had valuable contributions of a technical character to make to the program, but the Secretary could not always rely on them to carry out the Department's objectives, regardless of the political party in control of national administration. He feared that if the traditional pattern of relations with extension were followed, the power of the extension services and the American Farm Bureau Federation as interest groups would be vastly increased relative to his own. The evidence points to the further conclusion that Secretary Wallace's advocacy of new organs of local government was intended to strengthen national policy-planning and execution to the advantage of the Democrats and at the expense of the Republicans in Northern and Western States. Key has speculated that in these areas the Democrats have been unable normally to link their party organizations with state governments as effectively as the Republicans. "Hence, in considerable measure, the ends of the Democratic party become ends to be achieved through national action or not at all."[48]

At the end of 1936, the most important question about districts was whether state legislatures would understand and accept the idea that a new unit of local government was needed as the mechanism through which a cooperative federal-state program of land use adjustment could be focused.

[48] V. O. Key, *American States Politics: An Introduction* (New York: Alfred A. Knopf, 1956), p. 81; see pp. 74–84 *passim*.

CHAPTER 3

States Adopt the Standard Act

WHEN THE U.S. Department of Agriculture published the standard soil conservation districts law in May 1936, Secretary Henry A. Wallace apparently was ready to promote it. At the same time, the 1936 presidential campaign was just starting, and no one knew how much this model act might stir charges that the New Dealers were about to regiment farmers. Copies of the standard law were distributed, therefore, with some caution.[1] Wallace knew perfectly well that many of the leading administrators in the colleges of agriculture were actively opposing the "centralized" organization of the Soil Conservation Service (SCS), but he did not know whether they intended to oppose the soil conservation districts with equal vigor. He soon found that they did. In time, their opposition forced important modifications in the state laws authorizing districts, slowed the rate at which the laws were passed and districts created, and influenced several important modifications in district operations and responsibilities. The scope and intensity of this opposition varied considerably among the states, but it was never strong enough to prevent the steady growth of both SCS and conservation districts.

POPULARITY OF THE SCS PROGRAMS

In 1935, the Soil Conservation Service gained popularity with many farmers and congressmen because it was striving to curb erosion in the midst of overwhelming dust storms. In order to carry out its responsibilities SCS needed men, materials, and equipment. These were supplied in considerable measure by camps of the Civilian Conservation Corps (CCC), which were placed under the technical supervision of SCS. During the summer of 1935, SCS directed the work of approximately 500 CCC camps and thereby was able to be of service to

[1] National Archives, Record Group 114 (Soil Conservation Service, General Correspondence, 1936), T. L. Gaston to John M. Gaus, July 14, 1936. (National Archives records cited hereinafter usually will be identified by the abbreviation NARG.)

members of Congress who were likely to remember SCS when annual appropriations were voted.

On June 27, a normal day for that month according to the files, the Assistant Secretary of Agriculture, Milburn L. Wilson, replied to requests from eight members of Congress for CCC camps in their districts or states. All eight had compelling reasons why particular communities needed a CCC camp, as did many other representatives and senators. A file of correspondence from Representative Richard Kleberg of Texas—prepared by his secretary, Lyndon B. Johnson—is typical. One of Kleberg's supporters, Sam Fore, Jr., newspaper editor and "Democratic chairman for the fourteenth congressional district," complained that SCS had decided not to locate a CCC camp near Floresville, Texas, even after a conference in which Senator Tom Connally had participated. Fore, in his letter to Kleberg, said that the town had been "promised Free Water." He added that "our county has received no PWA loans. *This is our one project and we are sorely and sadly disappointed. . . .* We can't understand . . . Dick, after all you have done for the U.S. Department of Agriculture. . . . It does seem *very strange that a little CCC erosion camp can't be awarded— especially where it is so badly needed.*" Kleberg forwarded this appeal and made his own expectations clear to Paul H. Appleby, Assistant to the Secretary of Agriculture. "Someway, somehow, I am going to have to get another camp for Wilson county. I want you to have the Secretary arrange a conference for me with your Mr. LeCron and Director Bennett for some morning the early part of next week."[2]

This sort of demand created some stress for SCS, since all requests of this sort obviously could not be met to everyone's complete satisfaction, although apparently most were. Hugh H. Bennett and his assistants in SCS quickly discovered also that they could get into trouble if they chose a project without knowledge of its local political effects. In Nevada, for example, the State Soil Conservation Advisory Committee, established as recommended by the Secretary's Committee on Soil Conservation, approved a demonstration of range erosion control. Senator Key Pittman of Nevada complained about the project to Wallace who asked Bennett for an explanation. He found that the project was "proposed and sponsored by Director Creel and Mr. Witner, both of whom are reported as being political opponents of the Senator. . . . We knew nothing of any political antagonisms until the present discussion about the project developed."[3] C. W. Creel was the

---

[2] NARG 16 (Erosion File), Richard Kleberg to Paul H. Appleby, June 14, 1935.
[3] NARG 16 (Erosion File), Hugh H. Bennett to Paul H. Appleby, Dec. 11, 1935.

Director of Extension in Nevada and later a leader in the movement by the Association of Land-Grant Colleges and Universities to block districts and secure for the extension services a major share in administering soil conservation.

The operation and location of CCC camps, work projects, and SCS watershed demonstrations were bound to engage the attention of the members of Congress, since SCS was distributing sizable benefits to farmers. Senator Royal S. Copeland, Democrat of New York, asked Bennett how farmers liked the SCS program. Bennett replied that it was "very popular throughout the country." Copeland then wanted to know whether this attitude was due to the fact that the farmer "is getting something for nothing, or he is being hired, or is going to be given a job, or get some relief later?" Bennett denied that farmers had any such selfish motives and thought that they were simply delighted that their government "for the first time" was helping each to "save his land." Copeland then rather unkindly asked whether there was "a possibility that he has looked at the pictures of the Capitol and noticed that the dome is cylindrical and just like the top of a nursing bottle?"[4] Senator Richard B. Russell of Georgia, who has always warmly supported SCS on this subject, said a few years later that "the boys who really got the gravy are those who were fortunate enough to be in the intensive demonstration areas . . . in the early stages."[5]

In December 1935, Bennett sent Walter C. Lowdermilk, Associate Chief of SCS, to Arizona to carry out some important commitments which had been made to Senator Carl Hayden and Mrs. Isabella S. Greenway, Democratic representative from Arizona, who had both repeatedly requested conferences on SCS plans for flood control on the watershed of the Gila River and San Pedro River. Lowdermilk said he had "promised Senator Hayden that we should have this done by January 1, so that he might include flood control on the Gila as an amendment to the pending flood control bill in Congress." The problems were complex. Of particular importance was the fact that the Army Corps of Engineers had not reviewed the SCS plans for flood detention structures on the upstream watersheds of these two rivers. In order to make the SCS work on the upstream watershed effective, Lowdermilk also had to negotiate an agreement involving the Arizona Land Commissioner, the Grazing Service of the Interior Department, the Arizona Stockmen's Association, the Governor, and, of course, Sena-

    [4] *Department of Agriculture Appropriation Bill for 1937*, Senate, 74 Cong. 2 sess. (1936), pp. 390–401.
    [5] *Department of Agriculture Appropriation Bill for 1940*, 76 Cong. 1 sess. (1939), p. 551.

tor Hayden and Mrs. Greenway. The Soil Erosion Service (SES) had inaugurated the Gila River demonstration project at the request of governors of New Mexico and Arizona. At that time, it was understood that the U.S. Grazing Service would establish a federal grazing district on public lands in the watershed. The Arizona Land Commissioner and the Stockmen's Association opposed grazing districts to such a point that Lowdermilk told them that SCS would pull out of the area if the grazing district were not organized. The stockmen withdrew their opposition when they found that the Land Commissioner was opposing grazing districts because he feared that the Grazing Service would restrict his discretion to select sections of land for state administration. The stockmen had feared that the Land Commissioner would cancel or transfer their leases on state lands if they opposed him. At a conference at which all these interests were represented, a settlement was found and Lowdermilk hailed it as a "model for the Western States" as well as a "substitute AAA for the West." After further conferences with Mrs. Greenway, Lowdermilk reported to Bennett that she "appeared to be satisfied with the SCS and further asked if we needed additional funds to carry forward our work."[6]

Senator Hayden later cross-examined Captain Lucius Clay during appropriation hearings to show that the Corps of Engineers was moving more slowly than he wished. Clay said for the Corps that if SCS experimental work "were applied to flood control we believe that they would be erroneous as they now stand because the experiments have been conducted on a very small scale." Senator Russell pointedly asked Clay whether the Corps of Engineers had conducted experiments to determine the effects of siltation on navigable stream channels and reservoirs; Clay's reply was negative. Senator Hayden then told Clay and the Corps plainly that "I was responsible for the change in the law" authorizing the Department of Agriculture to conduct flood control investigations in upstream watersheds. Hayden wanted this testimony in the record to warn the Corps and let his colleagues in the House know that he did not approve of major cuts which they had made in the SCS appropriations for investigations and research.[7] The death of Representative James P. Buchanan of Texas had removed a staunch

[6] NARG 114 (General Correspondence, 1935), memorandum for H. H. Bennett from W. C. Lowdermilk, Dec. 4, 1935, for the statement of "commitments"; memorandum for files on subject of "Grazing Districts of Gila Project," signed by W. C. Lowdermilk and dated Tucson, Arizona, Dec. 23, 1935; another memorandum for files, signed by W. C. Lowdermilk, Dec. 21, 1935; and a telegram to H. H. Bennett from W. C. Lowdermilk, Jan. 9, 1936, from Phoenix, Arizona.

[7] *Department of Agriculture Appropriation Bill for 1938*, Senate, 75 Cong. 1 sess. (1937), pp. 628–29 and 714–16.

friend of SCS from control of the House Appropriations Committee.

The Soil Conservation Service had the support not only of Senator Hayden and Representative Greenway for upstream flood control work, but also of many other leaders in the Roosevelt Administration, including the Secretary of the Interior, Harold L. Ickes, and the President. In December 1935, Ickes had recommended to Roosevelt a report called *Little Waters,* which examined erosion control and detention structures on the small tributaries of rivers as measures contributing to flood control.[8] When spring floods, which were among the greatest ever recorded, washed over the Northeast and the Ohio and Mississippi Valleys, President Roosevelt insisted that Congress include the Department of Agriculture as one of the agencies authorized by the Flood Control Act of 1936 to make surveys and reports.[9] He said that there was a tendency to view flood control as "Army Engineering," ignoring the need for work on the headwaters. "You have got to think of it also from the point of view of the farmer's field up in the creek."[10] He told Morris L. Cooke, the first head of the Rural Electrification Administration (REA), who had been mainly responsible for preparation of the small, but influential, *Little Waters:* "We should all work to popularize those two words of yours 'upstream engineering.' I will try to use them in a speech."[11] Cooke, in turn, got Bennett's approval of a joint program of rural rehabilitation consisting of multiple-purpose county conservation demonstrations in which SCS would control erosion with land treatment measures and the REA would encourage the construction of small power projects.[12] None of these plans was carried out.

The standard district act made no mention of this upstream flood control problem. When the draft of the act was being discussed within

---

[8] H. S. Person, E. Johnston Coil, and Robert T. Beall for the Soil Conservation Service, Resettlement Administration, and Rural Electrification Administration, *Little Waters: A study of headwater streams & other little waters, their use and relations to the land* (Washington: Government Printing Office, 1936). See a discussion of this popular 82-page booklet in E. G. Nixon (ed.), *Franklin D. Roosevelt and Conservation, 1911–1945* (Hyde Park: General Services Administration, National Archives and Records Service, Franklin D. Roosevelt Library, 1957), Vol. I, p. 460.

[9] Nixon, *op. cit.,* Vol. I, pp. 503 and 506–08.

[10] *Ibid.,* Vol. I, pp. 510–11; see also pp. 512–18 and 559–73.

[11] *Ibid.,* Vol. I, pp. 519 and 532–33; Vol. II, pp. 54–55.

[12] NARG 114 (SCS, General Correspondence, 1935–36), exchange of correspondence, Morris L. Cooke and H. H. Bennett, March 5 and 18, 1936; May 8 and June 12, 1936. Bennett supported Cooke's proposal as author of an article entitled "Utilization of Small Water Powers," in *Soil Conservation,* Vol. I (October 1936), pp. 2–6. (This magazine is still published by SCS.)

SCS, one of Bennett's assistants, E. J. Utz, urged that it be broad enough in scope to include "such things as flood control, water conservation where it affects the water supply of municipalities, and possibly some basis of representation of municipalities where such conditions prevail."[13] Bennett did not encourage this view. When the chief engineer of the Oklahoma Conservation Commission asked Bennett whether the districts contemplated in the standard act were to have authority to conduct upstream watershed surveys and to prescribe preventive measures to comply with the Flood Control Act of 1936, Bennett said they "are not appropriate agencies" to carry on the sort of operations contemplated in that Act.[14] Instead, Bennett was convinced that SCS had to expand its demonstration projects so that no farmer had to travel more than fifty miles to see one. He was also willing to give aid to any special district organized under state law so long as it would create awareness of the erosion problem and could undertake preventive measures. Secretary Wallace, however, decided to restrain this "rather natural tendency . . . for the Soil Conservation Service to set up one demonstration after another, so that all farmers will be led to believe that if they can only exert enough pressure their lands will be brought within a demonstration project." He also decided not to sign agreements "with conservancy districts on anywhere from forty-eight different arrangements to an innumerable number." Consequently, he directed that a demonstration project was to be placed within one hundred miles of every farmer and that agreements with conservancy districts were to await the framing of the standard act.[15] In conformity with this policy, in September 1935, the Department disapproved a plan to include five projects under the supervision of SCS within the Muskingum Conservancy District in Ohio. Wallace departed from this policy, however, in May 1936, less than two weeks after the standard act was published. He authorized SCS to give technical assistance to nine wind erosion districts in the Panhandle area of Texas represented by Congressman Marvin Jones, Chairman of the House Committee on Agriculture.[16]

Pressures of the sort illustrated by the negotiations over the

[13] NARG 114 (SCS, General Correspondence, 1935), memorandum, E. J. Utz to D. S. Myer, Oct. 17, 1935.

[14] NARG 114 (SCS, General Correspondence, 1936), W. C. Burnham to H. H. Bennett, Oct. 19, 1936; reply by Bennett, Nov. 7, 1936. See also memorandums by Bennett to Solicitor and replies, Oct. 15 and 31, 1936; Nov. 5, 1936.

[15] NARG 16 (Erosion File), Paul H. Appleby to Hugh Calkins, SCS Coordinator, Albuquerque, New Mexico, Aug. 22, 1935.

[16] NARG 16 (Erosion File), R. G. Tugwell to Bryce Browning, Sept. 30, 1935.

location of projects and the extension of the SCS program to include upstream flood control surveys show clearly why the Department and some members of Congress wanted "centralized" administration of this program. The Soil Conservation Service needed protection in the states and localities from individuals or groups hostile to its program; it received support from the Roosevelt Administration and many incumbent congressmen. As early as 1935, some members of Congress were inclined to support SCS against claims from extension leaders for a greater share of responsibility for administering the program than the Secretary of Agriculture was willing to give. A decade later these differences mushroomed into a full-scale conflict between the two agencies and their respective congressional supporters.

EXTENSION SERVICES OPPOSE THE STANDARD ACT

Most of the issues over which SCS and the state extension services contended for many years were clearly fixed by the end of 1936. The principal spokesmen for the agricultural colleges opposed (1) the regional organization of SCS; (2) the alleged unwillingness of SCS to recognize adequately the role of the college staffs in specifying practices as part of a joint state-federal program of erosion control; and (3) the respective part of each in "conservation education." Many of these spokesmen questioned the wisdom of organizing soil conservation districts as provided in the standard act, especially on the basis of watersheds rather than county boundaries.

In December 1935, Dean I. O. Schaub of North Carolina State College complained to his congressman, William B. Umstead, that the regional organization of SCS had been devised to leave the state offices as powerless appendages. Umstead, a member of the House Committee on Appropriations, demanded an immediate explanation from SCS. Dillon S. Myer of the SCS assured Umstead that the state "coordinators" of SCS were to have an active role in reviewing plans and operations "and other important supervisory matters within the state." Umstead acknowledged Myer's explanation, but warned that he wanted to be "absolutely assured . . . that the control of the state office over the soil conservation work in the state will be maintained."[17] According to Myer, however, Chairman Marvin Jones of the House

[17] NARG 114 (SCS, General Correspondence, 1935), D. S. Myer to William B. Umstead, Dec. 19, 1935, and Umstead's reply to Myer, Dec. 27, 1935.

Committee on Agriculture insisted that SCS organize regionally and threatened to provide such organization by legislation, if it were not done by administrative action.[18]

In the following January, the Committee on Extension Organization and Policy of the Association of Land-Grant Colleges and Universities met with Myer and J. Phil Campbell of the Division of Cooperative Relations and Planning, SCS, in an attempt to smooth relations. The assembled directors complained that unsatisfactory relationships already existed between the SCS regional offices and extension "in some states." They recommended more frequent meetings of the state advisory committees to remedy the situation.[19]

The SCS regional offices were also unpopular with the directors of several state experiment stations. To develop its research program, SCS had worked during 1935 with the federal Director of Extension and Director of Research in the Department of Agriculture, to produce a memorandum of understanding with each of the state experiment stations. After working papers were presented to the station directors, C. B. Hutchison, Director of the California Experiment Station and Chairman of the Committee on Experiment Station Organization and Policy of the Association of Land-Grant Colleges, requested a conference. One was held in Chicago, May 4 and 5, 1936, but it was attended by representatives of the Department and the directors of only sixteen of the state experiment stations, although all had been invited.

At the conference, it was decided that SCS would have no officer bearing the title "Regional Director of Research" and that a research program was to be worked out for each state after direct negotiations between SCS and the director of the experiment station. The Soil Conservation Service was to retain authority to undertake its own projects involving "direct" research, as defined by the Secretary's interbureau committee of 1935. It was agreed that "all research within a State (or such research as may be mutually agreed upon) relative to soil erosion and its control, shall be jointly planned and conducted jointly or in a manner to be mutually agreed upon" by the two groups. Furthermore, SCS was to recognize that "so far as possible . . . projects with Planning Boards, Water Boards, State Departments of Conservation, and the like" were to be arranged through the director of

---

[18] Interview with Dillion S. Myer, Sept. 13, 1962.

[19] Committee on Extension Organization and Policy, *Proceedings of the Association of Land-Grant Colleges and Universities,* mimeographed (Washington, Jan. 24, 1936).

the state agricultural experiment station.[20] This second agreement was to block any tendency in SCS to build up state rivals to the experiment stations by arranging for lucrative federal research projects.

In June 1936, SCS made an overture to the agricultural colleges by suggesting that they jointly employ county and state extension conservationists and formulate state programs subject to joint review. These suggestions were in keeping with the recommendations made by the Secretary's Committee on Soil Conservation in 1935. For a while, SCS was willing to divert some of its funds to "extension work" to prepare for the organization of soil conservation districts. Conferences between extension directors and SCS representatives were held in the Southeast and the Dust Bowl states during the summer of 1936. The work plans of the states which evolved from these conferences varied in details, since some of the directors of extension had more funds available than others and, too, there were differing reactions to the proposed scheme. Both SCS and the extension services agreed that each would support half the cost of employing local workers—who would be called either assistant county agents or district conservation specialists—to work in conjunction with the SCS demonstration projects and CCC camp work areas. At a minimum, these workers were to organize county soil conservation associations and to aid in demonstrating SCS work on individual farms which they were to select. It was agreed they would report through the state extension soil conservationists, and copies of their reports were to be sent also to the regional and national offices of SCS. The location of CCC camp work areas would be determined by each state's soil conservation advisory committee, which would also formulate the state's erosion control program, subject to review by SCS.[21]

In October 1936, the Committee on Extension Organization and Policy met jointly with the Executive Committee of the Association of

[20] NARG 114 (SCS, Correspondence of the Central Office, 1936–1940), memorandum to Specialists in Research, Regional Conservators and State Coordinators, SCS, from H. H. Bennett, May 29, 1936. See also *50th Proceedings of the Association of Land-Grant Colleges and Universities, 1936* (Washington), p. 322.

[21] NARG 114 (Cooperation—Extension Service, 1936–41), memorandum, J. Phil Campbell to H. H. Bennett, June 15, 1936; and "Outline of Suggested Soil Conservation Procedures in Various States," Oct. 28, 1936. Also, NARG 33 (Soil Conservation File), memorandums exchanged between J. Phil Campbell, SCS, and C. W. Warburton, Director of Federal Extension Work, June 25, 1936, July 3, 1936, Aug. 31, 1936, and Oct. 12, 1936, and "Suggested Plan of Work (July 1, 1936 to June 30, 1937)," Nov. 11, 1936; *Report of the Chief of the Soil Conservation Service, 1936* (Washington: Government Printing Office), pp. 2, 45–46; and *ibid.,* *Report for 1938,* p. 32.

Land-Grant Colleges in Washington to review relations with SCS. The extension directors argued that there was no need for the regional offices of SCS which were "heavily manned with technical workers in practically every field of agriculture." They were incensed by the implications of SCS Field Memorandum No. 371, which reminded regional and state technical specialists they were responsible to the heads of the technical branches in the Division of Conservation Operations in Washington. In response to this memorandum, R. K. Bliss, Iowa's Director of Extension, and R. E. Buchanan, Director of the Iowa Experiment Station, had written to SCS, asserting that this memorandum was "not in accordance with the memorandum of the Secretary under which you are operating and will not, therefore, apply to the state of Iowa." They had claimed that the planning, extension, and technical phases of the SCS program were at variance with the June 1935 report of the Secretary's Committee, and they were dismayed at the assertion that there "are no state technical staffs recognized as such." The SCS regional technical staffs had a role to play, Bliss and Buchanan were willing to concede, but they refused to "permit a program in the state of Iowa which does not more effectively integrate the State Coordinator and the State Advisory Board with the technical direction of the program."

At this October meeting, the Committee on Extension Organization and Policy heard complaints from about a dozen states and then agreed that there ought to be no breakdown in the "traditional" interpretation of "extension work" or in channels for administering it. The regional offices of SCS should be retained merely for fiscal and procurement purposes. Only SCS work on public lands and under flood control legislation were to be excepted.

Members of the two Land-Grant College Association committees met with M. L. Wilson on October 8 in executive session with the hope that they could get him and Wallace to change their minds. They failed and, after deliberating on October 9, adopted a resolution saying they did "not approve or urge support of the suggested Soil Conservation Service District Law."[22] Apparently they thought that districts would be mere appendages to SCS to be opposed as heartily as the regional offices.

Members of the two Land-Grant College Association committees

[22] Committee on Extension Organization and Policy, *Proceedings of the Association of Land-Grant Colleges and Universities,* mimeographed (Washington, Oct. 8–10, 1936).

were so aroused over the SCS regional organization and the proposal to create soil conservation districts that they mimeographed portions of the proceedings and sent them to all of the agricultural colleges in time for staff members to prepare themselves thoroughly for the annual meeting of the Association at Houston, Texas, scheduled for November 16–18, 1936.

The resolution opposing the standard districts law came as no complete surprise, since Dean A. R. Mann of Cornell University, who was Chairman of the Executive Committee of the Association, already had written a pointed note to Bennett. "My attention has been drawn to a recent bulletin from the Soil Conservation Service entitled a Standard Soil Conservation Districts Law. I have not seen a copy . . . and we do not seem to have copies here. I am wondering whether I can obtain one or two from you." He followed with a warning which left no doubt about the character of the growing fight between the colleges and the SCS. "I may say that I have received some strong protests against the proposals set forth in this bulletin from certain of our land-grant institutions on the score that the proposals suggest the setting up of agencies wholly independent of the land-grant colleges, therefore involving conflicts of interest and activity. I should like to see the bulletin, after which I may wish to write to you more specifically."[23]

The whole festering conflict came to a head at the annual meeting of the Association of Land-Grant Colleges. According to at least one of the key participants in that meeting, the leaders of the Association had expected this annual convention to be a "hate Roosevelt session" organized by the "Texas A & M crowd." They had taken the *Literary Digest* poll seriously and planned to celebrate Roosevelt's defeat, but instead at meeting time they were flabbergasted at the results of the 1936 election.[24] At a private conference on November 14, Wallace,

[23] NARG 114 (SCS, General Correspondence, 1936), A. R. Mann to H. H. Bennett, Aug. 5, 1936 and Bennett's reply, Aug. 8, 1936. Dillon S. Myer testified later that the model act was drafted without consultation with members of the congressional committees on agriculture: "At least if it was, I do not know it." See *Department of Agriculture Appropriation Bill for 1940*, Senate, 76 Cong. 1 sess. (1939), p. 550.

[24] Interview with M. L. Wilson, Dec. 15, 1960. According to Wilson, relations of the Department of Agriculture with the extension services varied considerably among the states at that time. He said, for example, they were then "very bad" in California, Texas, Illinois, Indiana, and Kentucky; "not so good" in Wisconsin; and "quite good" in Iowa. Wilson thought that much of the disharmony stemmed from the pro-Republican sentiments of extension leaders, especially those in the Northeast and Midwest.

Wilson, Bennett, and others met with the Committee on Extension Organization and Policy but failed to settle their differences, and the conflict was brought into the open at the public sessions. Wallace tried to smooth off some of the rough edges by saying that he looked forward to greatly improved agricultural organization in the near future, since the current discontents were nothing more than "misunderstanding that can be corrected by improving the channels of communication."[25] It is doubtful that he convinced anyone, although Dean Mann agreed that Wallace had "ably" outlined the "organizational and relations problems" most in need of adjustment. Mann, therefore, appointed a special Committee on Land Problems under the chairmanship of F. D. Farrell, President of Kansas State College, to make the "executive and legislative" officers of the federal government see the need to eliminate "overlapping" in "promoting land conservation programs."[26]

Some idea of the bitterness of many of the extension leaders can be gleaned from a speech which Farrell made at the Association's November 1936 meeting. He listed twelve "obstacles" to land conservation. One of these was "fanaticism . . . in specialized areas of the field of land conservation. . . . It often retards progress by overemphasis. The extravagant, though doubtless honest, statements of over-zealous specialists may help in dramatizing a particular condition. But they also frequently hinder by first misleading the public then making it skeptical." An equally serious "obstacle," he said, was the "widespread confusion among public agencies in the administration of land conservation . . . and the fact that we are always in a dreadful hurry. . . . Land conservation is coming. . . . But . . . [it] will come slowly at best."[27]

According to some participants, Edward A. O'Neal, President of the American Farm Bureau Federation, made a strong attack in private on the Department for turning away from extension, and pressures were exerted by some state directors of extension to change the Department's commitment to soil conservation districts, but to no avail.[28] M. L.

[25] 50th Proceedings of the Association of Land-Grant Colleges and Universities, 1936, op. cit., p. 32.

[26] Ibid., pp. 35–36, and 89–90.

[27] Ibid., pp. 104–05.

[28] Interview with Dillion S. Myer, Sept. 13, 1962. Myer said that several of the extension directors got him in a room and called him a traitor to extension because he had accepted such a high position in SCS (he eventually became Assistant Chief). They implored him to work for a decision to turn the SCS program over to the colleges, but he refused on the ground that SCS was needed at the time.

Wilson gave the official reasons for the Department's support of the districts, and the Land-Grant College Association, in turn, bluntly refused to endorse the standard act.[29]

STRATEGY DECISIONS FOR 1937

When Wallace returned to Washington after the Land-Grant Colleges meeting, he discussed strategy with his assistants and decided to promote the standard act during 1937 in the face of apparently unanimous opposition from the colleges. He also had some opposition from Bennett. Even before the colleges officially announced their opposition to the soil conservation districts in November 1936, Bennett was convinced that it would be very difficult to organize them and that, if they were organized, it might prove equally difficult for SCS to work with them satisfactorily. Several times during the summer and early fall of 1936 he sent requests to the Solicitor or other officers of the Department, asking whether particular state laws authorizing special districts might be substituted for the standard act.[30] He was firmly turned down.

Wallace met with his principal staff members, except Bennett, in December and decided that he could counter extension opposition of districts with the popularity of SCS erosion control demonstrations, especially if he could tie them to widespread interest in flood control. On this occasion, Wallace decided upon a rather simple strategy. He chose to use the deadline of July 1, 1937, which his interbureau Committee on Soil Conservation had recommended two years earlier, to stimulate suitable state legislation. Specifically, he authorized SCS to differentiate the extent and kinds of assistance to states as a condition to be met before SCS would organize new demonstration projects. Those with suitable legislation on soil conservation districts would receive more than others. Also, SCS was authorized to recommend that this deadline be postponed for particular states or in general, depending upon the outcome of the 1937 legislative sessions in the states. Wallace also decided that the Department would ask the President and Congress to amend the Flood Control Act of 1936 so that the Secretary

[29] *50th Proceedings of the Association of Land-Grant Colleges and Universities, 1936, op. cit.,* pp. 187–90. See also Christiana M. Campbell, *The Farm Bureau and the New Deal* (Urbana: University of Illinois Press, 1962), pp. 158–59.

[30] NARG 16 (Office of the Solicitor), H. H. Bennett to Mastin White, Solicitor, and replies, Oct. 15, 31, and Nov. 5, 1936. Also W. C. Lowdermilk to Mastin White, Oct. 9, 1936; and reply, Nov. 7, 14, 1936.

of Agriculture could require state legislation "along the lines of" the standard act before the Department could undertake flood control investigations.[31] The Department prepared a draft of legislation enabling states to participate in the "watershed phases of the flood control program."[32]

Wallace also was persuaded by some of his subordinates to ask the President to use his enormous prestige in February 1937, to promote the districts law in the face of unyielding opposition from the leaders of the Association of Land-Grant Colleges and Universities. On January 12–15, 1937, the Association's Committee on Extension Organization and Policy met in Washington. It discussed SCS and soil conservation districts with Assistant Secretary Harry L. Brown, who had recently left his post as Director of the Georgia Extension Service, and received his promise that he would take up their views with Wallace.[33] According to oral reports, the extension leaders intended to ask Senator John H. Bankhead, 2d, of Alabama, Chairman of the Senate Committee on Agriculture, to offer an amendment to Public Law 46 which would establish the position of extension in the Department's erosion control program. They knew, however, that they would have trouble getting their bill through Congress if the Secretary did not approve of it, even if O'Neal of the Farm Bureau were from Senator Bankhead's home state. Brown arranged a meeting between Wallace, Creel of Nevada who represented extension, and others in the Department. After a spirited discussion during two days, Wallace decided not to support extension's proposed legislation.[34] Instead, he asked Roosevelt to go over the heads of the extension directors by appealing to all state governors to support state laws authorizing soil conservation districts organized on the basis of watersheds together with the amendments to the Flood Control Act of 1936. The President sent his message to the state governors on February 23, and two weeks later approved the Department's proposed change in the flood control legislation.[35]

This strategy produced only limited success in the state legislatures in 1937. Bennett learned in January that consideration of the standard

[31] NARG 114 (SCS, General Correspondence, 1936), memorandum, H. H. Bennett to the Secretary of Agriculture, Jan. 12, 1937. Bennett's memorandum presented his understanding of decisions made in his absence from this meeting.

[32] NARG 16 (Office of the Solicitor), Philip M. Glick to the Solicitor, Jan. 4, 1937.

[33] Committee on Extension Organization and Policy, *Proceedings of the Association of Land-Grant Colleges and Universities*, mimeographed (Washington, Jan. 12–15, 1937).

[34] Interview with Dillon S. Myer, Sept. 13, 1962.

[35] Nixon, *op. cit.*, Vol. II, pp. 21 ff. (esp. pp. 26–27 and 37–38); 50 *Stat.* 876.

act would be put off in Maine because the Association of Land-Grant Colleges had not approved it. No bill would be presented in Ohio during 1937 because only state legislation to carry out the program of the Agricultural Adjustment Administration would be considered. The state advisory committee in Kansas had rejected the standard act in favor of legislation incorporating soil conservation associations under an existing statute providing for county farm bureaus. In both Texas and Arkansas, there was opposition to land use regulations. The advisory committee in California was of the opinion that the standard act would not be "appropriate." The reports from several other states such as Iowa, Oregon, New York, Louisiana, Maryland, Montana, and the Dakotas indicated hopefully that the act might be more favorably received.[36] Some of these estimates, however, turned out to be much too optimistic.

ACTION IN THE MIDWEST, 1937

Dillon Myer said that it was difficult to secure consideration of the Standard Districts Act because it was long and contained some ideas of "doubtful value" in the minds of many who examined it. Even worse, not very many of the SCS state coordinators who were responsible for securing its enactment had shown much interest in working for its adoption.[37] Many of the coordinators were technicians or former research scientists who had neither the experience nor the aptitude for lobbying in state legislatures. Reports from the field also were often sketchy, so that Myer sometimes was not certain of either the status of the bills or their provisions in some states. Early in April, he sent the Minnesota State Coordinator copies of bills introduced in North Carolina and Utah with the remark that he thought they were the ones finally enacted, but "we are not positive." Ten states had adopted legislation by this time and it was his "understanding" that each "closely parallels the recommended Standard law with the exception of minor modifications."[38] By the time later in the month when a conference was held to discuss the matter, three more states had adopted legislation, but doubts of success were growing in the Department.

[36] NARG 114 (SCS, General Correspondence, 1936–40), memorandum for the Secretary from H. H. Bennett, Jan. 23, 1937.

[37] NARG 16 (Records of the Office of the Solicitor, 1937), unsigned memorandum which summarized the situation as of April 18, 1937.

[38] NARG 114 (SCS, General Correspondence, 1936–40), D. S. Myer to Herbert Flueck, April 2, 1937.

Wallace was again asked whether SCS should "go ahead with" the standard act and he answered in the affirmative. When he was asked: "Shall we stay with the July 1, 1937, requirement as to the inauguration of new projects?" he answered "affirmatively but not so definitely."[39]

Although the legislative situation was uncertain or unknown in many of the first states to adopt the standard act, its details were known well in Region 5 with headquarters in Des Moines, Iowa. Myer's office received copies of the bills introduced so that the Department was able to review and comment extensively on those introduced in Illinois, Iowa, Wisconsin, Minnesota, and Missouri. The Solicitor's office made numerous and detailed suggestions of changes in the bills in each of these states, and also sent representatives to aid in explaining its reasons for requiring those features of the act questioned by state legislators and interested citizens. Later these same practices were followed in reviewing the bills introduced in the states which were slow to consider the proposal.

With the support of the Southwestern Wisconsin Soil Conservation Association which met with the State Coordinator of SCS, M. F. Schweers, and drafted a bill, legislation was passed in Wisconsin after many modifications of the standard act were made. Opposition to the district as a governmental subdivision of the state rather than a nonprofit corporation was reported. In addition, the sponsors did not want mere "land occupiers" to vote in elections, since they did not want tenants to obligate landowners to carry out practices formulated by the district. The Department expressed doubts of the constitutionality of the proposed provision for adopting land use regulations and opposed authority in the State Soil Conservation Committee to approve such regulations. The State Coordinator replied that he would do all in his power to secure a bill embodying the Department's wishes. He warned, however, that "we may lose the support of the Southwestern Wisconsin Soil Conservation Association which was not only instrumental in having the bill introduced, but will also be very influential in having it passed and in carrying out the plan if it is adopted."[40] When compro-

[39] NARG 114 (SCS, General Correspondence, 1936–40), memorandum on "Considerations by Mr. Gaston of the Standard Soil Conservation Districts Law," April 26, 1937.

[40] NARG 114 (SCS, General Correspondence, 1937), M. F. Schweers to D. S. Myer, April 1, 1937. The file on the Wisconsin bill consists of letters and telegrams exchanged between Myer and Schweers on March 6, 11, 18, 22, 30, 31; April 1, 5, 6, 14; May 14; and June 2, 1937. Dean C. L. Christensen of the Wisconsin College of Agriculture was "pessimistic, but in favor of a bill of this type." The bill passed the lower House by a vote of 68 to nothing and Schweers thought there would be no trouble in the Senate.

mises were finally worked out at some points, Myer expressed his view
that the bill was quite comparable to the standard act. The bill was
passed, but two years later it was amended to make the agricultural
committee of each county governing board the supervisors of each soil
conservation district.

In Minnesota, SCS was aided by the Southeastern Minnesota Soil
Conservation Association and other similar groups organized in connection with CCC camps and work projects. The SCS State Coordinator, Herbert Flueck, reported that Governor Elmer Benson was initially
noncommittal and that F. W. Peck, Director of the Minnesota Extension
Service, would approach the proper legislative committees through his
own men. Flueck was "sure the University people will do all they can to
enact this legislation." Opposition to compulsory land use regulations
was strong and Flueck feared that, if he insisted upon them, he would
lose the bill. He wired an inquiry to Myer, asking whether provision for
compulsory regulations might be added after sufficient educational
work had created the proper climate of opinion for them. Myer replied
that such regulations would not be compulsory until they were voted
by land occupiers. "We feel this authority highly desirable." It would
be acceptable, however, to settle for a vote of two-thirds or even three-
fourths rather than a majority, if this concession would secure passage.
The upshot of the negotiations was a law which "may be impractical,"
according to Flueck.[41]

In Illinois, the college had bad relations with the Department, but
a modified version of the model law was eventually passed. F. A.
Fisher, the SCS State Coordinator, reported that Governor Henry
Horner placed the whole issue in the hands of his Director of Agriculture, Walter McLaughlin, who called a conference of representatives of
the College of Agriculture, the Illinois Grange, and the Illinois Agricultural Association. The Executive Board of the Agricultural Association
discussed the standard act and their secretary reported "they would go
on record as against the bill unless it was changed materially so far as
the voting status of the tenant is concerned. . . . They also felt that we
should demand a three-quarters vote of all those voting to establish

[41] NARG 114 (SCS, General Correspondence, 1937), letters and telegrams
exchanged between Dillon S. Myer and Herbert Flueck, March 11, 19, 20; April 1,
5, 11; and May 1, 1937. State Senator Galvin who sponsored the bill asked Dillon
Myer why there was objection to "having the counties of this state constitute the
districts and the County Board of Commissioners to constitute the Board of Supervisors. The County Board of Commissioners is the tax-levying body." Myer told
him to ask Flueck personally for "our comments." Senator Galvin to Myer, March
16, and Myer's reply, March 23, 1937.

land use regulations and that the absentee landowners be given the right to vote by proxy, or other delegated authority." E. A. Eckert, Master of the Illinois Grange, was reported to have said that "no corporation owning a large tract of land should be permitted to organize a district wherein they would benefit." This objection, Fisher was sure, "was to deal with coal companies who are ruining hundreds of acres by strip mining and then wanting aid to set the areas to trees." According to Fisher, Dean H. W. Mumford of the Illinois College of Agriculture and his representatives thought that a more extensive educational program was needed before compulsory land use regulations could be enacted. Representatives of these organizations wanted to know how "operations under the proposed law would fit in with the present programs of the Extension Service, the Soil Conservation Service, the Agricultural Adjustment Administration and related agencies." There was a feeling that it might produce just another agency privileged "to hire technical men and go into the field to promote a program. All present appreciate that very much demand would meet with failure for there is not enough competent and trained personnel to do this." Fisher said he understood that Governor Horner feared the "sweeping powers" in the standard act and "the size of this agency, if set up under a state law, that the federal government might withdraw its support and leave the state with a large financial responsibility." The Governor also questioned, Fisher said, the legality and wisdom of "giving the tenants so much power."[42]

A bill was prepared, despite the many sources of opposition to the standard act, but the modifications were so extensive as to prompt Dillon Myer to remark that the state might adopt it, but the Department might not cooperate under it since the bill had "serious defects." It had no provision to make the farmer pay costs, if the district performed work on his land in accord with district regulations, three-fourths of all landowners would be eligible to vote in land use referenda, the district was not a governmental subdivision of the state, and the expenses of hearings and other activities connected with forming a district were to be borne by the petitioners for a district. Although the bill passed the lower House by a vote of 97–12, it was

---

[42] NARG 114 (SCS, General Correspondence, 1937), F. A. Fisher to D. S. Myer, March 26, 1937. Fisher warned that he was reporting the substance of comments made verbally and not in writing. Other letters were exchanged on March 1, 4, 11, 13, 19, 24, 26 (telegram and letter), 31; April 10, 18, 21, 23; and June 8 and 17, 1937.

reported that the consideration came too late in the session to amend the bill to everyone's satisfaction.[43]

In mid-April, R. H. Musser, SCS Regional Conservator in Des Moines, reported his optimistic belief that satisfactory bills would pass in Illinois, Wisconsin, Minnesota, and Missouri, but not in Iowa. He asked whether he should be "more aggressive," although he volunteered his own judgment that "we should not necessarily appear to be the sponsors, but be in a position to help those who should be sponsoring" the standard act. He was told: "Just continue to follow your good judgment."[44]

The interested parties in Iowa were not ready to support the districts legislation. Philip M. Glick of the Solicitor's office was sent to a meeting with representatives of Iowa State College, the Farm Bureau, and other farm groups after Buchanan, Director of the Experiment Station, and Theodore W. Schultz, of the Department of Economics at Iowa State, had worked out a bill to provide a single state committee to "handle all Federal cooperative action programs." Bennett requested the Solicitor to prepare a draft of a bill to meet Iowa's constitutional requirements, the Department's policies, and the suggestions of the Iowa group. This was done but, in May 1937, Buchanan reported to Glick that no further effort would be made until 1939 to secure enactment of a bill because it would become a "political football between rival and partisan groups."[45]

The results of this campaign were even more discouraging in Missouri. The day after President Roosevelt sent his message to the governors, J. W. Burch, then assistant director of the Missouri Extension Service, told C. W. Warburton, the director of federal extension, that he and Professor M. F. Miller had gone over the standard act and had agreed that "a great deal more educational work had to be done in this state before such regulations could be applied without causing a tremendous kickback. It is, therefore, our understanding that no attempt will be made to pass such regulation at this session of the Missouri legislature."[46] K. G. Harmon, SCS state coordinator, remained

[43] NARG 114 (SCS, General Correspondence, 1937), D. S. Myer to F. A. Fisher, June 8, 1937; Myer to J. H. Lloyd, June 8, 1937; and Lloyd to Henry A. Wallace, June 11, 1937.

[44] NARG 114 (SCS, General Correspondence, 1937), R. H. Musser to D. S. Myer, April 13, 1937, and reply April 16, 1937.

[45] NARG 16 (Records of the Solicitor's Office, 1937), R. E. Buchanan to M. L. Wilson, Jan. 27, 1937; Buchanan to Philip M. Glick, May 5, 1937.

[46] NARG 114 (SCS, General Correspondence, 1937), J. W. Burch to C. W. Warburton, February 24, 1937. In a further letter to Warburton, on March 11,

optimistic, however, until late March 1937. After several meetings with representatives of the Missouri Extension Service, Harmon became discouraged upon finding Dean F. B. Mumford using several men "who are reactionary to any move unless it is in their direct charge." There also was opposition from a man whose "motives are quite questionable." In addition, there was a "storm of protest in the legislature" as controversy grew over the compulsory land use regulations. This opposition was brought about, Harmon reported, "through real estate and farm management groups from Kansas City and St. Louis." Dean Mumford, he said, was firm in his belief that the term "land occupier" in Missouri must mean landowner because the average tenure of tenants was two years. Regulations voted in by such a transient group, leaving obligations to be fulfilled by landowners, would be unfair in his view. In the end, the bill was impaled on charges that it was part of the Roosevelt Administration's plan for regimentation of agriculture—part and parcel of the "Court packing" plan![47]

Aside from the reported opposition of Missouri's College of Agriculture and the real estate and farm management interests, there was a split over the legislation between two major farm groups in the state. R. W. Brown of the Missouri Farm Bureau complained to a state senator, A. W. Lockridge, that "your committee" killed the bill even before there was an opportunity for the county farm bureaus to study the bill. It was, Brown charged, misrepresented as "regimentation," but there was no reason why Missouri should not have soil conservation districts provided by legislation which would have given the College of Agriculture more authority over the program.[48] Harmon reported to Myer that the most virulent opposition, however, came from "the leader" in the Missouri Farmers Association: "As you probably know, Mr. Hirth is absolutely against all measures which he thinks Secretary Wallace has approved."[49] The struggle ended for 1937 with a telegram from a staff

---

Burch elaborated this point by saying that the kickback would adversely affect "our entire agricultural educational program." Apparently, the Missouri Extension Service feared that its standing with farmers would be injured, if it were associated in any way with the proposal for creating districts with power to enact compulsory land use regulations. It is entirely possible that extension personnel in other states shared this fear and opposed the districts partly for this reason, since they were expected to advocate districts in their capacity as agents of the Department of Agriculture.

[47] NARG 114 (SCS, General Correspondence, 1937), exchange of letters between D. S. Myer and K. G. Harmon, March 9, 13, 18, 19, 23, 24, 25, 26, 31; April 5, 10, 24; and May 1, 1937.

[48] NARG 114 (SCS, General Correspondence, 1937), R. W. Brown to A. W. Lockridge, May 11, 1937.

[49] *Ibid.*, Harmon to Myer, April 10, 1937.

member of the Solicitor's Office who had been sent to Springfield,
Illinois, and Jefferson City, Missouri, to help draft the legislation. He
reported to Dillon Myer: "Tough fight here but progress made nothing
more I can do where do I go?"[50] Missouri refused to adopt a soil
conservation districts law until 1943.

SITUATION IN THE WEST, 1937

The legislative situation in the West at the end of the 1937 sessions
generally was not good. Colorado, Nevada, Utah, and Montana passed
laws. The Montana act was considered by the Department to be
unconstitutional and thoroughly unsatisfactory. The Utah statute contained one provision which the Department considered to be a serious
flaw. In California, a bill was prepared by two members of the staff of
the College of Agriculture and supported by the President of the state
Chamber of Commerce who, according to the Regional Conservator of
SCS, "has property in one of our southern projects."[51] The bill was
sponsored by the California Farm Bureau, but to the surprise of all the
interested parties, the Governor vetoed it. The only known explanation
was that the bill carried with it an appropriation of $5,000, and the
Governor refused to sign it along with nearly four hundred others like
it. In September 1937, the Board of Directors of the California Farm
Bureau discussed the standard districts law and directed its secretary
to have a bill prepared for the next session of the legislature. The SCS
State Coordinator reported, however, that he understood that there
had been "considerable criticism" by the "Washington office" of the
manner in which districts were to be created under the bill which had
been vetoed. He warned that both the Farm Bureau and college
spokesmen felt that "considerable opposition will be encountered if an
attempt is made to change the method of establishing districts as now
provided by many California acts." The county governing boards were
empowered to create such subdivisions as reclamation, irrigation, and
drainage districts and the "considerable opposition" would arise if a
soil conservation districts law were to "usurp these powers."[52] This fail-

[50] *Ibid.*, telegram, D. J. Sherbondy (in Solicitor's Office) to D. S. Myer, April 29,
1937.

[51] NARG 114 (SCS, General Correspondence, 107A, California), Harry E.
Riddick to D. S. Myer, Aug. 24, 1937; see also R. B. Couzzens to D. S. Myer, Aug.
23, 1937.

[52] NARG 114, *ibid.*, R. B. Couzzens to D. S. Myer, Sept. 20, 1937; Alex Johnson
to M. L. Wilson, Jan. 26, 1937 and reply, Feb. 23, 1937.

ure in 1937 was only the first stage of a protracted struggle over the law in California.

In New Mexico, as in California, animosities between extension and SCS flared as efforts were made to secure support for a districts law. The Soil Conservation Service enlisted the support of some members of the legislature early in the session of 1937, but the Resettlement Administration and the New Mexico extension service framed their own bill, authorizing wind erosion districts without power to control erosion caused by water. After efforts were made to reconcile some of the differences, a conference was held with Governor Clyde Tingley, who revealed that he "would veto any bill patterned after the Government sponsored standard act because he was not in accord with the work of the Soil Conservation Service." After some further delays in which he first charged SCS with objectionable "lobbying," and then claimed that he had never seen the districts bill, Tingley finally admitted to Wallace that the SCS employee who had referred the bill to members of the legislature "is entirely at odds with (the) state administration and what present status of this bill is now I do not know." Wallace considered the situation so serious that he sent Paul Appleby and M. L. Wilson to New Mexico to get a bill passed.[53]

An amended version of the model law was passed, but two years later it was altered to ease some of the tension between ranchers and farmers on irrigated lands, since the two were often at odds over control of the districts. This very important aspect of the organizational problem was raised in New Mexico, when a member of the State Soil Conservation Committee proposed that each district governing board, rather than the State Committee, be permitted to refuse to add to its existing territory. He argued that the law passed in 1937 permitted additions despite the fact that "such a change would be unacceptable to all the landowners of the original district." The SCS properly

[53] NARG 16 (Office of the Secretary; Soil 1), letters: D. S. Myer to W. C. Lowdermilk, Nov. 25, 1935; Stuart Moir, Assistant Regional Conservator, to D. S. Myer, Feb. 18 and 21, 1937; F. D. Matthews to H. G. Calkins, Regional Conservator, Feb. 24, 1937; telegrams: Gov. Clyde Tingley of New Mexico to Wallace, Feb. 19, 1937; reply Feb. 26, and further exchanges March 4, 5, 6, 8, 1937. One source of the Governor's opposition apparently was an undisguised struggle between the state's Extension Service and SCS, involving former county agents hired by SCS. The Director of the New Mexico Extension Service had asked SCS in 1935 to transfer out of the state at least two, and perhaps three, SCS employees who were former Extension personnel. The Extension Director charged them with working against him and warned that Senator Carl A. Hatch and maybe Senator Dennis Chavez would object to their continuance in the state. One of the three had been removed by the Governor from his position of county agent in Chavez County.

rejected this proposal which would have permitted a district board to concentrate the assistance provided by SCS upon only the lands of the operators of the original district, regardless of the need on others. This proposition, however, proved to be a portent of some of the problems which later arose between SCS and some district boards.[54]

In Texas, the legislative session was also disappointing in 1937. The Texas Extension Service prepared a bill and had it introduced in the legislature by State Senator E. M. Davis, Chairman of the Agriculture Committee, who had sponsored the wind erosion districts law in 1935. The bill made the county governing board and the county agent the supervisors of a district, with the agent to be the executive officer charged with supervising erosion control. All financial aid to the districts was to be channeled through the State Soil Conservation Board (the Regents of Texas A & M College). There was no provision for land use regulations. Since the act was to supersede all others inconsistent with it, Dillon Myer expressed fear that the wind erosion districts would be dissolved, if this act passed.[55]

Virtually all of the key leadership of SCS in Texas had been drawn from the Department of Vocational Agriculture at Texas A & M, a fact sufficient in itself to stimulate rivalry with the Texas Extension Service. In addition, SCS split the leadership of the Texas Farmers Association, an affiliate of the American Farm Bureau Federation, when V. C. Marshall, who was then the chairman of the legislative committee of the Texas Farmers Association, organized a "Committee of One Hundred," an informal association of ranchers, farmers, and others who formed the Texas Soil and Water Conservation Association.

Meanwhile, no single bill patterned upon the standard act was introduced in the Texas legislature during its regular session of 1937. Several watershed associations introduced bills for authority to create special districts, some with the expectation they might secure SCS assistance. The Trinity River Watershed and Flood Control Association, representing major urban interests in Fort Worth and Dallas, framed a bill which was submitted to the Department of Agriculture in December 1936, and judged to be adequate for this purpose.[56] This

[54] NARG 114 (SCS, General Correspondence, 1936–40), Mastin White to H. H. Bennett, Oct. 12, 1939; Frank H. Light, President, American National Bank, Silver City, New Mexico, to Hugh Calkins, Regional Conservator, SCS, Albuquerque, June 11, 1940.

[55] NARG 114 (SCS, General Correspondence, 1936–40), memorandum, D. S. Myer to Milton S. Eisenhower, Jan. 27, 1937.

[56] NARG 114 (SCS, General Correspondence, 1936–40), Paul H. Appleby to (illegible) Park, Ft. Worth, Texas, Dec. 31, 1936.

association had been formed chiefly to secure a municipal water supply and a canal to the Gulf Coast. Like many of the leaders of the other watershed associations, the Dallas–Fort Worth group were sold on the idea of upstream flood control. Reportedly, the Trinity River Association was told by the Army Corps of Engineers that it could do little to aid them unless runoff was controlled to reduce sedimentation. The Lower Colorado River Authority was interested in clear water in its lakes. The Nueces Watershed Association was interested in promoting water supply for industrialization.

The action of the watershed associations catalyzed a conflict. T. O. Walton, President of Texas A & M, told Bennett that these bills, patterned upon the standard act, had been introduced "under the guidance of your representatives in Texas and at the instance of Chambers of Commerce and other interested local citizens." As a result the state had been divided into two camps over what was called the "Chamber of Commerce Soil Conservation program based on the water shed district bills and a Farmers Soil Conservation program based on the principles of the Davis Bill." Walton could not understand why the Davis bill was unsatisfactory, since it was "patterned" on the wind erosion districts law and SCS was cooperating with these districts. He said he thought that all the difficulties had been ironed out in a conference in Washington with Representative Marvin Jones and that the Department would not "dictate the type of legislation the states must pass." The Department's objectives were understandable, he said, but the long view needed to be taken so that the program would not be pushed more rapidly than public support for it could be developed.[57]

Texas' Director of Extension told Bennett he thought that the Department's opposition to the extension bill was eliminated at the conference in Washington with Representative Jones, except that the Department insisted upon districts having authority to enforce land use regulations.[58]

Wallace told Walton that the principles of the standard act were sound, and there was certainly enough work for all agencies so as not to deprive any one of them of the opportunity to perform its proper functions. He thought there was no reason to be equivocal about the Davis bill which was "significantly different" from the wind erosion districts legislation and the standard act. Texas could enact any

[57] NARG 16 (Office of the Secretary; Soil 1), T. O. Walton to H. H. Bennett, Feb. 26, 1937.

[58] NARG 16 (Office of the Secretary; Soil 1), H. H. Williamson to H. H. Bennett, Feb. 26, 1937.

legislation it chose, but the Department of Agriculture was not obliged
to cooperate on such terms. "Frankly," Wallace charged to Walton, "we
do not feel that the Texas Extension Service has cooperated in this
program to the extent of providing the public with the truth about the
Standard Act or the Department's policies. On the contrary, we have
received repeated reports of opposition and misrepresentation."[59] Further conferences in Washington were fruitless. The Davis bill was
passed and vetoed by Governor James Allred upon the request of the
"Committee of One Hundred."

The sequel to these events, until a law was passed in 1939, illumi-
nates some of the reasons why Wallace and his associates favored
districts organized on watersheds. The Texas Soil and Water Conservation Association finally mustered the forces necessary to enact legislation, but only after State Senator Davis had given up his objection to
circumventing the county upon being convinced that soil conservation
districts organized on watersheds were needed to enable the Department of Agriculture to "coordinate" its flood control surveys with soil
conservation operations. He was reported to have said that he acted
upon the advice of Walton of Texas A & M, who convinced him that the
Department would not extend technical assistance in Texas without the
legislation it wanted.[60]

SUMMARY AND COMMENT

The documentary record of the interagency struggle to enact enabling legislation sharply etches the "political" character of the organizational problem.

First, the Department of Agriculture was extremely sensitive to any
proposal to vest in county governing bodies any great amount of
authority, either to create districts or to effect their operations in any
way. Wallace and his associates, as the struggle in Texas so clearly
demonstrated, wanted to insulate the administration of soil conservation from both the extension services and county political organizations
functioning as the "courthouse gang."

Second, the Department adopted the strategy of building support for
SCS by (a) co-optating local leadership previously committed to the

[59] NARG 16 (Office of the Secretary; Soil 1), Henry A. Wallace to T. O. Walton,
March 8, 1937.

[60] NARG 114 (SCS, General Correspondence, 1936–40), Henry A. Wallace to
Joseph Peacock, Executive Secretary, Texas Soil and Water Conservation Associa-
tion, Oct. 7, 1937; Peacock to M. L. Wilson, Nov. 13, 1937, and enclosures; and
Paul Walser, SCS State Coordinator for Texas, to T. L. Gaston, SCS, Nov. 3, 1939.

extension services, especially among county units of the American Farm Bureau Federation; and (b) combining erosion and flood control in a "package" program to attract urban supporters such as the Texas watershed groups to counter opposition from the extension services and their Farm Bureau allies.

Third, the Department found the strength of the links between extension and some state governors as well as members of Congress. To counter these ties, Wallace cultivated the friendship of congressmen not dependent upon extension for support at home in agricultural politics. In Texas, he even broke his rule that SCS would work only through soil conservation districts and permitted agreements to be made with the wind erosion districts in Representative Marvin Jones' district. The political power of the county judges turned out to be so strong in some of these districts that they were not replaced by soil conservation districts for twenty-five years or more.

Experience in the state legislatures during the 1937 sessions indicated on the whole, too, that the mixed responses of farm leaders and legislators would make operations through soil conservation districts much easier in some states than in others. Certainly, resistance to the districts as units having the power to enforce land use regulations was general, but not unanimous. If the Department could not gain acceptance of this key feature of the model law, any districts which might be organized would turn out to be quite different from those visualized by Wallace and Wilson in 1936.

When the legislative sessions of 1938 were added to those of 1937, twenty-seven states had passed district acts. However, only twelve states had enabling legislation which completely satisfied the Department's requirements. Four of these were in the Southeast, two in the Middle Atlantic states, two in the Middle West, one in the Northern Plains, and three in the Mountain West. Another nine states in scattered locations had laws which adhered substantially to the model act, but presented "administrative or legal problems." Six others had enacted laws which were unsatisfactory, since they failed to provide for effective enforcement of land use regulations.[61] Among these six, California did not actually pass a law in the strict legal sense until 1940

[61] NARG 114 (SCS, General Correspondence, 1936–40), memorandum for the Secretary from H. H. Bennett, Dec. 1, 1938. States in group I (with "satisfactory" laws) were Arkansas, Georgia, Indiana, Maryland, Michigan, Nevada, New Mexico, North Carolina, Pennsylvania, South Carolina, South Dakota, and Utah. Those in group II (which presented "problems") were Colorado, Florida, Illinois, Kansas, Louisiana, Mississippi, New Jersey, Oklahoma, and Wisconsin. And group III ("unsatisfactory" laws) consisted of California, Minnesota, Montana, North Dakota, Nebraska, and Virginia.

because of irregularities of legislative procedure in the 1938 act.[62] The
Governor of New York vetoed the law passed there in 1938, whether
because of extension opposition or for fiscal reasons is uncertain.[63] With
hindsight one can say that even among the dozen states which had
satisfactory laws, when measured by administrative and legal requirements, there were several in which districts were organized very slowly
against opposition from various sources. Among these were Pennsylvania, Michigan, Indiana, South Dakota, New Mexico, and Nevada.
The same was true in the other fifteen states, notably California and to
a lesser degree Colorado, Kansas, Florida, and Wisconsin.

Spokesmen for the Department optimistically made the most of this
record and claimed that districts were popular. It is doubtful that the
contending parties were deceived, for such major farm states as Texas,
Iowa, Oregon, Washington, New York, and Ohio had not acted. Except
for Vermont, most of the New England states did not pass districts
legislation until the early or middle forties. Tennessee and Kentucky
held off for a few more years, also. Under these circumstances, there
was a serious question whether the Department could adhere to its
decision to require that states authorize districts to be organized before
the Department would extend its erosion control operations through
demonstration projects. Also, there was already some question whether
any districts that might be organized would function as they were
visualized when the standard act was framed.

Districts had been conceived to focus upon "farms and watersheds"
all the research, technical, and informational facilities available
through the Department of Agriculture and the state colleges to which
it had been tied for two decades of cooperative relations. Unfortu-
nately, many of the colleges' principal administrators concluded that
the Secretary intended to make districts the administrative tool of the
Department's new special function agency, the Soil Conservation
Service. From 1937 to about 1942, when World War II obscured this

[62] NARG 114 (SCS, General Correspondence, 107A, California), Mastin White
to H. H. Bennett, May 28, 1938; J. Phil Campbell, SCS, to R. B. Couzzens, June 1,
1938; and (SCS, General Correspondence, 1940), Alex Johnson, California Farm
Bureau, to Henry A. Wallace, Feb. 19, 1940.

[63] At a meeting in December 1939, New York's Director of Extension L. R.
Simons was reported to have said that he did not want a districts law, but for two
years SCS had tried to sell the enabling legislation. He claimed that SCS had
inspired a resolution in the New York State Farm Bureau to consider the adequacy
of the existing state legislation in order to create pressure for districts legislation.
See NARG 16 (Cooperative Farm Forestry Program File), George R. Phillips,
Forest Service, to M. S. Eisenhower, Dec. 8, 1939.

issue, the major problem which the Secretary faced concerning districts was not only whether he could get the colleges to accept districts and work through them, but also whether he could persuade the bureaus of his own department to do the same thing.

CHAPTER 4

From Erosion Control to
Soil Conservation

THE PARTIAL SUCCESS of the Department of Agriculture in securing enactment of the standard soil conservation district law in some
form in twenty-two states during 1937 was rather remarkable in view of
the hostility of the Association of Land-Grant Colleges and Universities
toward organization of the districts. The Association's opposition is
understandable, however, since its leadership was convinced that the
Soil Conservation Service (SCS), more than any of the other new
"action" agencies created in the Department in the early thirties, was a
genuine threat to the long-run existence of the extension services. It is
usually assumed that administrators in the Department, the agricultural colleges, and the state extension services looked upon the Soil
Conservation Service as a special-function agency with the rather
narrow mission of carrying out an erosion control program. In fact,
however, the most perceptive of these men saw that SCS and the
districts were potentially multiple-purpose, or general-function, conservation agencies through which the Department could channel all of
its land use adjustment programs with, or without, cooperation from
the state extension services. Such a development would also cost some
of the Department's other agencies, such as the Forest Service and the
Agricultural Adjustment Administration (AAA), parts of their programs which would be transferred to SCS. In the thirties, this very real
fear was not publicized, but by 1948, when a campaign was under
way to organize every county in a soil conservation district and the
agencies were locked in a violent struggle for supremacy over the
Department's conservation programs, extension leaders openly charged
SCS with the intention of becoming a second extension service.

The Soil Conservation Service has insisted since the thirties that it
has a mandate to prepare and execute a comprehensive "national
plan for the sustained and beneficial use of soil and water resources."[1]

[1] National Archives, Record Group 114 (Soil Conservation Service, Policies and
Long-Range Program), "Report of the Committee on Long-Range Program and

In 1937, the Committee on Long-Range Program and Policies, appointed by Hugh H. Bennett, recommended a program which was then construed generally as a guide to the agency's policy—as it has been for the last twenty-five years. The Committee (known in SCS as the Watershed Committee) made five major recommendations.

First, it urged that for the next several years primary emphasis be placed on educating farmers to the need for conserving soil and the means of doing so. Once demonstration projects were gradually supplanted by soil conservation districts, their supervisors should take the initiative to secure land treatment on planned watersheds.

Second, the committee insisted that it was vitally important for SCS to develop a unified program based on both the soil conservation law of 1935 (Public Law 46) and the flood control laws so that all projects undertaken by SCS would be based on the "principle of watershed planning."

Third, SCS should recognize the competence of other agencies, but be prepared to assume its proper place along with them. It should, for example, recognize the technical competence of the Forest Service, but insist that its own responsibility was clear for all forestry work connected with erosion control and on "all farm woodlands where the farms are under cooperative agreement." Also, SCS would take the lead by assuming responsibility for regulating grazing on private lands adjacent to public lands administered by the Forest Service and the Grazing Service of the Department of Interior. And it should "advise and assist" the Agricultural Adjustment Administration to formulate and apply "sound conservation practices."

Fourth, the Committee insisted that "extension education" was not proper work for SCS, nor was it a "primary measure" for controlling erosion. Moreover, the Department was not legally dependent upon extension for "educational" services, since its organic legislation directed it to "develop and diffuse useful information pertaining to agriculture." The Committee recommended that SCS cooperate with all educational agencies, including the "vocational education system, State school boards and similar organizations." The Soil Conservation Service, therefore, should not divert any "considerable" amount of its funds to the state extension services for "educational" activities.

Fifth, the Committee recommended that demonstration projects be continued to promote erosion control, but it argued that in the long

---

Policies," mimeographed, April 1937, pp. 1–2. (National Archives records cited hereinafter usually will be identified by the abbreviation NARG.)

run there would be no sound substitute for "direct, close and intimate technical service to the individual landowner."[2]

One member of the Committee summed up its position, saying that, if the Soil Conservation Service were to function merely as an educational agency, it would be only a "question of time until our work will be merged with and absorbed by the Extension Service." Unless SCS were to become a service agency, it would lose its "identity and . . . standing as a unique governmental bureau with a distinct mission. The AAA has already stolen much of our thunder . . . and confused the issue with the public."[3]

These long-range goals which the Watershed Committee projected for SCS were, of course, alarming to rival agencies, but the Secretary of Agriculture disturbed them even more with his conception of the future of soil conservation districts. Henry Wallace urged Hugh Bennett to render maximum assistance to districts in December 1937, because they would be used to:

"unify a number of the different action programs which the Department is now authorized to carry on. Through local soil conservation districts the Department should be able to assist farmers in formulating and executing comprehensive plans for bringing about wise land use. I think it is important, therefore, that the districts should not come to be looked upon as having significance only for the program of the Soil Conservation Service,

[2] *Ibid.*, pp. 3 ff.
[3] NARG 114 (SCS, General Correspondence, 1936–40), John F. Preston, Woodland Section, to C. B. Manifold, Chief of Operations, SCS, March 10, 1937. The SCS has encouraged an "educational" program through teachers of vocational agriculture. On this subject, see John Dale Russell and Associates, *Vocational Education* (Washington: Government Printing Office, 1938), Staff Study No. 8, prepared for the Advisory Committee on Education, pp. 147–48. "As now administered the Future Farmers of America and the 4-H Clubs are to some extent competing organizations. . . . In some areas there has been considerable friction between extension workers and teachers of vocational agriculture. . . . In other areas there has been a high degree of cooperation." The authors warned: "The danger of using an organization such as the Future Farmers of America for purposes of publicity and propaganda on a national scale is far from imaginary. Organizations of this sort are almost inevitably guided by a limited group of adults, and if the adult leadership falls into the hands of persons who are not both wise and disinterested, more harm than good can easily result." In 1935, Bennett sent Wallace a plan for setting up a "conservation education" plan in the public schools. He said that some work was being done with adults through the vocational agriculture teachers and he had agreement from the U.S. Commissioner of Education that a program was "feasible." Bennett suggested that the Department set up "summer institutes and training courses" for these teachers. Wallace approved. See NARG 16 (Office of the Secretary), April 8, 1935.

but that they should be seen as local governmental units, organized demo-cratically . . . and possessing the necessary governmental power to . . . carry on well-rounded agricultural programs. The districts will need the help of most of the agencies of the Department and, in turn, they can help the Department as a whole to carry out more effectively the various programs it is administering."[4]

Wallace's desire to make this revolutionary change in agricultural administration apparently became known almost instantly, since Assistant Secretary of Agriculture Harry L. Brown gave his assurances immediately that the AAA county committees had no authority under, nor were they bound by, the standard soil conservation districts law.[5] Nevertheless, both AAA, and the Extension Service were so concerned that SCS and the conservation districts might be directed to assume responsibility for important phases of their programs that they took defensive steps. The Agricultural Adjustment Act of 1938 expressly directed that the county agent might be elected secretary, or serve ex officio, of the AAA committee organized in each county and also forbid these committees to include "more than one county or parts of different counties."[6] These provisions were clearly intended to prevent the Soil Conservation Service and districts organized on watersheds from becoming the primary or exclusive agencies for administering the Department's land use adjustment programs.

For the next decade or more the interested agencies struggled over this fundamental issue. For a brief time after a reorganization of the Department of Agriculture in 1938, the Soil Conservation Service appeared to be achieving the primacy postulated by the Committee on Long-Range Program and Policies. But, in the meantime, the districts failed to serve some of the basic objectives which they were originally created to achieve. In part, these failures are attributable to both the Extension Service and the SCS.

[4] NARG 114 (SCS General Correspondence, 1937, 107A), "Memorandum for Mr. H. H. Bennett," from the Secretary of Agriculture, Henry A. Wallace, December 10, 1937. Two years later, when the Department was still trying to get Arizona to adopt districts legislation, Under-Secretary Milburn L. Wilson told Alfred Atkinson, President of the University of Arizona, that there "is already reason to believe that soil conservation districts may become the nucleus of future efforts to bring about sorely needed adjustments in our land use and agricultural economy." NARG 114 (SCS General Correspondence, 1939, 107A), letter, Wilson to Atkinson, February 7, 1939.

[5] NARG 16 (Office of the Secretary), Harry L. Brown to E. E. Scholl, Acting Director of the Oklahoma Agricultural Extension Service, Dec. 20, 1937.

[6] 52 *Stat.* 31, Sec. 8.

A LIMITED COMMITMENT TO DISTRICTS

In December 1936, when Secretary Henry A. Wallace decided that states would have to enact district enabling legislation if the Soil Conservation Service were to locate new demonstration projects within them after July 1, 1937, he also agreed that this deadline might be postponed under some circumstances. On this basis, in June 1937, Bennett told Wallace that he would like to organize immediately about two-thirds of the 350 demonstration projects SCS planned to have established by 1944. Bennett did not want the Department to require that a district be organized around each project, but if the Secretary insisted, Bennett wanted to enter into working agreements directly with farmers and not with district governing boards. He also wanted district boards, if he had to work with them, to understand clearly that "demonstration work" would receive priority over "district work." Bennett contended that the demonstrations would give farmers time "to recognize the need for erosion control work" and to "desire a soil conservation district." He thought that demonstrations would also convert the personnel of the state experiment stations and extension services to the SCS program. The interval would provide time for the state soil conservation committees to "perfect" their policies and procedures for organizing districts. He recommended, therefore, that the Secretary permit SCS to continue to establish its complete demonstration projects in any state, so long as the projects were justified by erosion problems. He would make an exception, however, by not setting up any new demonstration projects in those states which had passed enabling legislation which was not satisfactory to the Department. He wanted authority, also, to set up a demonstration project even within each soil conservation district which might be organized during the next few years.

Bennett was deeply concerned by the sentiment against land use regulations evident in several of the state legislatures during the recent sessions. He recommended to Wallace, therefore, that for at least the "next two or three years" no soil conservation district should be required to enact land use regulations as a condition of receiving SCS assistance. The idea of regulation, he said, ought to be introduced "gradually" and not "abruptly" to the nation's farmers. He confessed, too, that the Soil Conservation Service needed more information about sound practices before recommending regulations which would stand any tests. He was afraid that regulations based on the information then available would apply "low standards" and make later "deviation difficult."

In order to satisfy Wallace's directive that he promote soil conservation districts, Bennett recommended that SCS provide districts with technical service and supply the transportation, equipment, space, and clerical assistance for SCS personnel. In a "very few cases," SCS would also provide funds to purchase supplies, equipment, labor, and new or uncommon planting stock for demonstration purposes. The districts would be expected to provide "all other" equipment, labor, supplies, and space requirements. Since they were not expected to have the power to tax and the state governments were not interested in assuming new obligations to finance this program, it was evident that the districts would either go without supplemental aid or they would have to bargain for it from one other likely source—the Works Progress Administration (WPA). As governmental subdivisions of the states, the districts would be eligible for grants and, eventually, 69 districts in 18 states received WPA grants.

Wallace approved Bennett's recommendations just a week before the July 1 deadline on the condition that new demonstration projects were to be held to the "lowest possible minimum in order that there may be as great as possible a transfer of funds and activities from demonstration projects to cooperative activities with soil conservation districts."[7]

In December 1937, when Wallace told Bennett that SCS should accelerate its work through districts so they might in the future unify a number of action programs, he was disturbed because the memorandum of understanding proposed by the SCS to be signed by the Department and districts implied that the former was not yet "ready and willing to offer them substantial assistance." He thought that this situation would discourage the organization of districts. He wanted it made clear, instead, that the Department was ready to offer "substantial" aid, possibly more to those districts initially organized than to those established when their number became large. For this reason, he urged Bennett to find ways to see how quickly existing demonstration projects could be put on a maintenance basis to "release a rapidly increasing share of the annual appropriations . . . for expenditure through the districts." He expressed, also, his concern for the "confusion" in states already having districts laws, since bad adminis-

---

[7] NARG 114 (SCS, General Correspondence, 1936–40), memorandum to the Secretary from H. H. Bennett, June 9, 1937; memorandum, Chief of SCS, to Regional and State Offices, June 25, 1937.

tration through the improper location of boundaries, inadequate public discussion, and unwise district programs might set back the whole program.[8]

Bennett took Wallace's mild admonition and pondered the problem with his staff for five months before making a counterproposal. He finally recommended a refinement of the principle that the amount of SCS assistance to districts be differentiated, depending on the degree to which each state's enabling legislation conformed to the Department's requirements. Under existing state legislation, this policy would enable Bennett to devote the bulk of SCS manpower, funds, equipment, and other resources to a variety of projects other than districts. Bennett proposed to Wallace, therefore, that the maximum of SCS service, equipment, materials, labor, and other assistance be given to districts only in those states whose enabling acts "fully meet the Departmental standards." To districts in those states whose legislation "departs substantially" from the standard act "assistance may be made available prior to such time as the basic statute can be adequately amended but on a limited scale only."[9] After two more months of discussion, Wallace approved Bennett's proposal on June 7, 1938. By the end of the year, only twelve states had enacted legislation which "fully" met the Department's standards.

In December 1938, Wallace forced Bennett to extend full assistance to all districts, except those in states which had not granted them effective power to enforce land use regulations. Districts in this class received only SCS technical assistance, but not camp labor and equipment from the Civilian Conservation Corps (CCC), or SCS equipment and materials.[10]

Bennett formulated this classification scheme to limit the amount of assistance which SCS would be required to give to districts while it concentrated on demonstrations and several other types of operations. Secretary Wallace accepted it because he wanted to induce states to enact the sort of legislation the Department wanted as a condition for receiving SCS assistance. It may have been effective, although several states passed laws which did not conform in major respects to the standard act and a few others amended their original acts in ways not calculated to please the Department. In 1939, Virginia was shifted from group III—states with "unsatisfactory laws"—to group II—those which

[8] NARG 114 (SCS, General Correspondence, 1936–40), Henry A. Wallace to H. H. Bennett, Dec. 10, 1937.

[9] NARG 114 (SCS, General Correspondence, 1936–40), memorandum for the Secretary from H. H. Bennett, April 29, 1938.

[10] NARG 114 (SCS, General Correspondence, 1936–40), memorandum for the Secretary from H. H. Bennett, Dec. 1, 1938.

presented "problems"—when assurances were given the Department that efforts would be made at the earliest opportunity to secure the amendments considered appropriate. North Dakota was shifted from group III in 1938 to group I—states with "satisfactory" district laws—in 1939, when its statute was amended to provide for the enactment and enforcement of land use regulations on all lands, and not just those of landowners who signed cooperative agreements with districts. Although Vermont in 1939 enacted a law which made it doubtful that land use regulations could be enforced, it enacted changes satisfactory to the Department in 1941. Montana enacted a particularly unsatisfactory law in 1937, but made desired changes in 1939. Iowa enacted its first districts legislation in 1939, and was promptly put in group III for failure to provide adequately for land use regulations. Arizona and Ohio did the same thing in 1941. In Indiana, during the 1941 session of the legislature, farmers from existing districts appeared and "furthered the prevention of an amendment . . . which . . . would have seriously limited the authority of the districts."[11] The New Jersey law was "questionable" because it permitted district supervisors to submit land use regulations to the state soil conservation committee for approval. The Wisconsin law was extensively amended in 1939 to permit a county board of supervisors to organize a district, using county boundaries and not a watershed as the basis of organization. It vested in the agricultural committee of the county governing board the authority of soil conservation district supervisors, and provided that only the county board might enact land use regulations, if two-thirds of the land occupiers affected approved in a referendum. Even these material deviations from the standard act were not considered by SCS to be serious enough to make it impossible to cooperate in Wisconsin under this statute.[12] Despite these experiences, Bennett concluded that the classification scheme had "materially influenced amendments in several states . . . simplifying administrative procedures and adding authority to districts."[13]

Although Bennett was successful in limiting the SCS commitment

[11] NARG 16 (Soil 1, Cooperation 3), memorandum, H. H. Bennett to the Secretary of Agriculture, June 24, 1942. At this time, states still classed in Group III (those with "unsatisfactory" district laws) were: Arizona, Idaho, Iowa, Maine, Minnesota, Nebraska, New York, and Ohio.

[12] NARG 114 (SCS, General Correspondence, 1936–40), memorandum, "Analysis of Wisconsin Soil Conservation Districts Law," Nov. 25, 1939.

[13] NARG 16 (Soil 1, Cooperation 3), memorandum, H. H. Bennett to the Secretary of Agriculture, June 24, 1942. This classification scheme was still in use in 1947. See NARG 16 (Soil 1), T. L. Gaston to Secretary Clinton P. Anderson, with attached legal analysis by E. M. Shulman, Acting Solicitor, Aug. 18, 1947, approved Aug. 26, 1947.

to districts for several years, he did not reach his goal of 350 demonstration projects by 1944; the total never exceeded 182.[14] The demonstrations undoubtedly were important in promoting SCS work, since
most of the earlier districts were organized near or around them.
During the five years of operations between 1937 and 1942, SCS
devoted most of its manpower and other resources to a variety of
projects administratively organized to carry out its various responsibilities. In 1940, for example, CCC camp areas covered over 30 million
acres; water facilities areas included over 107.5 million acres; land
utilization projects included over 61 million acres; and forestry projects
embraced over 15.5 million acres. There were also 96 "soil conservation
areas" in 1940, and 233 soil conservation districts encompassed a total
area of a little more than 146 million acres. One year later, SCS
reported 462 soil conservation districts covering a gross acreage of
almost 262 million acres, and the area reported as part of demonstrations had dropped sharply to less than 13.5 million. Surveys under
the flood control acts were also under way.[15]

EFFECTS OF SCS-EXTENSION RELATIONS
ON DISTRICTS

The effects of disharmony between SCS and the leading administrators in the land-grant colleges on the growth and operations of soil
conservation districts are not easy to gauge since so many people were
involved. A decentralized pattern of conflict or cooperation developed
early and has tended to remain relatively stable over more than two
decades. It is difficult to determine how much the extension services'
criticism of SCS operations and procedures limited the operations of
districts, since SCS was not fully committed to work through them for
several years. The little documentary evidence available indicates that
often it was both. While the state directors of extension did not
welcome SCS or soil conservation districts, most permitted them to be
organized. A few fought them relentlessly and demonstrated what all
the state services could have done to thwart the Department had they
chosen to do so. County agents acted the same way.

[14] *Department of Agriculture Appropriation Bill for 1943*, House, 77 Cong. 2 sess.
(1942), pp. 945–47.
[15] *Report of the Chief of the Soil Conservation Service* (Washington: Government Printing Office, for the years 1936–51); see especially the reports for *1937*, p.
17; *1938*, p. 23; *1940*, p. 8; and *1941*, pp. 5 and 70.

In 1942, Dillon S. Myer testified for SCS that county agents had conducted much of the educational work to organize about 90 per cent of the 700 districts which had been organized. He said that the "service in a great many places is not too intensive, but it has been helpful."[16] In 1944, J. C. Dykes, Assistant Chief of SCS, testified that there were still "some sizable areas in the country where the district program has not yet taken hold." Alabama and South Carolina were completely organized into districts, but "we find, particularly in the West, quite large areas where the soil conservation districts are rather scattered at the present time." In most areas where there had been demonstration projects, people soon organized districts, but there were others in which these projects had not had this desired effect. "For example here . . . is the big Central Valley of California, which is one problem area . . . the farmers have not yet seen fit to go very strong for soil conservation districts." Senator Richard B. Russell remarked that northern Georgia contained an area not yet organized and recalled that, although the land needed treatment very much, there had been "two or three" elections at which the proposal to organize a district had been voted down. It was his recollection, he said, that a report was circulated to the effect that, if one were organized, "the Government took over your land . . . it definitely influenced a considerable number of votes."[17]

Myer and Bennett had testified a few years earlier that most referenda had resulted in favorable votes (78 per cent by January 1941), and that several voted down the first time had later been approved. Bennett cited several of the stock arguments used on the hustings to stymie organization of districts. "They speak of regimentation, and 'the Federal Government is going to take over your land.'" Myer identified some of the opposition as urban. Some of the insurance companies and other absentee landlords possessing large holdings supported districts and others "have gone more cautiously on it." There were not many instances where districts had been voted in only because of an interested minority, Myer said, unless one measured this proportion of interest by "the whole eligible voting list."[18] Bennett once testified that some state soil conservation committees had refused to organize districts because SCS did not provide enough personnel to

---

[16] *Department of Agriculture Appropriation Bill for 1943*, House, *op. cit.*, p. 236.

[17] *Department of Agriculture Appropriation Bill for 1945*, Senate, 78 Cong. 2 sess. (1944), pp. 401–03. The Georgia area referred to by Senator Richard B. Russell was in Tennessee Valley Authority territory.

[18] *Department of Agriculture Appropriation Bill for 1942*, House, 77 Cong. 1 sess. (1941), Pt. 1, pp. 1030–31.

serve them. He did not hint that his own deliberate policy might have had a bearing on their inactivity.[19]

Congressmen familiar with agricultural politics understood that friction was common. The President of the American Farm Bureau Federation rather regularly testified in favor of reduced appropriations for SCS. In 1942, for example, Edward A. O'Neal urged the Senate to cut "at least" $5 million from the SCS estimates. It "has done some splendid work . . . [but] it has been extremely costly." He claimed that the "soil and moisture conservation and land use operations, demonstrations and information" are the "same type of work carried on by other agencies in the States for many years, except that the Soil Conservation Service carries out this work on a much more elaborate and expensive scale."[20] O'Neal's listeners understood that the "other" agencies were the extension services, financed in part by the AAA, which "recognized" their technical competence.

For several years, the two parties wrangled over their respective roles in conservation "education." In 1938, SCS decided to terminate its financial support of conservation assistants to county agents in states where it had worked out such agreements. It did continue to assist extension services to employ state extension conservationists, although the total employed never reached the goals projected in the middle thirties.[21] In 1940, Bennett and M. L. Wilson, who had become Director of Extension in the Department, issued a joint statement which recognized the Extension Service as the "primary channel" for disseminating agricultural information. The Soil Conservation Service agreed to work through Extension to the extent that both parties might "mutually agree"; SCS promised to let Extension clear information about SCS programs in each state provided that Extension "agrees at the same time to share responsibility for carrying forward adequate information" on SCS's program.[22] This announcement was prompted by the refusal of directors of extension in some states, especially Texas and Wisconsin,

[19] *Department of Agriculture Appropriation Bill for 1945*, House, 78 Cong. 2 sess. (1944), pp. 1041–42. See also *Department of Agriculture Appropriation Bill for 1947*, House, 79 Cong. 2 sess. (1946), pp. 1016–17.

[20] *Department of Agriculture Appropriation Bill for 1943*, Senate, 77 Cong. 2 sess. (1942), p. 730.

[21] NARG 114 (Soil Conservation Service, General, 107A), memorandum, I. L. Hobson to J. Phil Campbell, Division of Cooperative Relations and Planning, SCS, June 30, 1938; *Report of the Chief of the Soil Conservation Service, 1944* (Washington: Government Printing Office), pp. 32–33.

[22] Quoted in W. Robert Parks, *Soil Conservation Districts in Action* (Ames: The Iowa State College Press, 1952), pp. 200–01.

to use promotional materials prepared by SCS.[23] Five years later, the situation had not improved in Minnesota where extension leaders wanted SCS to agree that all "educational" work preparatory to organizing a district should be handled by the extension service. In fact, they insisted that all inquiries from farmers who wanted a district organized would have to be cleared by the state director of extension and not the state soil conservation committee. All conservation plans were to be based on a program of "soil management" recommended by the University of Minnesota. And SCS could present a program of "soil conservation" only at the end of the "educational phase." The Minnesota leaders insisted that they were operating according to the 1935 report of Secretary Wallace's interbureau committee, as if no changes had been made in the Department's policies since then.[24]

A district supervisor from North Carolina explained to members of the House of Representatives that the ideal procedure was for someone to observe the need for a district and to ask the appropriate county agent to hold an "educational meeting in such-and-such a community." The agent would hold the meeting to explain the services which all agricultural agencies, including SCS, could render through a district. "But it does not work like that all the time. Here is what happens lots of times . . . we have county agents that will not even tell a group of farmers that there is a soil-conservation technician back there that can help them. He just will not do it. Then we have soil-conservation people, perhaps, who will not tell what the Extension Service is doing." A soil conservation district supervisor from Texas explained that "is what we first understood was their prerogative . . . to promote that district by educating the people to see the necessity of a district." The county agent "can more effectively do educational work, working with the organized bodies, than he can if he does action work." Some county agents, he added, had carried out "splendid" educational programs for districts.[25]

Both parties to this squabble understood that the Extension Service claimed the right to make initial and exclusive contacts with

[23] NARG 33 (Federal Extension Service, General Correspondence, 1936–45), W. W. Clark, Associate Director of Extension, to Reuben Brigham, Assistant Director, Federal Extension Service, Feb. 4, 1939.

[24] NARG 33 (Federal Extension Service, General Correspondence, 1936–45), Paul M. Burson to J. L. Boatman, Aug. 14, 1944; and memorandums, Boatman to Hobson, Aug. 30, 1944; Hobson to Boatman, Aug. 31, 1944; and B. F. Rusy to Reuben Brigham, May 22, 1944.

[25] *Soil Conservation,* Hearings on H.R. 4150, H.R. 4151, and H.R. 4417, House Committee on Agriculture, 80 Cong. 2 sess. (1948), pp. 203, 205.

farmers in order to maintain its identity and to advocate practices which sometimes differed from those of SCS. A determined state extension service could keep SCS from operating through districts in the state if only the Secretary of Agriculture could be induced to follow literally the recommendations by his interbureau Committee on Soil Conservation. Minnesota was one of the states in which few districts were organized until after 1945. This issue was not settled generally until changes were made in the composition of state soil conservation committees, starting about 1945. These committees then began to function as the Department originally expected. They encouraged farmers to organize districts with or without help from the extension services.

Some of the state directors of extension were not pleased that the Department developed detailed procedures for district operations. Administrative guidance of this sort presumably was to be left to each state soil conservation committee, since the standard act anticipated that the state committee would provide assistance to districts after it had encouraged their organization. Instead, the Department prepared detailed procedures and forms of agreements to cover the whole range of district operations. Each district was required to formulate a long-range program in order to receive technical assistance from the Department. It was next required to make a work plan with the assistance of all agricultural agencies and farm organizations within the district before submitting the plan to the state soil conservation committee for approval. Each district's program and work plans were then reviewed within the Department by the Solicitor, the Coordinator of Land Use Planning, and the SCS. The Department supplied forms for a memorandum of agreement with each district, a supplemental memorandum of understanding between SCS and each district, and an agreement between the district and each farmer who desired to become a "cooperator" eligible to receive assistance. The Department's Solicitor prepared detailed forms which could be used by state soil conservation committees to perform their various functions. Also, the Solicitor gave legal opinions of proposed state legislation, but with the understanding that they were not substitutes for the opinions of duly constituted state officers.[26]

In January 1939, the Committee on Extension Organization and Policy of the Association of Land-Grant Colleges met in Washington to discuss the procedures for organizing districts, developing their work

[26] NARG 114 (SCS, General, 107A), memorandum, H. H. Bennett to staff, Sept. 13, 1937.

plans, and coordinating their educational procedures and operations once districts were formed. There was considerable grumbling about the detailed and lengthy requirements which the Department had imposed. The Wisconsin director of extension complained that the Department was simply trying to control each district's program. Wallace answered that this charge was not true. The Department did not force any district to accept its assistance, but once the district did so it would have to conform because the Department was charged with "the responsibility of seeking permanent results from the expenditure of federal funds for erosion control." Wallace informed the Wisconsin soil conservation committee, of which the director of extension was the chairman, that it rejected a memorandum of understanding with one of the districts in Wisconsin because it refused to require a written agreement with each farmer who received assistance.[27] The Department insisted that a complete farm plan was the necessary foundation for sound operations. The agreement between the district and the farmer was to insure that the farmer would use his land properly after he and the technician had worked out the plan. The agreement was intended to insure that the plan would be executed progressively, but not necessarily at once, under the supervision of the district governing board. Similar difficulties had to be worked out with a conservation district in New Jersey where supervisors complained that the agreement with the Department made them responsible for collecting liquidated damages from any farmer who failed to live up to his agreement. M. L. Wilson emphasized that a decision of this sort was properly one for the supervisors to make, but that only the supervisors were in a position to determine cases of such a failure.[28]

In 1940, SCS and extension representatives held another conference. A spokesman for the Committee on Extension Organization and Policy specified several points of friction, although he also had praise for some of the work of SCS. He indicated that extension did not object to official SCS policy so much, but that some of the SCS regional officers did not carry out the spirit of these directives. Echoing the objections of the Wisconsin Extension Service in 1939, the committee recommended that SCS no longer require each district to prepare a long-range program and a plan of work to execute it. Many extension workers objected to the "excessive" detail of both documents which were to be prepared before the Department entered into a memorandum of

[27] NARG 16 (Soil 1.3), Henry A. Wallace to Noble Clark, Chairman, Wisconsin State Soil Conservation Committee, March 23, 1939.
[28] NARG 16, *ibid.*, M. L. Wilson to Henry W. Herbert, March 25, 1939.

understanding with a district. It was argued that the detailed technical information required was beyond the competence of the supervisors and they ought not, therefore, to be required to write out a technical report which inevitably would be prepared by SCS personnel. For example, the program and work plan for one Wisconsin district totaled sixty-five pages and was required to be made in seven copies. Some extension representatives thought that supervisors could spend their time more effectively by promoting soil conservation among their neighbors. The SCS procedures for making farm plans were so detailed, they said, that each technician could make only about twenty-five a year.

The extension spokesmen also requested that SCS assign its personnel to each district board immediately after it was organized to assist in preparing the program and plans. This recommendation was based on complaints from several states that SCS did not have, or failed to provide, sufficient staff to service districts. The Extension Committee also recommended that after each district had prepared its program, it execute a "general" memorandum of agreement with the Department and follow it with separate agreements with each federal and state agency so as to specify the services each would render to the district. This last recommendation would, for example, enable each state extension service to agree with each district on the extent to which it would use its personnel for district "educational work." There was a feeling among some of the extension representatives that the Department and SCS were asking it to devote too much of the county agent's time to this activity, so long as the Department was unwilling to use SCS funds generously for the joint employment of assistant county agents. Several directors hoped that all the SCS regional offices would permit the SCS state coordinators to develop state programs, since specialists at the colleges were still convinced that SCS failed to consult them in working out technical recommendations.[29]

The discussions in March and April 1940 stimulated a "major staff conference" on the part of SCS during the first week in May to achieve "greater accomplishments in connection with the districts program." Philip M. Glick, who had continued to work closely with SCS in matters relating to the districts, reported the techniques of the demon-

[29] NARG 114 (SCS, General, 104), F. A. Anderson, Director of Extension for Colorado and Chairman of a Special Subcommittee of the Committee on Extension Organization and Policy, to C. E. Brehm, Director of Extension, University of Tennessee, and Chairman of the Committee on Extension Organization and Policy, March 29, 1940.

stration projects were being used erroneously in districts and "limiting progress." He found that district supervisors were not prepared to function effectively in their new positions and the instructions from Washington to the field had been "too many and too detailed . . . resulting in confusion."[30]

C. B. Manifold, SCS Chief of Operations, said that SCS was responsible for several of the criticisms offered by the extension directors. Manifold called for better coordination between the Operations Division and the Division of Cooperative Relations and Planning which was responsible for securing the adoption of necessary state legislation, the organization of districts and interagency relations, especially with the Extension Service. Manifold wanted it clear that the Operations Division was responsible in districts once they were organized. He also said bluntly that before a district was organized and made a program and work plan it was essential that farmers understand the reasons for such an organization. "Interests other than conservation of the soil and water sometimes are the real reasons." District programs should express the "honest objectives of the supervisors." They should not be "dressed up for clearance, with possible hesitation later on doing the things included which were required for Departmental cooperation." He warned that the employment of "Service personnel to stimulate interest in the formation of new districts should be used with caution and discretion at the present, where it appears that more districts are being formed than can be given attention with present resources without jeopardizing the effectiveness of the Service program."

Manifold had other criticisms, too. The time of the technicians of his Operations Division should not be used for "educational" activities, especially after districts were organized, certainly not without the full consideration and approval of his Division. More research was "greatly needed . . . [to] give adequate information on the validity of our field work, but the experimental work should not be mixed with any district's program." Supervisors agreed to work with SCS, he observed, because they believed that "we know what to do and will do it in line with a correct work plan developed by the supervisors." In working with farmers, SCS personnel had to realize that each has his own problems and conservation plans must meet their needs. The SCS farm planner needed to make an "honest plan without being hampered by rigid restrictions and procedures, and still stay within the limits of

[30] NARG 114 (SCS, General Correspondence, 1940), memorandum from C. B. Manifold to H. H. Bennett, May 22, 1940.

sound quality and practical applications." Supervision should consist of "suitable criteria" produced by staff technicians and review of plans by higher offices should be only for general control.

In working with district supervisors, Manifold continued, SCS personnel needed to develop harmonious relations so that the governing boards could "be led gradually from an incorrect concept of conservation and suitable practices to accept a more nearly correct understanding and thereby revise or amend district programs and work plans to proper objectives. Technicians must avoid open resistance to the supervisors but tactfully assist them to make use of established facts; otherwise, successful continuation of district program is doubtful." Manifold also argued that the "agreement docket," consisting of the farm plan and the "contract form," should be modified by eliminating the latter. The "farmer-district agreement" which had been a bone of contention with the Wisconsin and New Jersey extension services, should be replaced by assurance from districts "of getting the work done by the supervisors in such simple form as they desire." The Department would be protected by the memorandum it signed with each district. The "contract" type of agreement had been carried over from the demonstration projects where it was necessary to protect federal investment of service, labor, equipment, and materials used on the private lands of farmers for their direct benefit and a presumed public good.[31]

The outcome of these conferences and SCS self-examination was a statement by the new Secretary of Agriculture, Claude R. Wickard, on September 21, 1940, making significant changes in district operations. Under the new policy, the Department would enter into a memorandum of agreement with each district as soon as a program had been prepared and any agency of the Department was ready to assist it. "The Department will assume no responsibility for the soundness of the program and will neither approve nor disapprove it. Each agency will be free, upon examining the program, to determine the degree to which it will assist toward carrying out that program." For this reason, Departmental clearance of the district programs and work plans was terminated. "Each agency" working with districts would be free to consult with district boards to secure modifications it considered necessary, but this would no longer be a departmental matter. Each agency would formalize its intentions to assist districts by entering into supplemental agreements with them. It was expressly provided, also,

[31] *Ibid.*

that if any agency of the Department were to assist a district, the state coordinator of SCS would be responsible for "arranging discussions . . . between agency representatives."[32] One of the most significant implications of this directive was that it terminated, in effect, joint preparation by the SCS and extension services of district programs and left the technical operations, including planning, wholly in SCS hands. It implied with equal force that SCS would work with any district regardless of whether the latter had "cleared" its plans with the state soil conservation committee. This last change, undoubtedly, was a major objective of the directive.

Four years later, Representative Clarence Cannon of Missouri introduced a proviso to the departmental appropriation bill for 1945 which would have required that all agreements between districts and agencies of the Department have the approval of the state soil conservation committee prior to execution. The Department opposed this "Cannon Amendment," saying that it would reinstate procedures tried during the earlier years and abandoned because they had "discouraged district boards from taking the initiative or responsibility by the farmers directly concerned so that districts did not function as independent, self-governing units of local government as desired and contemplated in the state laws."[33] Senator Russell put his finger on the raw issues by remarking that: "If you had an antagonistic State body, either to a certain area within the State or to a Federal operation, it seems to me they could practically prevent the work of an independent soil conservation district within the State." An SCS spokesman agreed.[34] The Senate Appropriations Committee, observing that comity peculiar

[32] "A Statement by the Secretary of Agriculture Concerning Departmental Cooperation with Soil Conservation Districts," mimeographed, Sept. 21, 1940, 7 pages. This memorandum reflected a nearly complete breakdown in the Department's relations with the land-grant colleges and the American Farm Bureau Federation in 1940. The future of the Soil Conservation Service and districts in agricultural administration was a major issue, but the colleges and the American Farm Bureau Federation were also disturbed somewhat by the Agricultural Adjustment Administration and the Farm Security Administration. The Farm Bureau broke with the Department of Agriculture on the ground that the Department had become "too partisan" by the time of the presidential election of 1940. One competent observer has concluded that by this time the President of the Farm Bureau, Edward A. O'Neal, intended to place the state extension services under the domination of his organization, as his plans for reorganizing the U.S. Department of Agriculture in 1940 indicated. O'Neal charged that the U.S. Department of Agriculture was not "representative" of farmers. See Christiana M. Campbell, *The Farm Bureau and the New Deal* (Urbana: University of Illinois Press, 1962), pp. 156–78.

[33] *Department of Agriculture Appropriation Bill for 1945*, Senate, *op. cit.*, p. 25.

[34] *Ibid.*, p. 407.

to legislative bodies, amended the "Cannon Amendment" to apply only in Mr. Cannon's home state of Missouri.

The 1940 directive made another significant change by altering the character of the district-farmer agreement under criticism not only from extension leaders but also the Operations Division of SCS. The new form of memorandum of understanding between the Department and districts provided that each district would enter into agreements with owners and operators, "fixing the responsibilities of the parties in carrying out . . . [conservation] plans." The district was to be responsible "for determining the kind and amount of erosion control and soil conservation work to be performed by it on individual farms and ranches, and for seeing that provisions of agreements it enters into with owners and operators of land are carried out." To implement this policy, Bennett directed the field forces of SCS to discuss appropriate forms of the agreement with district boards without distributing his memorandum to them. To protect the federal investment the "district *may* at its option terminate the agreement and be reimbursed by the farmer for the value of the labor and materials, and the rental value of the equipment made available to him." Therefore, in the event of careless or intentional misuse of materials or equipment, the form of the agreement "may provide that . . . such assistance . . . is to be disposed of and used only as provided in the plan."[35] Coupled with a directive of May 22, 1939, which had allowed SCS technicians to provide plans for farmers who had not signed agreements with the district boards, this shift of responsibility was undoubtedly intended to smooth SCS relations with farmers and supervisors and to reduce one of the critical sources of disagreement with the extension services in several states. More important, this dilution of each district board's responsibility for securing permanent results from farmers changed fundamentally and permanently the character of soil conservation districts. It may not have been pure coincidence that Wickard issued his directive in September 1940, just as another presidential campaign was under way.

SCS PROGRAM EXPANDED

At the time Bennett was transferred to the Soil Erosion Service in the Department of the Interior he was convinced that the nation needed a

[35] Soil Conservation Service, memorandum D.C. 21, mimeographed, Feb. 1, 1941.

permanent program of erosion control directed by a single unified agency. This idea was also widely accepted by many leaders in resources policy planning, including Secretary of Agriculture Wallace. One difficulty which arose soon after responsibility was transferred to the Department of Agriculture and the Soil Conservation Service established, however, stemmed from the way Bennett and his associates in SCS interpreted Public Law 46 of 1935. They believed it vested in SCS alone a mandate and a unique mission to plan and execute a national program of soil and water conservation. But Secretary Wallace interpreted the legislation to mean what it said; the Secretary and the Department of Agriculture were given authority to carry out a coordinated program. When Wallace reorganized the Department in 1938, however, he used provisions of Public Law 46 to transfer several functions from other agencies of the Department to SCS. By so doing, he probably contributed to the belief that the responsibility of the Department and one of its bureaus was identical. To a degree, no doubt, this belief was reinforced when Wallace occasionally expressed his hope that on the level of local government the full range of the Department's soil conservation operations could be "focused on farms and watersheds" through soil conservation districts.

The responsibilities which Wallace assigned to SCS, starting in 1937 and 1938, included the Water Facilities Act of 1937 (known also as the Pope-Jones Act) to permit the construction of water facilities, to develop conservation plans, and to provide technical service on farms and ranches. Secretary of Interior Harold L. Ickes complained that this legislation permitted the Department of Agriculture to set up "another Bureau of Reclamation" in the West. It was "slipped over" on him, he complained, because his "own people were asleep at the switch."[36] Also in 1938, SCS was given responsibilities for administering the programs for land utilization and retirement of submarginal lands under Title III of the Bankhead Jones Farm Tenant Act of 1937 and the farm forestry program provided by the Cooperative Farm Forestry Act of 1937 (the Norris-Doxey Act).[37] During the following year, the work of the Division of Irrigation and Drainage of the Bureau of Agricultural

[36] E. G. Nixon (ed.), *Franklin D. Roosevelt and Conservation, 1911–1945* (Hyde Park: General Services Administration, National Archives and Records Service, Franklin D. Roosevelt Library, 1957), Vol. II, p. 375. The Water Facilities Act, 50 *Stat.* 869, like Public Law 46 of 1935 and the Flood Control Act of 1936, amended, permitted the Secretary of Agriculture to require appropriate state legislation to enable the Department of Agriculture to carry out the purpose of the Act.

[37] Secretary of Agriculture, memorandums 785 and 790, 1938.

Engineering was transferred by the Secretary of Agriculture to the Soil Conservation Service, and a few months later 38 CCC drainage camps formerly under the Bureau of Agricultural Engineering were transferred to SCS for administration.[38] Responsibility for flood control operations on agricultural lands was also vested in SCS by the Secretary.[39] When these additional duties had been laid upon SCS by the 1938 reorganization of the Department of Agriculture, Secretary Wallace explained to a displeased representative of the Association of State Foresters that SCS was "no longer concerned only with soil erosion control . . . The reorganization of the Department placed in the Soil Conservation Service responsibility for carrying out five major activities of the Department: namely, erosion control, flood control work relating to the land in agricultural areas, development of water facilities in arid and semi-arid regions, submarginal land acquisition, development and administration, and farm forestry."[40]

The farm forestry program authorized by the Act of 1937 was the outgrowth of rather distinct trends in public policy. One was to set aside land under public ownership, regulation, and management to reserve forests for an adequate supply of timber, to preserve the aesthetic beauties of woodlands for recreational purposes, and to conserve soil and water resources by protecting watersheds from erosion and destructive flooding. Programs of afforestation were recommended to provide cover and erosion control on croplands no longer suited for production. A second trend, started in the Clarke-McNary Act in 1924, was to improve management in small private forests with public educational programs and better fire protection. This Act vested responsibility for educational work in the agricultural extension services. The U.S. Forest Service and state forestry agencies were made responsibile for research, improved fire protection, and direct assistance to landowners. This program was extended by the Cooperative Farm Forestry Act of 1937, which provided also for demonstration farm forests.[41]

The Committee on Extension Organization and Policy of the Association of Land-Grant Colleges held numerous meetings with Secretary

[38] Secretary of Agriculture, memorandum 799, 1938. Also *Department of Agriculture Appropriation Bill for 1939*, Senate, 75 Cong. 3 sess. (1938), pp. 534–35.
[39] Secretary of Agriculture, memorandum, Nov. 27, 1939.
[40] NARG 16 (Cooperative Farm Forestry), Henry A. Wallace to Fred C. Pederson, Executive Committee of the Association of State Foresters, Jan. 6, 1939.
[41] See Charles H. Stoddard, *The Small Private Forest in the United States* (Washington: Resources for the Future, Inc., 1961), esp. pp. 1–2, 58–75.

Wallace and his staff to secure an allocation of funds authorized by the Forestry Act of 1937 so that they might expand extension forestry with increased federal grants. When Wallace decided to administer this program through the Soil Conservation Service in order to channel all land use programs through a single agency, he precipitated a three-way contest for control of this program.

The dismay of leaders in the extension services upon discovering that the Soil Conservation Service would administer the farm forestry demonstrations was matched by that of the state forestry agencies and the U.S. Forest Service. The belief of both forestry groups that the demonstrations should be administered by the state forestry agencies was based on the principle that only their own professional foresters were competent to give direct technical assistance to farmers. Their direction of the Norris-Doxey cooperative farm forestry program would be, they argued, a logical extension of the cooperative responsibilities which had been established by the Clarke-McNary Act of 1924. The U.S. Forest Service would continue its research and cooperative programs for fire protection and the production of planting stock, while the state foresters would give direct technical assistance to the farmers who agreed to join in the program.

In a few states, such as Florida, there was tension between the state's forestry agency and the Extension Service. The Department, therefore, was forced to intervene in support of the forestry agency before agreeing to support financially the employment of an extension forester.[42] In some states, such as New York, there were different situations. The Extension Service, which had already developed some work in forestry, sided with the state's department of conservation in blocking the SCS efforts to require acceptance of the complete farm plan and the signing of a 5-year agreement as conditions for the farmer's receiving technical assistance in farm forestry management. L. R. Simons, Director of Extension, told Milton S. Eisenhower, the Department's Land Use Coordinator, that the issue was "the old controversial question of the complete 5-year Soil Conservation Plan." The state extension services and forestry agencies wanted only one group of cooperators in the forestry demonstrations—those who indicated that they wanted technical assistance from the extension forester or the state forester only for forestry practices and not for soil

[42] NARG 16 (Cooperative Farm Forestry), H. L. Brown to H. L. Baker, Florida Forest and Park Service, May 13, 1938; H. A. Wallace to Wilmon Newell, Director of Florida Extension Service, May 13, 1938.

conservation in general. The Department of Agriculture acceded to this demand, and thereby undercut the effort of the Soil Conservation Service to insure results with some degree of permanence.[43] In Virginia, for example, a state forestry plan was developed to provide that it "will be the policy of the Service to direct the Farm Forestry Program through (soil conservation) districts wherever they exist." The chairman of the Department of Agriculture's Farm Forestry Committee informed H. H. Bennett that the Virginia plan was "unacceptable" because of this feature. The Soil Conservation Service might properly "give adequate attention to the possibility" of establishing projects within soil conservation districts, but "it is not the policy of that Service at this time to direct the Farm Forestry Program through" districts. The Virginia program, therefore, would require modification in keeping with this policy to be acceptable to "all other cooperating parties."[44] At about this same time, Milton Eisenhower informed the Chairman of the Association of State Foresters that the Department would permit a memorandum of agreement to be negotiated in each state. The practical question to be settled was whether a state forestry agency could provide its own foresters by matching federal funds.[45]

The upshot of these developments was that SCS officially administered the farm forestry program for the Department, but actually was forced to negotiate agreements which, in effect, turned the program over to the forestry agencies in such states as New York and Virginia which had vigorous state agencies with staffs large enough to meet the demands of the program.

The difference at issue was put very bluntly by spokesmen for the Association of State Foresters. Fred C. Pederson, chief of the Virginia Division of Forestry and a member of the Executive Committee of the Association of State Foresters, wanted to know how federal funds would be distributed to the states and whether foresters added by these funds would be under the direction of SCS and "not an integral part of the state forester's organization," or would traditional lines of direct "cooperation" with the state agencies prevail?[46] When the Foresters met with the Committee on Extension Organization and Policy to discuss this program, Wisconsin's State Forester was reported

[43] NARG 16, *ibid.*, M. S. Eisenhower to L. R. Simons, Director of Extension, Nov. 10, 1939, in reply to Simon's letter of Nov. 8, 1939.
[44] NARG 16, *ibid.*, George R. Phillips to H. H. Bennett, Nov. 9, 1939.
[45] NARG 16, *ibid.*, M. S. Eisenhower to C. P. Wilber, Oct. 18, 1939.
[46] NARG 16, *ibid.*, Fred C. Pederson to H. A. Wallace, Jan. 9, 1939; Wallace to Pederson, Feb. 8, 1939.

to have said that administration by SCS was "simply an extension of Federal bureaucracy." He claimed that the Association of State Foresters had been "doublecrossed on its understanding with the Forest Service."[47]

Finally, in 1944, after continuous lobbying by the Committee on Extension Organization and Policy and the Association of State Foresters had failed to divest SCS of responsibility for this program, Lyle F. Watts, Chief of the U.S. Forest Service, recommended to the Secretary of Agriculture that his agency be given responsibility for its administration. He argued that SCS administration had failed to achieve the original purpose of having only one agency deal directly with the farmer in carrying out the Department's land use programs. Instead, the result had been overlapping, duplication, and confusion for the farmer. Other considerations which Watts mentioned, however, indicate that the proposal to expand the forestry programs to include compulsory regulation of some forestry practices, was a major element in the decision to relieve SCS of its responsibilities for this program. This change did not, however, relieve the Soil Conservation Service of all concern for farm forestry practices, since it continued the SCS staff of foresters and advised farmers to use their woodlands as part of the complete farm plans made for cooperators with soil conservation districts. Indeed, the Chief of the Forest Service insisted that, since the SCS had a "vital" responsibility for controlling and preventing erosion, it should continue to render assistance to farmers in managing their woodlands within soil conservation districts and where woodlands were a "minor" part of the farm economy.[48] His advice was finally accepted and the program was transferred from SCS on July 1, 1945.

The real reason why Secretary of Agriculture Wickard finally agreed to relieve SCS of responsibility for farm forestry after eight years of interagency bickering is not clear. In any event, several state extension services, state forestry agencies, and the U.S. Forest Service were largely responsible for frustrating Wallace's plan to administer the forestry phases of land use adjustment and service programs through soil conservation districts. Their opposition postponed into an indefinite future any likelihood that soil conservation districts would become multiple-purpose conservation units.

[47] NARG 16, *ibid.*, memorandum, George R. Phillips to M. S. Eisenhower, Nov. 9, 1939.

[48] NARG 16, *ibid.*, Lyle F. Watts, Chief, Forest Service, to Claude R. Wickard, Secretary of Agriculture, Jan. 4, 1944.

COMMITMENT TO EXPAND
SOIL CONSERVATION DISTRICTS

Shortly before the Secretary of Agriculture divested SCS of responsibility for farm forestry, Congress authorized a program for treating the upstream watersheds of eleven relatively minor rivers in the Flood Control Act of 1944. According to testimony which Bennett gave in support of SCS plans for the Trinity River in Texas, this work was intended to consist of land treatment, and not "works of improvement." Until then, only one project of any notable size (that on the Los Angeles River in 1938) had been authorized since President Roosevelt and Senator Carl Hayden insisted in 1936 that "upstream engineering" by the Department of Agriculture be added to the work of the Corps of Engineers as flood control.[49] There is no evidence that Wallace ever effectively used his authority to require a state to have suitable districts legislation before the Department would undertake upstream surveys. Since he had no projects to undertake, he had no leverage on state legislatures. After Wallace left the Department in 1940, plans were soon altered by the war crisis and flood control operations were curtailed. When the Flood Control Act of 1944 authorized the Department of Agriculture's operations on watersheds, it did not require that they be performed through, or in conjunction with, soil conservation districts. It is not clear, therefore, whether Secretary Wickard intended to use this authority as leverage to secure suitable state legislation, or whether it was simply to permit the Department, and SCS, to operate in areas where there were no districts. It is quite possible that it was no more than part of the Roosevelt Administration's program for "reconversion" after World War II. And there is also the fact that 1944 was a presidential election year. It is doubtful, however, that extension of such activities merely happened, for this expansion occurred simultaneously with an apparent decision to push for soil conservation districts organized on all agricultural lands, including many on which erosion was not a problem in any form. It is not clear who made this decision, just as it was never clear whether Wallace and Wilson expected to promote districts only on eroded lands, but the evidence indicates that such a decision was made. The steady expansion of the SCS program from 1938 onward to include drainage and irrigation

[49] *Agriculture Department Appropriation Bill for 1943,* House, *op. cit.,* Pt. 1, pp. 955–58. See also *Flood Control Plans and New Projects,* House Committee on Flood Control, 78 Cong. 2 sess. (1944), pp. 459–75.

operations was evidence that erosion was not the only problem intended to be solved with technical assistance. Of course, related
problems of soil and water management often occur in the same area.
For example, irrigated croplands are often organized within special
districts having boundaries which do not include the surrounding
uplands from which flood waters and silt may flow. The boundaries of
an existing district could not always usefully be extended to include
such areas, since state enabling legislation usually did not authorize
these districts to undertake erosion control operations and to perform
other related functions considered by the Department of Agriculture
to be necessary.

Issues of political power and strategy also were involved. Irrigation
and reclamation districts in the seventeen Western states had much the
same relationship to the Bureau of Reclamation in the Interior Department as soil conservation districts had to the SCS. The Bureau of
Reclamation, in turn, had agreements with state extension services to
provide technical assistance on reclamation projects to farmers needing
to solve "water management" problems. It was reasonable to expect
that the governing boards of these existing districts would maintain
their traditional agreements with the extension services rather than
turn to SCS for technical assistance. If soil conservation districts were
organized in these areas, SCS would have access to farmers and could
provide assistance to those who had not yet been served or who had
problems that had not been solved. It was possible, too, that SCS might
offer a wider range of valuable services and equipment.

In a broader perspective, erosion control was not an obvious problem
in much of the politically influential North Central area, on the Gulf
and Atlantic coastal plains, and the flood plains of the Mississippi
Valley. Erosion control on private range lands in the West had been
almost a complete failure, if measured by the area organized in
districts. The northeastern portion of the country was almost virgin
territory for organization; indeed, in late 1943, when Bennett acted on
this evident decision, Connecticut, Massachusetts, and New Hampshire
did not have district enabling laws, and Delaware and Rhode Island
had just passed their statutes. Moreover, by June 30, 1943, the WPA
and the CCC were terminated. Their demise deprived SCS of the labor
corps which had been at its disposal and eliminated federal financial
assistance to districts.

At least one other important consideration probably prompted the
decision to commit SCS wholeheartedly to working through soil conservation districts. The SCS had found that the demonstration projects

were only partially successful. In 1937, Bennett's timetable called for ending the first, or demonstration, phase of his program in 1944. His long-range policy committee predicted that educational techniques alone would never secure the permanent national results desired by the Department. Demonstrations could be used to get farmers to adopt some soil conservation practices, but SCS found that most farmers could not make out their own plans and execute them. Most could not lay out terraces, contours, outlets, gully plugs, and other more or less permanent measures. What SCS needed was a representative who could walk over a man's farm with him, lay out a conservation plan, and come back to help him install the more difficult practices and structures. The Department had been committed for many years to the principle that this kind of assistance would be offered only through districts, unless SCS technicians "bootlegged" service to farmers living outside districts—and this very often happened unofficially. Despite the fact that he had resisted this policy for seven years, Bennett was finally convinced that SCS should promote districts widely and work primarily through them once they were organized. Moreover, despite the fact that SCS spokesmen occasionally complained that the extension services did not promote districts with sufficient energy, at the end of 1944 there were 1,235 districts in the United States, about 40 per cent of the total in 1964.

In December 1943, Bennett wrote to M. L. Wilson, the Department's Director of Extension, proposing that SCS and the Extension Service make a joint study of existing soil conservation districts. Bennett said that it was time to find out "the extent to which district governing bodies are taking advantage of the opportunities opened to them." Such a study would "no doubt bring out many related facts, such as post-war opportunities for employment in soil conservation work of men returning from the armed forces." Bennett thought that someone from Wilson's office should work with SCS and that his proposal should be laid before the Committee on Extension Organization and Policy of the Land-Grant College Association.[50] Wilson took up the proposal with J. W. Burch, the Chairman of the Committee and also director of the Missouri Extension Service. Burch soon complained that the members of his Committee had expected the state directors of extension to be consulted in making the plan for this survey, but that had not been done. He told Wilson bluntly that "our committee would be consider-

[50] NARG 33 (Federal Extension Service, General Correspondence, 1936–45), H. H. Bennett to M. L. Wilson, Dec. 28, 1943.

ably disgusted if a superficial survey should be made simply to drum up more business for soil districts and SCS generally."[51]

Eventually nineteen states participated in the study in spite of suspicions about Bennett's motives which were shared by a number of directors of extension.[52] Director D. W. Watkins of South Carolina was reluctant, but said that he was willing to go along. He did not like the "tendency" for local publicity to feature prominently the name of the local SCS employee and the district supervisors in reports of accomplishments under the soil conservation program. Supervisors were credited with performing duties which they did not in fact do, since they "are busy farm operators." He was especially incensed that these reports "never" mentioned the contributions to soil conservation of "other agricultural workers and agencies." Unless this matter of publicity were looked into and corrected as a result of the proposed joint survey, Watkins feared that "the study will only be used to promote additional appropriations for the Soil Conservation Service."[53]

An extension director in one state declined outright to participate because he thought that the survey would not answer the pointed questions which he raised in correspondence with M. L. Wilson. Extension personnel in his state thought that the tendency was for district supervisors to become little more than rubber stamps for SCS, giving *pro forma* approval to activities which they were too busy to supervise personally. Because farmers and ranchers wanted assistance

[51] NARG 33, *ibid.*, J. W. Burch to M. L. Wilson, Feb. 29, 1944. Wilson appended a note to this letter, telling his assistant, J. L. Boatman, to reply to Burch over Boatman's signature, explaining that Wilson was out of town.

[52] NARG 33, *ibid.*, M. L. Wilson to Aubrey Gates, May 5, 1944. Wilson explained to J. L. Boatman (memorandum, July 20, 1944) that he had discussed the survey with Gates personally and found him to be "critical of some basic relationships between Extension and Soil Conservation Service which get into policies which are planned and inaugurated by Congress and administrative officers on a higher level than either the Chief of the Soil Conservation Service or ourselves." Nixon, *op. cit.*, Vol. II, p. 582, reveals a mystifying note in the Roosevelt papers, dated Jan. 31, 1944. Morris L. Cooke had suggested that Roosevelt write to Bennett, congratulating him on the great progress he had made. Roosevelt penned a memorandum to Stephen T. Early asking him: "Do you think I should see Bennett and tell him that I think what he is doing is all to the good but not send him any written communication? I do want to tell him about some ideas I have anyway, outside of this." According to the editor's note, neither an appointment with Bennett nor a letter to him followed this suggestion. Charles M. Hardin, *The Politics of Agriculture* (Glencoe, Ill.: The Free Press, 1952), attributed a shift of SCS services to the Corn Belt region after World War II to what he called "an apparent shift to Republicanism." This interpretation appears to be too simple in light of the evidence in this chapter and Chapter 5.

[53] NARG 33, *ibid.*, D. W. Watkins to J. V. Webb, copy to M. L. Wilson, May 25, 1944.

for irrigation, weed control, and drainage as well as soil erosion, it was possible to justify almost any activity as "soil conservation." The SCS was staffed with "capable and ambitious" men who were "eager to take on anything that might come their way" and had, therefore, done educational work in just about every extension field but 4-H and home demonstration. It was possible for SCS to do this because it could shift its staff to outnumber the county agent ten to one wherever it was advantageous to do so. In addition, there appeared to be a move under way to get all of the state organized in districts regardless of whether there was a serious erosion problem or not. County agents were hearing all over the state from farmers who wanted districts organized, despite the fact that the state conservationist of SCS had instructed his men to let the county agents know of farmer interest "before any promotional work is done." This director of extension could not but wonder whether the drive to organize districts, regardless of the lack of erosion in particular areas, was not the start of a post-World War II program for SCS, since there would be need for much new heavy equipment and many men to do the job.[54]

The hearings on agricultural appropriations conducted by House and Senate subcommittees early in 1944 provided answers to the suspicions of these extension directors. In a lengthy hearing, Representative Malcolm C. Tarver, Democrat of Georgia, exchanged his usual role of friend of the SCS for that of an apparent critic. The estimates for SCS proposed an increase of more than $8 million for the fiscal year 1945. Tarver wanted to know first of all why any of it was to be spent on further demonstration projects, since the services provided within them were greater than those provided farmers in districts. Wasn't the federal government "playing favorites in helping farmers who are not trying to help themselves?" J. C. Dykes, who was testifying for SCS, replied that SCS had only been trying to get its program "as fairly spread as possible over the entire country." Tarver wanted to know: "What was wrong with my statement?" Dykes replied: "Well, not a whole lot." Tarver asked Bennett what was wrong with telling farmers that they had state laws which allowed them to organize districts, just as other farmers had done; if they wanted SCS services, they should be told "you must organize a soil conservation district."[55] Bennett replied that, since there had been generally a high correlation between the locations of demonstrations and districts, SCS would like

[54] NARG 33, *ibid.*, letter marked confidential to M. L. Wilson.

[55] *Department of Agriculture Appropriation Bill for 1945*, House, *op. cit.*, pp. 1068–69.

to have a "little leeway or life" for a few demonstration projects. Tarver shot back: "This discrimination ought to stop." He asked whether there was any reason why all demonstrations should not be terminated in the very near future.[56]

When the appropriation bill cleared the House it included a provision that no funds could be used for demonstration projects after June 30, 1945. Officially the Department opposed this provision before the Senate subcommittee, but Dykes, who testified in Bennett's place, said that SCS did not "feel very strongly" about this restriction. Farmers would "get busy and organize soil conservation districts." Senator Russell graciously assisted in making the record by quoting a telegram from farmers in a demonstration project in Mississippi who opposed the rider. Dykes remarked that Mississippi "has a districts law, Senator, and if the farmers want to organize a district covering this project the authority is available in the State to do it."[57]

This whole proceeding bears the marks of complex maneuvers in the Department and Congress. The Soil Conservation Service needed a congressional command just as it had needed President Roosevelt's support in the earlier fight with the land-grant colleges. In 1944, Congress directed SCS to administer its program only through soil conservation districts. Few people knew or noticed the fact that this policy was precisely the one which Secretary Wallace earlier had tried in vain to force upon Bennett.

The muscle for the drive was supplied by a greatly expanded program of "water conservation" and new SCS heavy equipment. Two million dollars were earmarked for increased drainage and irrigation work which was attractive in the coastal areas of the Atlantic and Gulf states, large areas of Iowa, Illinois, Indiana, and on the alluvial plains of the Mississippi Valley. Also, SCS would provide new services to 938,600 acres of irrigated land in the West. Tarver asked whether this was a "new venture you are proposing?" Dykes' reply was: "That is correct."[58] After he explained the program in further detail, Tarver remarked: "You think you are offering them such a good proposition that they are bound to accept it?" Dykes responded: "We are reasonably sure they will do so."[59]

[56] *Ibid.*, pp. 1073–74.

[57] *Department of Agriculture Appropriation Bill for 1945*, Senate, *op. cit.*, p. 405.

[58] *Department of Agriculture Appropriation Bill for 1945*, House, *op. cit.*, p. 1082. See also *Department of Agriculture Appropriation Bill for 1944*, Senate, 78 Cong. 1 sess. (1943), pp. 447–49.

[59] *Department of Agriculture Appropriation Bill for 1945*, House, *op. cit.*, pp. 1083–84.

Heavy construction equipment to use on these projects was essential. Senators Carl A. Hatch, Democrat of New Mexico, and Representative W. R. Poage, Democrat of Texas, had already introduced legislation to permit the Secretary of Agriculture to acquire surplus military equipment to be lent to districts. At 1944 hearings, Senator Gerald P. Nye, Republican of North Dakota, warmly supported the drainage work of SCS. In the Red River Valley of his state, five districts had just been formed so that work could be undertaken to repair damages caused by heavy floods in 1943. He told Dykes that, although loans of surplus equipment to districts might create fears of unfair competition among the manufacturers of machinery, such grants would eventually create a greater market for sales through private retail outlets. Dykes agreed that this plan "will enable these districts to purchase new equipment from the manufacturers in the future." Nye applauded Dykes' answer as "well put."[60]

With authorization for the eleven watersheds an apparent directive to operate through districts and a vast increase in regular appropriations, Bennett enthusiastically predicted that within six to eight months after the end of hostilities 100,000 men could be put to work "improving and protecting the nation's farm plant." The SCS "is preparing plans for advancing the work on a thoroughly practical basis." In the second year, "3,000 qualified technicians could be recruited and trained to augment the present technical staff of the Soil Conservation Service." This staff could "efficiently utilize" the labor of at least 500,000 man-years, and with sufficient heavy equipment complete 40 per cent of the "heavy duty job" in three or four years.[61] The very thought made personnel in rival agencies shudder.

SUMMARY AND COMMENT

From the time the Soil Conservation Service was created, Hugh H. Bennett and his associates claimed that their agency had a unique and exclusive mission to plan and execute a national program of soil and water conservation. Secretary Henry A. Wallace, however, interpreted his authority, granted by Public Law 46 of 1935, to mean that the whole

---

[60] *Department of Agriculture Appropriation Bill for 1945*, Senate, *op. cit.*, p. 400.

[61] *Report of the Chief of the Soil Conservation Service, 1944*, *op. cit.*, p. 4. A year later, he said that there "is a 15 to 20 year job ahead of us in soil and water conservation." This startling change was brought about by the new SCS timetable, the "Soil and Water Conservation Needs Estimates for the United States." *Report*, 1945, p. 7.

Department of Agriculture had this responsibility, and the Soil Conservation Service was but one of several bureaus equipped to contribute to a departmental program. For two or three years, Wallace apparently thought that he wanted all departmental agencies to work eventually through districts, but from the outset he placed on SCS all responsibility for launching district operations. Bennett agreed to this policy very reluctantly because he wanted to concentrate on the "educational phase" of a two-stage program by massing technical assistance, equipment, and materials on demonstration watersheds. His opponents in the land-grant colleges were generally aware of the situation and charged that work done by SCS "duplicated" their own. Insofar as SCS worked through districts, most extension personnel viewed the districts as, at best, thinly disguised "fronts" for SCS.

In 1937 and 1938, Wallace shifted major responsibility for various phases of water conservation to SCS, and assigned the demonstration farm forestry program to it. These assignments strengthened the view that the conservation responsibilities of the Department and SCS were nearly identical, although the conservation program of the Agricultural Adjustment Administration (and its successors) was administered separately. Foresters in other agencies proved strong enough in 1945 to divest SCS of most forestry responsibilities. But at the same time, SCS retained and broadened its program with upstream flood control projects, and greatly increased technical assistance for drainage and irrigation in areas where it had not previously functioned when erosion control was its primary emphasis.

In 1940, Wallace resigned as Secretary, M. L. Wilson became Director of the Extension Service, and Claude R. Wickard moved up through the AAA to become Secretary. Wickard immediately terminated the joint SCS-Extension review of district programs and plans on the state level. He made district governing bodies solely responsible for determining the amount and kind of assistance to be given farmers cooperating with them and for deciding whether the assistance was used properly. Thus, the district was not required to give the Department assurances that aid which it supplied to farmers was used to achieve permanent results in the public interest. Since there were then only about 200 districts, few of those now in existence had occasion to know of these events or become aware that at one time the districts were expected to perform this important public function. Secretary Wickard also cemented the identity of districts with SCS in 1940 by making the latter responsible for all "discussions" between a district board and any agency of the Department of Agriculture.

Resistance from SCS and the colleges, alike, contributed to the slow growth of districts in the 1940's and to the virtual demise of their planning function—a function which most supervisors probably never understood to begin with. These failures to achieve the objectives for which districts were conceived in 1935–36 were combined with the unwillingness of many state legislatures to accept key features of the standard districts law, especially the power to make and enforce land use regulations. The result of all this was that the contemplated function of districts was radically altered by the time World War II began. Moreover, since only about 600 of the districts were organized, they never really served to "focus on farms and watersheds" the whole of the Department's programs for land use adjustment. The county committees of the AAA and the state and county land use planning committees under the joint guidance of the Extension Service and the Department's Bureau of Agricultural Economics tended to perform this function—although by 1942 even the planning committees were moribund.[62]

During the War, districts increased at a rate of better than 200 a year. Late in 1943 a decision apparently was made to promote them in areas where they had not yet been organized. To the Department's appropriation act for 1945, Congress attached a rider forbidding use of any of the greatly increased appropriations for SCS to be used for demonstration projects. Key members of Congress and officers of the Department of Agriculture understood this rider to mean that in the future SCS would give technical assistance only to farmers living within districts. When SCS proposed to the Extension Service a joint review of district operations, several directors construed it to be a signal that SCS was finally aiming to organize a district in every county in the nation.

---

[62] Hardin, *op. cit.*, p. 135, charged that the AAA joined SCS to terminate the land use planning program. Grant McConnell, *The Decline of Agrarian Democracy* (Berkeley: University of California Press, 1953), p. 136, concluded that the committees fell victim to opposition from the extension services and the American Farm Bureau Federation. See also Campbell, *op. cit.*, p. 177. M. L. Wilson in an interview Sept. 13, 1962, said he agreed with McConnell. See also John M. Gaus and Leon O. Wolcott, *Public Administration and the United States Department of Agriculture* (Chicago: Public Administration Clearing Service, 1940), pp. 154–59; and John M. Gaus, "Agricultural Policy and Administration in the American Federal System," in Arthur W. MacMahon (ed.), *Federalism, Mature and Emergent* (Garden City, N.Y.: Doubleday and Co., 1955), p. 296. See also U.S. Department of Agriculture, *Agricultural Statistics of the United States, 1941* (Washington: Government Printing Office), p. 656; *ibid., 1943*, p. 48; and *ibid., 1945*, p. 664.

Because the documentary evidence is too fragmentary to permit even a guess at the answer, one question must remain a mystery for the present: Was the congressional rider forbidding SCS to operate demonstration watersheds intended to help it reach a far broader clientele than in the past, or to restrict its operations largely to areas where it was already functioning in 1944? If the purpose was to restrict SCS, despite the increase in appropriations, was it intended to help the AAA mount a drive to "absorb" SCS and finally provide the Secretary of Agriculture with a single agency to administer the Department's soil conservation program of cash subsidies and technical assistance?

For the next seven years or more, extension leaders acted on the theory that the rider was intended to help SCS organize districts throughout the country. The SCS personnel acted on the same interpretation. The Production and Marketing Administration (under which the AAA functions had been placed in August 1945) acted as if its leaders believed they could "absorb" SCS. Both organizations were convinced that the leadership of the Soil Conservation Service was determined to make the agency into the Department's exclusive and unique, general-purpose land use organization, which would eventually absorb agencies from the Department of the Interior as well.

CHAPTER 5

Organizational Issues Reopened

WHEN HENRY A. WALLACE decided, in February 1936, to let the Agricultural Adjustment Administration (AAA) and the Soil Conservation Service (SCS) administer their soil conservation programs through separate organizations, he had no other practical choice. For nearly the next decade, the two agencies got along well enough, although there was not much formal cooperation between them. In 1937, Wallace's Economic Adviser, Mordecai Ezekiel, told Howard R. Tolley, the Administrator of AAA, that SCS was achieving results very slowly because a farm planner could work on only an average of ten to twenty farms per year. One specialist had estimated that it would take "70 years to cover the areas of the state that were badly in need of attention through the methods now being used."[1] In 1948, Norris E. Dodd, Under Secretary of Agriculture who had earlier been Chief of the Agricultural Adjustment Agency, said: "I know of many counties . . . where there might be 25 or 30 or 50 farms being surveyed and planned by the Soil Conservation Service, and I wish it were 500 or 1,000. There might be 2,000 farms in that county carrying out good conservation practices under the triple A program."[2]

The administrators of SCS and AAA assured Congress that their work did not overlap. In 1937, Tolley said that they were working toward the "same general objective, but not trying to do the same thing."[3] Hugh H. Bennett, for SCS, testified that AAA made conservation payments for "soil-building" practices and SCS advised AAA on whether it ought to make payments for certain practices such as terracing. Even then, Bennett was putting up a good front for a situation he did not like. As Senator Richard B. Russell of Georgia

[1] National Archives, Record Group 114 (Soil Conservation Service, General Correspondence), Mordecai Ezekiel to H. R. Tolley, Sept. 15, 1937. (National Archives records cited hereinafter usually will be identified by the abbreviation NARG.)

[2] *Department of Agriculture Appropriation Bill for 1949*, 80 Cong. 2 sess. (1948), Pt. 2, p. 199.

[3] *Department of Agriculture Appropriation Bill for 1938*, Senate, 75 Cong. 1 sess. (1937), pp. 335 and 341.

pointed out, the AAA "land policy was predicated solely on recommendations made to them by the various State extension services." Dillon S.
Myer assured the Senator that SCS was responsible for some recommendations on practices, although procedures varied around the
country.[4] The AAA did not develop its own staff of agronomists,
foresters, agricultural engineers, and other specialists, but relied chiefly
on the agricultural colleges to recommend practices and specifications.[5]
Naturally, this division of labor tended to ally the AAA with the
colleges, rather than with SCS, but the relations of AAA and the
extension services were not always harmonious where other AAA
programs were concerned.

These general attitudes and relationships applied during the various
reorganizations of the AAA: when it was changed from an "Administration" to an "Agency" to become a part of the Agricultural Conservation
and Adjustment Administration early in 1942; and later, in 1945, when
the AAA's functions were consolidated within the new Production and
Marketing Administration (PMA).

PMA-SCS RELATIONS

Hearings on the 1950 agricultural appropriation bill offer a clue to
the fluid state of PMA-SCS relations by the late forties. H. L. Manwar-
ing, testifying for the Production and Marketing Administration at the
hearings in 1949, said that in each state the Soil Conservation Service
had representation on the technical advisory committee which recommended to the state PMA committee the practices for which PMA
conservation payments would be made. On the county level, he said,
PMA kept a file of farms having SCS farm plans and agreements with
the districts and "to the extent possible" used the SCS technicians for
laying out practices requiring construction. He explained that because
there were not enough SCS technicians to serve all farmers in most
counties, and many district boards "required" SCS personnel to work
only on the farms of cooperators, the remaining farmers were served by
personnel of PMA, the extension services, or private contractors. "But
we do use those technicians where they are available and can give the
assistance." When Representative Jamie L. Whitten, Democrat of
Mississippi, remarked that "we have complaints from time to time . . .

[4] *Ibid.*, pp. 616–17 and 640–52.
[5] *Department of Agriculture Appropriation Bill for 1946*, House, 79 Cong. 1 sess.
(1945), p. 312; *Bill for 1948*, House, 80 Cong. 1 sess. (1947), Pt. 1, p. 1239.

that the Soil Conservation Service goes one way and PMA goes another," J. C. Dykes answered that SCS was represented on the technical advisory committees "in nearly all States . . . although I am not absolutely sure of that, but nearly all." These representatives "present the results of our research work to the State PMA committees. What they do with it from that point on, Mr. Whitten, is, of course out of our hands." Whitten commented that it might be out of "your hands, but it is not beyond your interest, is it?" Hugh Bennett hastily assured him that it was not.[6]

During 1943, AAA placed new emphasis on practices to improve the use of water supplies and range and pasture forage and to prevent land and water erosion. In 1944, it stressed permanent practices requiring construction by bringing contractors and farmers together to accelerate work. These efforts were so successful that AAA's Southern Region overspent its allocation of funds for conservation payments, and Congressman Whitten accused Norris E. Dodd, AAA Chief, of deliberately trying to overspend the money which Congress had appropriated. In previous years, AAA had not utilized all of these funds.[7] Whitten did not mention the fact that this drive took place during a presidential election year. There may have been no necessary connection, however, since PMA reported in 1946 that county committees continued to promote construction of ponds by bringing farmers and contractors together. "The committees sought out farmers who wanted dams built . . . and encouraged local contractors to purchase heavy dirt-moving equipment to do the work. Enough dams were laid out to make the venture profitable for the parties."[8]

In turn, SCS placed new emphasis upon finding neighborhood leaders who would help to convert farmers in their communities to proper methods of land use and erosion control.[9] For 1946, Bennett ordered that the number of agreements between farmers and districts be doubled. He kept abreast of this development, calling for monthly reports when he found that only 1,114 had been signed in January, whereas the total "should have been 2,228, or at least 1,900."[10] R. M.

[6] *Department of Agriculture Appropriation Bill for 1950,* House, 81 Cong. 1 sess. (1949), Pt. 2, pp. 84–85 and 362.

[7] *Department of Agriculture Appropriation Bill for 1946, op. cit.,* pp. 338–40.

[8] *Report of the Administrator of the Production and Marketing Administration, 1946* (Washington: Government Printing Office), p. 18.

[9] NARG 33 (Federal Extension Service), M. L. Wilson to H. H. Bennett, and reply, Nov. 21, 1944. See also *Report of the Chief of the Soil Conservation Service, 1944* (Washington: Government Printing Office), p. 31.

[10] NARG 114 (SCS, General Correspondence), memorandum, H. H. Bennett, Chief of SCS, to All Regional Conservators, "Doubling Production," March 18, 1946.

Musser, SCS Regional Conservator in Milwaukee, adviscd his field forces to reach returning veterans and community leaders such as "businessmen, professional men, civic leaders. . . . Be sure that the children in the community know more about soil conservation and its benefits. . . . There is a lot of good conservation on the back 40's. There should be much more on the 'front 80' along the highways and railroads."[11] During the fiscal years 1945 and 1947, Congress appropriated money for SCS to purchase surplus heavy equipment which could be lent or granted to soil conservation districts. This appropriation, however, was only a limited substitute for an authorization sought by SCS to provide districts with this kind of assistance on a large scale.[12]

In competition with PMA, SCS suffered a disadvantage, since PMA had a farmer committee in every county together with several community committees. According to official data submitted to Congress, PMA had 9,090 county committeemen and 99,305 community committeemen. The county committeemen received per diem compensation for their services, averaging $5.65 per day for 38 days worked a year in 1944. Community committeemen received an average per diem of $5.06, working an average of 6.3 days a year. At the end of 1944, there were 1,235 soil conservation districts with approximately 3,700 supervisors. In the Southern states, however, many soil conservation districts were organized on a multi-county basis, so that in a single county there was normally only one district supervisor to advocate the work of SCS in contrast to three county committeemen and nine to ten community committeemen to support the PMA conservation program. In the Western states, the soil conservation districts were so few and scattered—many organized for areas smaller than county units—as to make it impossible for the supervisors to reach all farmers and ranchers. The supervisors received no pay for their work, although some received expenses. Organized districts included approximately 446 million acres of farm land and ranch land, both public and private. Over 615 million acres of land outside of districts were said to need "conservation work."[13] That is, the land area which had not been organized in districts for SCS conservation operations was almost one and one-half times greater than that within districts.

The drives by PMA to expand applications of permanent practices

[11] NARG 114 (SCS, General Correspondence), memorandum, R. M. Musser to field staff, Jan. 23, 1946.

[12] *Department of Agriculture Appropriation Bill for 1947*, House, 79 Cong. 2 sess. (1946), pp. 1050–52; *Bill for 1948*, House, *op. cit.*, Pt. 1, pp. 1048–50.

[13] *Department of Agriculture Appropriation Bill for 1946*, House, *op. cit.*, pp. 172 and 316–18.

and by SCS to organize soil conservation districts covering all agricultural counties coincided in 1945 with rumors that the new Secretary of Agriculture, Clinton P. Anderson, would reorganize the Department. He soon deprived SCS of the farm forestry program, but nothing more. In 1947, Anderson cautiously called for a "consolidation" of the conservation programs of the two agencies and a "strengthening" of the PMA county committees which would make them responsible for conservation planning and for requests for departmental "assistance" from the soil conservation districts.[14] This announcement alarmed the officers of the recently created National Association of Soil Conservation Districts (NASCD). Kent Leavitt, President of the organization, both wrote to Anderson and visited him to discuss the subject. Anderson told him that PMA had provided the "incentive for phenomenal increases in many desirable practices. Because of the program much more rapid progress has been made in putting into practice conservation plans developed by the Soil Conservation Service. . . . At the same time the Soil Conservation Service program has caused more desirable practices to be carried out and carried out more effectively . . . the two programs have materially strengthened each other." His recommendation to Congress would "strengthen" the work of the districts, he reassured Leavitt.[15] Anderson assured Senator Carl Hayden of Arizona, a long-standing friend of SCS, that the heads of both PMA and SCS had "concluded that these activities can and should be consolidated."[16]

In 1947, the Administration proposed to Congress that PMA appropriations be increased to raise the salaries of county office personnel by about $4.3 million for 1948, since there had been no increase since 1943 and the rates of pay in force were below prevailing local wages. Representative Everett M. Dirksen, Republican of Illinois, as Chairman of the Subcommittee on Agricultural Appropriations, conducted hearings on both the PMA and SCS appropriations. At the SCS hearings, he put Hugh Bennett through the most unfriendly hearing Bennett had ever experienced before the Appropriations Subcommittee.[17] Showing

[14] Assistant Secretary Charles F. Brannan (who became Secretary in 1948) confessed to Representative Harold D. Cooley that "we are not prepared to say in detail just exactly how we want to recommend it." See *Soil Conservation*, Hearings on H.R. 4150, H.R. 4151, and H.R. 4417, House Committee on Agriculture, 80 Cong. 2 sess. (1948), p. 43.

[15] NARG 16 (Soil—1), Secretary Clinton P. Anderson to Kent Leavitt, Feb. 18, 1948.

[16] NARG 16 (Soil—1), Secretary Clinton P. Anderson to Senator Carl Hayden, April 8, 1948.

[17] *Department of Agriculture Appropriation Bill for 1948*, House, *op. cit.*, Pt. 1, pp. 987–1027 (see esp. pp. 990 and 1009).

keen interest, also, in the way in which PMA funds were being spent, he found that 3.6 million farmers had received payments during the last program year, and that 80 per cent of the participants received only 40 per cent of the conservation payments, while 20 per cent received the remaining 60 per cent of the funds. Dirksen also delved into the exact wording by which the Department expressed its commitment to make payments to farmers.[18]

This strife over the agencies and their programs in an election year resulted in a reduction in funds for the PMA program. The cuts hurt, too, for Under Secretary Dodd testified that the average community committeeman (and there were 99,300 of them) was employed only 3.8 days in 1948 in contrast with 6.3 days in 1944. Some supporters of SCS were among the warmest advocates of this reduction in funds. The California Association of Soil Conservation Districts wanted the PMA program eliminated and $10 million added to the SCS appropriations for 1949 to "save $290,000,000 in federal funds and . . . get some real conservation work done."[19] In 1948, Ollie Fink, Executive Secretary of the Friends of the Land—an organization friendly to Bennett's views—passed on to Secretary Anderson a warning he had from the Mississippi Valley Association, another group supporting SCS. Fink said the report was that President Harry S. Truman intended to issue an order for the PMA-SCS consolidation before January 31, 1948, but that Friends of the Land opposed it because PMA subsidized "poor farming" and was subject to political pressure based on "regional greediness."[20]

The Committee on Extension Organization and Policy of the Association of Land-Grant Colleges and Universities opposed consolidation of PMA and SCS because it would enable the new organization largely to deprive the extension services of any effective role in the soil conservation programs. The consolidated PMA-SCS would combine the fiscal resources of the former with the technical skills of the latter so that PMA would no longer rely upon the agricultural colleges for recommendations and specifications on practices. There would no longer be any need for PMA to use county agents to lay out any of the construction practices in the several hundred counties in which no districts had been organized and SCS was not providing technical services.

---

[18] *Ibid.*, pp. 1233–39 and 1248–51.

[19] *Department of Agriculture Appropriation Bill for 1949*, House, *op. cit.*, Pt. 2, pp. 206–07 and 210.

[20] NARG 16 (Soil—1), Ollie Fink to Clinton P. Anderson, Jan. 22, 1948; reply Feb. 18, 1948.

Secretary Anderson's proposal to merge SCS and PMA stung the extension services and the American Farm Bureau Federation into an attack on SCS, but not on the PMA's conservation program. The reasons are not difficult to understand. The first has already been stated: the SCS-PMA combination would probably displace the extension services from most of the federal soil conservation programs. The second was that PMA provided the Department with a more reliable political channel to the farmers than the extension services could be. When Henry A. Wallace built up "line" organizations to administer the Department of Agriculture's new action conservation programs, his political objective was to create an administrative system insulated from control by local extension leadership. He wanted "the people," instead, to participate directly in planning and administration. If SCS were grafted onto PMA, the administration in power in Washington would control a treasure of funds, materials, services, and technical advice to be distributed to farmers through this direct channel which would by-pass the extension and Farm Bureau leadership.

THE AMERICAN FARM BUREAU FEDERATION AND SCS

The sweeping Republican victory in the 1946 congressional elections and the widespread assumption that President Truman could not be re-elected in 1948 encouraged the foes of the consolidation of PMA and SCS. They believed that for the first time in twelve years there was a good opportunity to control the federal soil conservation programs. Action in the executive branch and Congress in 1948 demonstrated a pattern which was becoming increasingly clear. The rhythmic intensification of conflict between the extension services and SCS over administrative reorganization seemed to reach an apogee in presidential election years.

Secretary Anderson's threat to consolidate PMA and SCS was not the first sign of danger to the extension services. Ever since Bennett had proposed in 1944 the joint study of districts by SCS and the extension services, SCS had waged an aggressive campaign to expand the range of its activities and to promote districts. Some of the techniques it used have been mentioned, but there were other specific actions which aroused the ire of extension leaders.

In Oklahoma, for example, a state association of districts was organized. A conflict followed in the Legislature in 1945 over a change in the composition of the State Soil Conservation Committee. The representatives of the college of agriculture were eliminated and the "Committee" was replaced by a "Board" consisting of five farmers who were required to be district supervisors appointed by the Governor from five areas described by the enabling act. In the same year, a similar change was made in Georgia where the heads of seven agencies were retained in an "advisory capacity," but the State Committee on Soil Conservation was altered to consist of five district supervisors appointed by the Governor. It soon became clear that changes in the composition of the state committees and the organization of state associations of district supervisors were the beginning of a significant trend.[21]

Measures to improve the working climate and the production records of the districts were supported by maneuvers to undercut the American Farm Bureau Federation as the national spokesmen for the extension leaders seeking "decentralization" of SCS. In April 1944, the SCS State Conservationist in Tennessee reported to the Regional Director that during a recent conversation the Executive Secretary of the Tennessee Farm Bureau had revealed that the "district" approach to soil conservation "would receive the wholehearted support of the Farm Bureau." He indicated that he saw no reason why county farm bureaus in Tennessee could not be instrumental in establishing districts and cooperate "wholeheartedly" with their boards. Since this expression of viewpoint was materially different from any voiced in the past by the President of the Tennessee Farm Bureau, it was called especially to the attention of J. C. Dykes, Assistant Chief of the SCS.[22]

Whether this information stimulated the move or other considerations entered the picture is not certain, but soon various farm organizations, including state farm bureaus, passed resolutions at their 1944 and 1945 annual meetings expressing support for either the SCS or districts, and in many cases for both. The National Grange added its support and in a few states the National Farmers Union followed suit.[23] The precise total of favorable resolutions by state farm bureaus is not certain, but key congressmen were aware that Allan B. Kline, President of the American Farm Bureau Federation, was speaking for a divided federation when he advocated the decentralization of soil conservation activities to the state extension services. At hearings held by the House

[21] W. Robert Parks, *Soil Conservation Districts in Action* (Ames: The Iowa State College Press, 1952), pp. 211–13.

[22] NARG 114 (SCS, General Correspondence), T. S. Buie to J. C. Dykes, May 3, 1944.

[23] *Department of Agriculture Appropriation Bill for 1946*, House, *op. cit.*, Pt. 2, pp. 149–50; *Bill for 1947*, House, *op. cit.*, pp. 1046–50. See also *Soil Conservation*, House Hearings, *op. cit.*, p. 154.

Committee on Agriculture in February 1948, Representative John Flannagan of Virginia, ranking Democrat on the Committee, asked H. E. Slusher of the Missouri Farm Bureau whether his attitudes were those of the American Farm Bureau. Slusher answered "yes," insofar as the extension service would be the "main county agency" and that "co-ordination" of conservation would mean to "have our soil conservation engineers and technicians, but let them be working under and through the county agents' offices." Flannagan then wanted to know: "Well, are all the State organizations together on that?" Slusher replied: "I would not say that." Flannagan wanted to know whether there was "opposition among the State organizations or not with respect to service through the Extension Service?" Mr. Slusher replied that the "majority of the Farm Bureaus of the States, and that includes all of them, feel that the work should be better coordinated."[24] Slusher had made the best case he could for "decentralization," but it was very damaging to the American Farm Bureau's claim that it spoke for all the state farm bureaus on this issue.

The lack of unanimity among the state farm bureaus was matched to a degree by differences among the agricultural colleges. John A. Hannah, President of Michigan State College, supported the American Farm Bureau's bills, although he denied that the Association of Land-Grant Colleges and Universities (of which he was President in 1948–49) had initiated them or participated in their preparation. He admitted only that there had been discussion of the "principles" involved. The Association, he said, was neither for nor against the proposed legislation so long as it coincided with the "sound basic principles . . . [of] the Smith-Lever Act and the . . . memorandum of understanding of 1914."[25] The Virginia Director of Extension, J. R. Hutcheson, spoke for the "moderates" within the Association of Land-Grant Colleges when he wrote to an Assistant to the Secretary, William A. Minor, saying that there had been cooperation in his state so long as SCS abided by the 1914 Agreement, but that ever since the SCS Regional Office had laid great stress on farm planning there had been "duplication" and a lack of "cooperation" between the Virginia Extension Service and SCS. Director Hutcheson said that he was "not at present numbered among those who wish to see all of the activities of the Soil Conservation Service turned over to the Extension Service. However, if the Soil Conservation Service continues to do educational

---

[24] "Farm Program of the American Farm Bureau Federation," Committee on Agriculture, House of Representatives, 80 Cong., 2 sess. (1948), p. 22.
[25] *Soil Conservation*, House Hearings, *op. cit.*, pp. 121–22.

work in the whole field of farm management, I will soon be found in this group." He argued that the SCS regional offices ought to be abolished and soil conservation work be achieved through districts with assistance from the colleges, as in the area of the Tennessee Valley Authority (TVA). He suggested as an alternative that the extension services be authorized to give the "educational recommendations in each farm plan" as a joint endeavor with SCS. He also repeated the frequently made demand that the state offices of SCS be located at the land-grant college to achieve maximum coordination of effort so long as the two agencies remained separate.[26]

H. C. Saunders, Louisiana's Director of Extension, was among the militants demanding that the extension services be allowed to take over most of the work of SCS. In testimony before the House Committee on Agriculture, he revealed in specific detail the new source of friction between the SCS and the extension services. Saunders said that he and the county agents of Louisiana had actively promoted the organization of soil conservation districts, to the extent that all referendums had carried by an overwhelming vote, except in one parish—Placquemines. Like the extension services of other southeastern states, Louisiana's had sponsored the Department's soil conservation program in the thirties, despite Saunders' opinion that Secretary Wallace "was a great exponent of a planned economy" and thought that the leaders of the "land-grant colleges were static, reactionary and without vision. College personnel would not jump when he cracked the whip. Not wishing to antagonize them too much, he invented the term 'action program' and reserved for Federal agencies all activity in that field." He felt that SCS had "wasted" a lot of time writing farm plans which were "long, complicated and . . . [in] much detail." Many had been written for farms "now growing broom sedge and pine saplings." Meanwhile, SCS had received funds to do engineering work in the Delta area and that had attracted "a lot of attention." Actually only "good practical engineers are needed in our Delta areas. The county agents were well-informed agronomists." Duplication and overlapping resulted, Saunders claimed, because the SCS personnel "as I see it, have no place to stop. You go out on a farm and you make a plan. We are also making farm plans in Extension. They recommended that this particular field be planted in grass. All right, shall he stop there with that recommendation or shall he say, 'Here is the kind of grass you shall plant and here is the amount of lime and fertilizer that will be needed,' and so on. All right, that

[26] NARG 16 (Soil—1), J. R. Hutcheson to W. A. Minor, Nov. 20, 1947.

leads into utilization. What are you going to do with the grass? It is either going to be cut for hay or for livestock. If you go into livestock, what kind are you going to get? So there is just no end to where your Soil Conservation Service program begins or ends. You have both groups out there working on pasture, both groups working on winter legumes, both groups working on summer legumes, on lime, phosphate."[27]

SOURCES OF SPECIAL DIFFICULTY

The militant extension leaders were demanding the "decentralization" of SCS because its complete farm plans challenged extension's complete farm plans. The SCS plans did not stop at recommending merely "engineering" practices such as drainage, improved irrigation, terraces, and similar activities. Its plans also included recommendations for "agronomic" practices and management to cover most of the range of operations carried out on any farm—and thereby challenged the very base of the extension services' standing with farmers. In some instances, the recommendations of SCS and the extension service differed with respect to the same practices. Apparently in Missouri such a difference—in this case, over terracing—was one root cause of bitter hostilities. There Hugh Bennett and members of the staff at the agricultural college had come to verbal blows because the Missouri people disregarded, or disagreed, with Bennett's distrust of terraces unless they were constructed according to the specifications developed by SCS. The Missouri Extension Service established a powerful working alliance with the Missouri Terracing and Contractors Association. The University of Missouri sponsored annual short courses for members of the Contractors Association, and provided them with its own specifications for terraces. This alliance, together with other complicating elements involving farm management companies in the state, was powerful enough to delay legislation for soil conservation.[28]

[27] *Soil Conservation*, House Hearings, *op. cit.*, pp. 125–34.

[28] *Ibid.*, pp. 68–70 and 202. See also *Department of Agriculture Appropriation Bill for 1948*, House, *op. cit.*, Pt. 1, pp. 1009 and 1020–21; and *National Land Policy Act*, Hearings on H.R. 6054, House Committee on Agriculture, 80 Cong. 2 sess. (1948), p. 539. The Doane Agricultural Service in St. Louis was also active in opposing SCS. Howard Doane was a member of the Task Force on Agriculture of the First Hoover Commission in 1948–49. There are numerous letters in the files—NARG 16 (Reorganization)—dealing with this relationship during the proposed reorganization of 1953, including: C. W. McIntyre to Secretary Ezra Taft

The Missouri Extension Service also organized so-called Balanced Farming Rings, which were patterned after terracing associations in preference to soil conservation districts. These were publicized by the Extension Service of the Department of Agriculture in 1948 as examples of means for furthering soil conservation. This system originated in Carroll County where, it was claimed, the first terraces were built west of the Mississippi River in 1915. Stimulated by the county agent, there was growing interest in the program which was "emphasizing good rotations, fertilizers, and terracing." During 1941–42, the Missouri College of Agriculture developed plans for a state-wide balanced farming system to be effected through the county agents, who organized farmers into groups of 50. Each of these farmers paid $50 to help meet the cost of an assistant county agent. By 1948, there were 32 such groups, and clinics were being held to stimulate them in 108 of Missouri's 114 counties. The idea spread into Tennessee and North Carolina, and 38 of the other state extension services sent representatives to Missouri during 1948 to study this system. Apparently it was being given careful scrutiny by most of the state extension services as an alternative—or addition—to districts, in case the American Farm Bureau's recommendations for "decentralization" were put into effect.[29]

Other immediate points of conflict based on different supporting groups could be found in 1948, and many of these are continuing in the 1960's. For example, in the arid states of the West several governmental agencies have provided various types of technical assistance to farmers operating on irrigated lands. They have designed and supervised the construction of works requiring engineering advice, such as drains installed to correct the excess salinity which results from the application of more water than highly impervious soils will absorb. Cropland has been graded or planed to improve the distribution of water over entire fields and to reduce runoff. Bench and contour irrigation systems have been installed. Canals and laterals for delivery of water have been lined to reduce water losses—losses which have been estimated by responsible authorities to amount to 70 to 75 per cent of the total delivered from impoundments.

For many years, the Bureau of Reclamation of the Department of

Benson, and letters to and from J. Earl Coke, Assistant Secretary of Agriculture, March 12 and 16 and April 21, 1953. The Missouri Terracing and Contractors Association, the St. Joseph Chamber of Commerce, and the St. Joseph Livestock Exchange strongly supported "decentralization."

[29] U.S. Department of Agriculture, *Report of the Cooperative Extension Work in Agriculture and Home Economics* (Washington: Government Printing Office, 1948), pp. 7–8.

the Interior provided funds to state extension services of the Western states for assisting farmers in irrigated areas with services based on research at the state experiment stations. In some cases, it also helped to train county agents to lay out and supervise practices for increasing efficiency in the use of water. California developed a large staff in its Extension Service devoted to the solution of the problems of irrigated agriculture. Utah and Montana created agencies in addition to the colleges of agriculture. Before 1945, SCS intensified its research into these problems at its centers in Berkeley, California, and in Logan, Utah. In the early forties, SCS succeeded in establishing several soil conservation districts which included existing irrigation districts and surrounding land surfaces contributing silt or flood waters to the croplands. Then, in 1945, SCS scored a minor triumph in California in its struggle with the Extension Service by entering into a memorandum of understanding with the Imperial Irrigation District. The SCS technicians had developed a particularly effective drainage system to cope with some unusual problems of salinity caused by the use of water from the Colorado River which carries 2,200 pounds of salt per acre-foot.[30]

The relative competitive positions of the Bureau of Reclamation, the SCS, and the Western states' extension services were disturbed in 1947 by threats of alterations in their "historic" relationships. The growing success of SCS in reaching farmers on irrigated lands, especially by concentrating staff and equipment in the soil conservation districts, threatened to upset the older arrangement whereby the extension services were, in effect, allied with the Bureau of Reclamation in serving farmers. Suddenly, the Bureau of Reclamation terminated its financial support of the Western extension services. This move evoked a demand from the Committee on Extension Organization and Policy of the Association of Land-Grant Colleges that the Department of Agriculture take defensive mesures for their protection. It also stimulated a resolution from the National Reclamation Association commending the Department of Agriculture and "associated state agencies" for their services to irrigation farmers together with requests for additional funds for research and other activities.[31] This situation broadened and

---

[30] *Department of Agriculture Appropriation Bill for 1951*, House, 81 Cong. 2 sess. (1950), Pt. 4, pp. 1156–57.

[31] *Department of Agriculture Appropriation Bill for 1949*, Senate, 80 Cong. 2 sess. (1948), pp. 577–91. See also NARG 16 (Reports), Charles F. Brannan (Assistant Secretary), to M. L. Wilson, March 10, 1947. Gladwin Young reported a meeting in Denver where all the directors of extension in the Western states bemoaned the termination of the "historical relation" they had enjoyed with the Bureau of Reclamation by means of funds which it allocated to them.

intensified the area of competition between SCS and the extension services nationally by ranging the Western extension services in support of the American Farm Bureau's proposals for reorganization, even in states where the ties between the extension services and farm bureaus were weak.

Still another region in which competition between the extension services and SCS had been intensified since the end of World War II was the Tennessee Valley. This was related to the manner in which the TVA Board of Directors allocated funds to the land-grant colleges by entering into contracts for the performance of most of the TVA's agricultural program using the "educational" methods of the extension services. The Tennessee Valley Authority regionally, very much like the AAA-PMA nationally, accepted the colleges' emphasis on the "soil management" version of soil conservation. This approach was especially convenient because it provided an outlet for the TVA's fertilizer production. Enough funds were provided for assistant county agents for soil conservation. Voluntary soil conservation associations, rather than soil conservation districts, provided the county sponsorship which typifies most agricultural programs. Essentially, these associations were the sort advocated by the pro-extension group under Milton S. Eisenhower in the Department of Agriculture in 1935–36 and called Balanced Farming Rings in Missouri.

This system, called "grass roots democracy," was staunchly defended by TVA, whose spokesmen claimed that the Department of Agriculture had consented to it in the report made by the Secretary's Committee on Soil Conservation in 1935. By 1941, however, soil conservation districts had been formed in northern Alabama, and in 1942 the Secretary of Agriculture abrogated the Department's agreement to keep SCS out of the Valley. Negotiations continued with little organizing of districts until SCS discovered in 1944 that the Tennessee Farm Bureau was no longer unanimous in its opposition to the organization of districts in the Valley counties. From that time until the issue was aired before the Senate Public Works Committee in 1948, an organizing drive to satisfy farmer demands for SCS service through districts was covertly waged. When high-level support of SCS, in the form of a cut in appropriations for assistant county agents in the Valley counties was made by the Bureau of the Budget, the extension services became desperate.[32]

[32] A summary account of these relations, somewhat favorable to the TVA-extension position, can be found in Norman L. Wengert, *Valley of Tomorrow* (Knoxville: Bureau of Public Administration, University of Tennessee, 1952), pp. 98–112.

It was evident to knowledgeable persons that the drive to organize districts and the counter efforts to keep them out of the TVA area were complicated by the movements to create similar authorities in the Colorado, Columbia, and Missouri basins. The colleges of the Valley states were forcing TVA to fight their battle with the Department of Agriculture and SCS. The TVA Board was placed in a dangerous position which invited a renewal of the issue of public *versus* private power by the private electric utility interests. The TVA had more to lose than to gain from its championship of the colleges' position, for it was placed in the position of the boy whose little brother urges him to fight the neighborhood strong boy. Even many of the extension leaders came to realize that their resistance to SCS was hurting them with many farmers.[33]

Circumstances such as these account for variations among the positions of the state extension leaders and the state farm bureaus in respect to the organizational issues. The same phenomenon can be observed within states where variations occur among counties. For example, during the hearings before the House Committee on Agriculture in 1948 the presidents of several state farm bureaus supported the American Farm Bureau proposals. Yet other witnesses from the same states testified against these proposals and were careful to identify themselves as presidents or other leaders in their county farm bureaus. There was testimony that county farm bureaus had been active in many states in organizing soil conservation districts. Undoubtedly, these identifications were planned for effect, but they served the useful purpose of allowing congressmen to gauge the degree of support which the American Farm Bureau had from within its own ranks on this issue.

The major general farm organizations were also ranged against each other. The National Grange and the National Farmers Union supported SCS against the American Farm Bureau. Russell Smith, testifying for the Farmers Union, was frank to say that his organization opposed the Hill and Cooley bills (bills containing the Farm Bureau proposals) because they would greatly strengthen the power of the Farm Bureau-extension alliance—although the alliance was not of uniform strength in all states. He quoted California's Director of Extension who, in 1940, chided his colleagues for thinking that the alliance—with the extension

---

[33] See Philip Selznick, *TVA and the Grass Roots* (Berkeley: University of California Press, 1948), p. 175. The opposition of politically potent forces kept the TVA counties of Virginia, for example, free of TVA electricity. Starting in 1951, the TVA counties were rather quickly organized as soil conservation districts. The public *versus* private power struggle has also delayed the appropriation of funds for the Potomac Watershed.

services acting as "chore boy" for both the Farm Bureau and the Department of Agriculture—would bring extension "vast resources and power." Smith claimed that in no state did either the Grange or the Farmers Union exert influence over extension for achieving its organization aims. Other organizations joined the chorus of opposition out of the same considerations of expediency. E. T. Winter spoke for the Mississippi Valley Association, an organization whose members, he said, "go to sleep every night hating the Tennessee Valley Authority idea."[34] Members of his Association supported SCS in its hour of need because they expected the Department effectively to block a Missouri Valley Authority and SCS could play a pivotal role in the maneuvers.

Similar dispositions on the part of key congressmen toward the agencies are evident from the record, although motives are normally both hidden and complex. As is usual with respect to national policy issues, formal party affiliation does not dictate patterns of congressional behavior toward the programs of service agencies in the executive branch. During this period of agitation in interagency relationships, members of the same political party serving on the same standing committee displayed differing attitudes toward a single agency. An example is the case of Representative William S. Hill, a Republican, and Representative Harold D. Cooley, a Democrat, who sponsored the Farm Bureau's proposals in twin bills. One clue to Hill's support of the extension services was his revelation that he had once served in the Colorado Extension Service. Cooley's home state, North Carolina, was strongly organized by the Farm Bureau as a model for the South in the late thirties; by 1948, it had become a battleground because of SCS efforts to organize districts in the western fourth of the state which constituted a TVA sanctuary for the North Carolina Extension Service. Representative Clifford R. Hope of Kansas, Republican Chairman of the Committee on Agriculture, like W. R. Poage, then a senior Democrat from Texas, was warm in his support of SCS. Both of these gentlemen had special regard for the flood control potential of the SCS small watershed program then emerging as a possible alternative in the minds of many of their constituents to the large dams of the Army Corps of Engineers. Hope was from the wheat-producing areas of western Kansas where the Farmers Union is a major farm organization.

[34] *Soil Conservation*, House Hearings, *op. cit.*, pp. 106–11 and 141–42. Representative W. R. Poage remarked (*ibid.*, p. 10) that in Texas the boundaries of soil conservation districts did not necessarily coincide with those of counties, but were set up along drainage lines. "It seems to me that when you resort to political lines rather than the lines that God Almighty made, you are breaking down one of the things that has made the Soil Conservation Service such a success."

Representative Orville Zimmerman, a Democrat from Missouri, was instrumental in revealing the American Farm Bureau's objectives, while critically questioning the President of his own state's Farm Bureau.

Representative Dirksen, Republican from Illinois, who used the 1947 hearings of the Subcommittee on Agricultural Appropriations as a forum for airing extension's case against SCS, had forced the House Committee on Agriculture to hold hearings on the Farm Bureau's bills, to the distaste of several members regardless of party.[35] His colleague, Republican H. Carl Andersen of Minnesota, was during his service one of the most uncritical supporters of SCS in Congress—matched, perhaps, only by Republican Ben F. Jensen of Iowa, a former senior member of the parent House Committee on Appropriations. Andersen's support of SCS was particularly interesting, since most of Minnesota was not then organized in soil conservation districts. However, his own congressional district was composed in part of the "pothole" prairies lying in western Minnesota and eastern South Dakota where farmers wanted aid to drain wet lands.[36] Whitten, who since 1949 has been Chairman of the Subcommittee on Agricultural Appropriations when Democratic majorities have prevailed in the House, has supported both SCS and PMA. Representing a poor hill country constituency in Mississippi, he has favored the conservation payments program as an aid to the operators of small general farms, which are still numerous in

[35] *Ibid.*, pp. 1–2. Representative Orville Zimmerman, Democrat from Missouri, grumbled; and Representative John Flannagan of Virginia, the ranking Democrat, said: "I resent another committee of this House trying to usurp the authority of this committee by the back-door method of withholding appropriations to carry out legislation recommended by this committee and passed by the House." Representative Charles B. Hoeven, Republican from Iowa, said (p. 13) he understood that the Hill and Cooley bills had been introduced to mollify the House Appropriations Committee. "The fact that they did name such objections has resulted in decreased appropriations."

[36] *Department of Agriculture Appropriation Bill for 1959*, House, 85 Cong. 2 sess. (1958), Pt. 3, pp. 1609–10. Representative H. Carl Andersen put remarks "deliberately in the record" in 1958. The Fish and Wildlife Service, supported by wildlife interests seeking to protect waterfowl breeding grounds by maintaining the potholes, opposed the Ten Mile Creek watershed project in Andersen's district. Andersen complained that SCS had talked the local interests into a compromise which would leave about 10 per cent of the potholes, but that the Fish and Wildlife Service sabotaged it. Its obstinacy, he said, had made the farmers in his area "lose $1 million in Federal aid." See further, *Drainage of Wet Lands*, Hearing before the Subcommittee on Conservation and Credit, House Committee on Agriculture, 87 Cong. 1 sess. (Aug. 18, 1961), Serial Q. For a recent example of a Jensen panegyric in praise of SCS, see *Department of Agriculture Appropriation Bill for 1961*, House, 86 Cong. 2 sess. (1960), Pt. 4, pp. 388–93. Jensen's failure to be re-elected in 1964 may have interesting effects for SCS.

the Southeast. His support of SCS has not always been viewed within the agency or by some of its supporters as that of a warm friend. But his attitudes and behavior, reflecting the interests of his constituents, were crucial in the maneuvering to secure legislative authorization for the Small Watershed Program. Whitten differed somewhat from Andersen, since they reflected two contending points of view within SCS.[37] Representative Clarence Cannon, the Missouri Democrat who, except for brief interruptions, long served as chairman of the parent House Committee on Appropriations, supported SCS in important ways when its organizational integrity was threatened.[38]

In the Senate, George D. Aiken of Vermont, ranking Republican on the Committee on Agriculture and Forestry, was rated a strong supporter of extension and the PMA program. The SCS has had a friend, however, in the Senate Subcommittee on Agricultural Appropriations, which with only slight interruption for nearly twenty years, has consisted, in practice, of Senator Russell of Georgia. He is backed up on the parent Committee on Appropriations by Senator Hayden of Arizona, who for reasons already described was one of the first supporters of SCS in Congress.

The National Association of Soil Conservation Districts did not play an important role in blocking the American Farm Bureau's plan of reorganization in 1948. It was still quite new, having been created at a meeting in Chicago held late in July 1946. The Chicago meeting took place after an *ad hoc* group had met in Washington at the suggestion of Governor Robert S. Kerr of Oklahoma to lobby for a bill sponsored by Representative Poage of Texas to make surplus military equipment available to the districts. The bill was not passed, despite a conference with President Truman and members of Congress, so that it was necessary to persuade the Appropriations Committee to approve an item in the budget for 1947 permitting SCS to purchase surplus heavy

[37] Their differences stemmed partly from aligning themselves with either of the two major groups in SCS (those for or against the Watershed Program) and partly over the Agricultural Conservation Program payments in relation to SCS. See *Department of Agriculture Appropriation Bill for 1948,* House, *op. cit.,* pp. 1044–46; *Bill for 1950,* House, *op. cit.,* Pt. 2, pp. 27–30. Representative Jamie L. Whitten's Second Congressional District in Mississippi had the highest percentage of employment of farm workers in the United States (41.7 per cent) according to the census in 1960. Representative Andersen's former district ranked ninth; and Representative Ben Jensen's Seventh Congressional District in Iowa ranked twelfth, according to Congressional Quarterly Service, *CQ Census Analysis: Congressional Districts of the United States* (Washington, August 21, 1964), p. 1820.

[38] NARG 16 (SCS, Organization 1), Clarence Cannon to True D. Morse, Under Secretary of Agriculture, Sept. 19, 1953.

equipment and lend or grant it to districts. Even before World War II, several state associations of district supervisors had been formed and discussions held on the need for regional organizations. Since few states were organized when the war broke out, it was decided to concentrate first on organizing the states. According to former employees of SCS, there were intermittent discussions for several years of the wisdom of encouraging the creation of a national association and two of the regional directors—T. S. Buie in Spartanburg, S. C., and Lewis Merrill in Ft. Worth, Texas—particularly favored creation of the association. Although the need for a national organization was discussed and decided affirmatively, it was not until the meeting of July 1946 that the NASCD was officially organized. It was generally understood by informed parties that SCS initiated this move to give it a support group organized at the "grass roots."[39]

PROPOSALS FOR REORGANIZATION

The goal of the American Farm Bureau Federation was made clear when H. E. Slusher of the Missouri Farm Bureau testified that: "We would retain the technical help out of the Soil Conservation Department" and "what I would call the AAA machinery out there for carrying out the administration of every other part of the program." Representative Zimmerman asked whether the American Farm Bureau would

[39] Because many of my sources of information about these events are persons who are still active in public affairs, their names must remain confidential. One letter does not make a court record of evidence, but it is sometimes helpful. An example is NARG 16 (SCS, Organization 1), George W. Thatcher to Ezra Benson, Oct. 14, 1953. The writer was an employee who had served with the organization practically from its beginning until his retirement. He claimed in 1953 that "about six years ago" the International Harvester Company put on a big machinery show in Albuquerque with SCS personnel from four states in the region and the Washington office present. At a dinner, J. C. Dykes, Assistant Chief of SCS, was reported to have "made quite a talk. He . . . threw caution to the wind. He talked at length about the various other USDA agencies having their gangster organizations to fight their battles, such as the Farm Bureau for Extension and Farmers Union for FHA, etc. To meet this type of problem he says 'We decided we had to have our gang too, so we started the Soil Conservation Districts. These we are building into a national organization which should put up a pretty good battle against all comers.' "

After the NASCD was formed in 1947, its New England Director reported: "Upon returning from our July meeting, my job was to contact chairmen of districts in this region. From State Conservationists of the Soil Conservation Service, I received complete lists of supervisors from each state in the district, *in many cases with key men designated.*" [Emphasis added.] National Association of Soil Conservation Districts, "Minutes," Feb. 25–26, 1947, mimeographed, p. 9.

"abolish the Soil Conservation Service . . . and place the benefits of the Soil Conservation Service in the Extension Service?" Slusher replied: "You would consolidate it with the Extension program, you would merge it. You would probably . . . have a man who is responsible at the Washington level to assist the various State organizations. . . . That probably would eliminate the districts and regional offices. I don't think they are necessary." Representative Cooley added: "I understand the things we should maintain are the agricultural colleges, the experiment stations and the Extension Service; but at the county level we should bring together all soil conservation activities."[40]

The key features of the Farm Bureau-extension plan of reorganization for SCS were plain. A small staff to give "advice" would be retained in the Department of Agriculture in Washington; this would be similar to the federal Extension Service. The regional, state, and district offices of SCS would be abolished. The work of technicians on the county level would be "decentralized" to the state extension services and placed under the county agents, as in the TVA area. The PMA would be continued to administer the conservation payments through the county committees. Under this plan, the extension services, at long last, would be solely responsible for all the technical phases of soil conservation and would no longer have to battle with SCS over practices and specifications. The final wisdom in matters agricultural would again filter down to the American farmer exclusively from his land-grant agricultural college. The landscape of rural America would once again be at peace.

The hearings on the Farm Bureau's bills to reorganize work of SCS exposed to the public the fact that "decentralization" would vastly increase the political power of extension in the states. Allan Kline, who was President of the Iowa Farm Bureau before becoming head of the American Farm Bureau Federation, frankly said that ever since 1936, and especially since 1940, when it broke politically with the Department, the Farm Bureau had recommended an end to the "overlapping and duplication" of the Department's agencies. The Farm Bureau's experience had been that "sooner or later straight-line agencies get into partisan politics . . . regardless of the party in power." In addition, farmers "who become associated with the administration of one of these straight-line programs frequently become intensely partisan toward it. . . . It is accurate to say that the personnel of straight-line agencies sometimes encourage this." Kline contended specifically that

[40] "Farm Program of the American Farm Bureau Federation," *op. cit.*, pp. 23 and 36–38.

PMA "has been used politically in the State of Iowa. . . . It is a Republican State and people who operate PMA are Democrats." In his opinion: "You are always suffering a delusion when you think you have control of a program but you do not have control of the personnel, their terms of employment, length of tenure in office or anything else."

When Representative Hope asked Kline how much of SCS would be left under the Farm Bureau's proposed reorganization, Kline answered that he could not draw a fixed line for everyone because political activity by SCS was one of its characteristics as a "straight-line agency." Representative Poage, Democrat of Texas, agreed and added: "I think we can say further that all of the jointly supported agencies, the state and Federal government, have done it, too . . . and I do not think there is a bit of question but what you say is perfectly true about . . . the Soil Conservation Service . . . [and] the Extension Service and any other Government agency that engages in a fight either to swallow somebody or to keep from being swallowed." Kline then innocently stated that he had "not known the Extension Service in our area to get into this kind of thing." There were exceptions, he admitted, so far as some states "are Democratic and some . . . are Republican and you just cannot get this straight-line activity in a certain particular direction." Poage asked if he did not mean simply that "you get political activity in the States, and that in Texas it is going to be Democratic and in Wisconsin it is going to be Republican, whereas, if you have a straight-line agency it is going to be whatever the national administration is?" Kline agreed that Poage's characterization of the situation was "accurate."[41]

Representative Hill of Colorado also had this problem on his mind when he asked E. T. Winter of the Mississippi Valley Association how he would like it if he had a "fine technician" in a soil conservation district and the "boys in Washington could come out and take that man away leaving the local organization with practically no recourse." If, on the other hand, it were an organization like an extension service, "any time they do not want that county agent, they will just simply notify the Extension agent to remove the man. In order to put a county agent in a county you have to have the support, in my State . . . of the county commissioners. We have no Farm Bureau control of the county agent in my State. If a county agent gets mixed up in politics and goes out and campaigns for the wrong commissioner, he is definitely going to be moved."[42]

[41] *Soil Conservation,* House Hearings, *op. cit.,* pp. 87–89.
[42] *Ibid.,* p. 114.

Meanwhile, SCS was busy defending its program in a variety of ways which did not escape the attention of members of Congress. During 1948, SCS put on more than forty demonstrations at which farms were planned for conservation and the equipment manufacturers supplied the heavy equipment needed to convert land to proper practices.[43] These demonstrations were quite dramatic displays and were given widespread publicity to insure maximum effects. For example, when the Columbia Broadcasting System wanted to cover a demonstration in Westmoreland County, Pennsylvania, Wesley McCune, an aide to Secretary Charles F. Brannan, suggested that it would probably be better for "your program" to cover one scheduled for October 12 and 13, 1948, in Chester County. He did not elaborate, but Chester was the home county of a key member of the Pennsylvania Legislature—where SCS needed supporters at the time—and also of the Brandywine Valley Association which had supported the program for small watersheds advocated by SCS. Also, CBS was informed that there would be a "conservation caravan" in the Southeast between September 22 and October 19. A demonstration had been held earlier at Winder, Georgia, the home of Senator Russell. McCune gave CBS the names of SCS regional information officers who would assist in planning coverage of the demonstrations.[44]

It is possible that it was only a coincidence that these demonstrations were being held in 1948—a presidential election year in which the Farm Bureau was waging such a determined fight to "decentralize" the SCS program. In 1950, however, Joseph J. Lawlor, Assistant Postmaster General, wrote to Secretary Brannan, on behalf of Congressman Harry P. O'Neill, Democrat from Pennsylvania, and requested a conservation demonstration in Lackawanna County, Pennsylvania. Lawlor said he could give several reasons, but "to me the most important reason is the coming election. Of course, I know you are personally acquainted with Senator Francis J. Myers [Democrat from Pennsylvania] who goes up for re-election this year. Such a program, if approved, would have great effects on the coming campaign." Lawlor said he understood that "your Department has approved such demonstrations in other parts of Pennsylvania and you personally attended one or more." Brannan replied that all the demonstrations, which were being organized by PMA, had been planned and that he could not accommodate O'Neill.[45]

[43] NARG 16 (Soil—1), Charles F. Brannan (Secretary of Agriculture) to Frank Bronson, Sept. 3, 1948.

[44] NARG 16 (Soil—1), Wesley McCune to William Stewart, Sept. 7, 1948.

[45] NARG 16 (Soil—1), Joseph L. Lawlor to Charles F. Brannan, May 3, 1950, and reply.

Representative Hope decided in the spring of 1948 that the Farm Bureau attack on the SCS had gone far enough and he offered his own bill to counter the bipartisan Hill and Cooley bills. The Hope bill was an omnibus piece of legislation which proposed the consolidation of a whole host of agencies in both the Interior and Agriculture departments into a single agricultural resources administration.[46] It is reasonable to guess that he knew the bill would not even be reported out of committee for a variety of reasons. Since the "organizational" issue had been raised as a talking point in the presidential campaign of 1948 to influence the farm vote, the Farm Bureau and extension services opposed his bill.

The Hope bill was an easy target for critics also because it provided for such drastic alterations in the administration of the public lands and in certain activities of the Bureau of Reclamation that spokesmen for the livestock and irrigated agriculture interests of the Western states came to Washington in haste to protest any change which would disturb their relations with the Department of the Interior. In addition, Hope could safely guess that the Department of Agriculture would oppose it, since he included provisions which would have the effect of destroying PMA's control of the distribution of conservation benefits through the county committees. In place of the existing arrangements, Hope proposed to administer payments through the soil conservation districts and, incidentally, to expand greatly the role of SCS—although under a new name. By introducing this bill, Hope knew that as Chairman of the House Committee on Agriculture he could use the hearings as the occasion for producing a record of testimony that would counter—and confuse—the criticisms of SCS and the districts given in support of the PMA and the Farm Bureau's proposals. He could anticipate, also, the possibility (which eventuated) that this testimony could be used to blunt recommendations which the First Hoover Commission was expected to make in 1949. He could, therefore, confidently expect no action in an election year.

The provisions in his bill dealing with SCS-PMA relations were crucial, and it is interesting that he baited Under Secretary Norris E. Dodd into revealing the exact source of the Secretary's distaste for relinquishing PMA's administrative control over the conservation payments to the districts. Hope proposed two schedules of payments: "A" payments for permanent or semipermanent practices to be subsidized at rates not to exceed one-third of their cost; and "B" payments for

[46] *National Land Policy Act,* House Hearings, *op. cit.,* May 4–26, 1948, p. 316.

"recurring practices that will retard depletion of soil productivity" which could be made annually at rates not exceeding one-fourth of the cost. The "A" payments could be made only once for a given practice and only if they were based on a cooperator's agreement (and SCS farm plan) with a soil conservation district. These payments could be made only through soil conservation district boards, acting as agents of the PMA. The "B" practices were also to be "certified" through the district boards, but where there was no board, the PMA "local" committees could perform this function.

Clearly, these provisions of Hope's bill would have aided materially in speeding the organization of soil conservation districts in the areas of remaining indifference or hostility. They also would have put an end to the PMA system of providing "technical" services through its county committees. Dodd testified that although the Department was wholly in accord with the objectives of this bill, it could not incorporate the soil conservation districts into its "administrative mechanism to the degree necessary to insure the proper execution of programs for which Congress holds the Department responsible" because they were instrumentalities of the state governments. Thus had the wheel turned. Almost ten years to the day after Secretary Henry Wallace had committed the Department to promoting soil conservation districts because it could not work reliably with the state extension services, new leaders in the Department rejected districts as appropriate instrumentalities for carrying out its soil conservation work.

The transparency of this argument was revealed when Dodd remarked that the proposed bill would channel through the districts certain programs (farm forestry) "now administered through State forestry agencies." Disturbing the existing relations in forestry, he counseled, would be "very unwise."[47] No member of the House Committee on Agriculture saw fit to challenge Dodd's—or the Department's—inconsistency. No one asked the obvious questions because they were unnecessary. Representative Pace, Democrat of Georgia, put the issue quite simply when he complained bitterly to Hope that under the provisions of his bill 70 per cent of the farmers in Georgia would get nothing out of such a program.[48] No one saw fit to point out the fact that

---

[47] *Ibid.*, pp. 14–15.

[48] *Ibid.*, p. 34. Testimony by interested farmers on the Hope proposals reveal sources of the intense loyalties of some farmers to PMA and others to SCS. Often, if not always, they are founded on bread-and-butter issues and fashioned out of personal experience. See, for example, *ibid.*, pp. 128–30, 153–58, 163, and 237. Ben Glover, who identified himself as a soil conservation district supervisor, Farm

district supervisors, although chosen without formal party designation, might be "politically unreliable" in many areas of the country.

If the Hope proposals were any indication of the sort of reorganization Congress might enact, the Department wanted no part of it. It would be bad enough for a Democratic administration to take the blame for depriving farmers of their conservation payments and services, but it would also be equally disastrous to deprive temporarily, at least, a varied host of construction contractors and the suppliers and purveyors of equipment and materials to the rural market which the conservation payments supported. At a minimum, it would anger such supporters of PMA as the county committeemen who not only sat in judgment approving practices, but also, in some cases, supplemented their incomes by contracting to provide goods and services subsidized by their own agency. Any reduction or expansion of conservation payments is a bread-and-butter issue which has to be taken seriously in an election year.

In 1948, the uncertainty of the outcome of the impending presidential election was so very great that the parties to the controversy were disposed to await the results before pushing the reorganization issue further. At the congressional hearings in 1948, several witnesses mentioned the fact that the First Hoover Commission was due to report in 1949. They took refuge in suggesting that reorganization be laid over until after the Commission had reported. Had the Republicans won the presidential election, a new Administration might have been inclined to use some of the Commission's recommendations. The report of the Commission's Task Force on Agriculture would have provided a blueprint for a reorganization favored by the extension leaders. In essence, the Task Force repeated the recommendations which Secretary Wallace's Interbureau Committee made in June 1935. The Task Force urged that all forestry and soil conservation "education" be placed within the Extension Service; that the technical services provided for farmers by SCS be transferred to Extension which would be authorized to add conservation assistants for county agents. These services would be rendered through soil conser-

---

Bureau leader, state senator, and member of the legislative committee of the Alabama Association of Soil Conservation Districts, testified with amazing frankness (*ibid.*, p. 257) that he had planted "2 acres of alfalfa last year as an experiment . . . and considering the fact that I was a member of the Senate, through politics, I finally got [a] thousand pounds of potash. That is just how it is." In contrast to the supporters of SCS, the proponents of PMA at this particular hearing were generally smaller operators who had an unusually direct stake in the continuation of PMA's program.

vation districts or directly to farmers. The SCS would be left to "coordinate" and "plan" a national conservation plan, to include upstream flood control investigations.[49] The Task Force was heavily staffed with supporters of the Extension Service, but the Commission represented a far broader spectrum of interests. It is not surprising, therefore, that the Commission greatly toned down the Task Force's recommendations.

POLITICS AND ORGANIZATION

The congressional hearings of 1948 provided a good picture of the extreme complexity of the struggle for political power in rural America. Although the focus was on efforts to control the soil conservation programs, the struggle was far broader. One other revealing aspect, for example, was the attempt by the National Farmers Union, the National Grange, and certain members of Congress to publicize, criticize, and, if possible, eliminate the legal ties between the extension services and state farm bureaus where this arrangement prevailed. It is easy to slip into the error of thinking that the extension services' ties with the state farm bureaus were universal and uniform, but they were not and never had been. The tie existed in some states which were normally Republican and others which were durably Democratic. When issues of this sort were aired in a national forum there was a tendency to overlook the variety and complexity of the extension services' relations with support groups in the states. Indeed, witnesses to the hearings on this issue differed over a fundamental issue. Some thought that the American Farm Bureau was the master of the Extension Service and others thought that the reverse situation prevailed. They opposed the legal tie in either case.[50]

The hearings on soil conservation showed that the organizational question was not a party-line issue, nor was it clearly sectional, except that there were no key supporters of SCS from the Northeast. Moreover, it was not an issue producing any clear-cut cleavage between the Administration and Congress. It is true that the Department of Agriculture under Democratic control developed the "action" agencies in or-

[49] The Commission on Organization of the Executive Branch of the Government, *Task Force Report on Agricultural Activities, Appendix* (Washington: Government Printing Office, 1949), pp. 34–41. The full Hoover Commission rejected the idea that the Extension Service should provide the technical services to farmers in place of SCS.

[50] William J. Block, *The Separation of the Farm Bureau and the Extension Service* (Urbana: University of Illinois Press, 1960), pp. 102–06.

der to have direct lines of communication and administration to farmers instead of helping the Extension Service to expand to serve these purposes. It was understandable, therefore, that the more militant extension leaders expected the Republicans, if they captured the presidency in 1948, to organize the soil conservation activities to suit the Extension Service. This would occur, they assumed, because the leadership in the American Farm Bureau and the agricultural colleges of the Northern states provided some of the most likely prospects available to the Republicans to staff the Department. These men understood that they could make common cause over the organization of soil conservation despite the almost infinite variety of relationships which the state extension services had with their political friends and enemies at home.

Members of Congress indicated at the hearings in 1948 that they understood this situation very well. Republican Representatives Jensen of Iowa, Andersen of Minnesota, and Hope of Kansas had no difficulty agreeing with Democrats Whitten of Mississippi, Flannagan of Virginia, and Poage of Texas in supporting SCS. In all five of their states, the extension services had fought SCS at home and in Congress with varying degrees of success; in three of the states, its tie to the Farm Bureau was legal and formal, and in the other two it was informal but strong. These key members of the House and others in Congress had much to gain by the Department's administering action programs through line agencies which had to run the annual gauntlet of the appropriations process and to secure their legislative authorizations from committees dominated by veterans such as those just mentioned. Members of Congress know perfectly well that they can exercise influence over policy-making and administration in the Department in attempts to satisfy the needs and wishes of their constituents. They know equally well that they could not oversee the administration of federal funds by fifty extension services. By overseeing federal administration, congressmen defend themselves against the attacks of rivals from home. It is an elementary rule that one of the sure ways to build political power is to perform favors for voters, and SCS had built a good reputation with many members of Congress. It had been extremely fortunate—almost lucky—to be able to offer a program which particularly suited the needs of key members of Congress from districts having quite diverse soil or water conservation problems. Members of Congress were too shrewd to vote federal money to their states and districts where the money would be distributed as goods, services, or advice by administrators beyond their control.

Obviously, there were exceptions among congressmen. Extension had its friends there, too; and some, like Senator Aiken of Vermont, were key members. Their motives in supporting the extension services, rather than SCS, are not wholly clear. Aiken, for example, came from a state which had few erosion problems. Such factors cloud analysis of the problem, but the basic consideration to every congressman who was much concerned with this issue probably was, first, the extent and character of the services which SCS had performed for satisfied interests in his constituency. The second consideration, although not necessarily in importance, was the degree to which actions by the extension service and its supporters in his constituency would influence the outcome of his election. In some areas, this factor did not exist; in others, a well-established member of Congress had no need to fear this sort of opposition; and, in still others, the congressman either used the extension system to his advantage or had to respect its political power.

Whatever the mixture of motives among the interested members of Congress in 1948, none of the reorganization bills left the House Committee on Agriculture and, once again as so often in its earlier history, SCS was well served by strategically situated members of the House. Even when the Republicans finally gained control of the executive branch in 1953, and the Extension Service and its friends staffed the Department of Agriculture, they had very limited success in reorganizing SCS. To a considerable degree, they were thwarted because Representative Whitten and Secretary of Agriculture Brannan imposed a two-step settlement on PMA and SCS in such a fashion that neither one "swallowed" the other, and the Extension Service and its supporters lost all hope of "decentralizing" the SCS program.

SUMMARY AND COMMENT

Once Secretary of Agriculture Henry A. Wallace decided in February 1936 that the Department would administer cash subsidies for soil conservation through the Agricultural Adjustment Administration and give farmers technical assistance through the Soil Conservation Service, his successors accepted this arrangement until 1947. There was actually a close tie, however, between the AAA and the state agricultural colleges, since the AAA depended chiefly upon their extension services, rather than upon SCS, for technical guidance. Consequently, even after their relations were strained over other issues, AAA and the Extension Service maintained close operating relations in administering the

Agricultural Conservation Program. In the meantime, SCS went its own way, assisting only farmers who were "cooperators" with soil conservation districts. The AAA conservation program was open to virtually all farmers, whether they were "cooperators" with districts or not.

When the dimensions of the SCS drive to organize conservation districts and expand its program to be serviceable in all areas of the United States threatened the Extension Service's farm management planning operations, the American Farm Bureau Federation threw the organizational issue into the legislative hopper, thereby acting on threats which it had been making against SCS since 1936. A Western Republican and a Southern Democrat introduced in the House in 1948 bills to "decentralize" administration of the SCS program to the state extension services. Bipartisan opposition from key members of Congress stopped these proposals, which were essentially the same as the recommendations of the Secretary of Agriculture's interbureau Committee on Soil Conservation of June 5, 1935. Representatives friendly to SCS forced officers of the American Farm Bureau Federation to admit that they spoke for only some, not all, of their state organizations; and state organizations did not always speak for county farm bureaus on this issue.

During public hearings, Representative Hope maneuvered Under Secretary of Agriculture Norris E. Dodd to admit that the Department opposed any new legislation which would make soil conservation districts responsible for funds appropriated to the Department by Congress. In short, he revealed that the Department could not hold districts responsible for accomplishing its conservation objectives any more effectively than it would control the state extension services when the districts were conceived in 1935–36. He confirmed the truth that was implicit in Secretary Claude Wickard's memorandum of September 21, 1940.

Key members of Congress refused to act on any of the reorganization proposals put before it in 1948 because many of them preferred the political effects of national administration regardless of the party in control, to those of state administration by the extension services. Congress showed that, where the Soil Conservation Service was concerned, it was no more capable during an election year of pushing "coordination" or "decentralization" than the Secretary of Agriculture. A settlement was postponed until PMA and SCS became locked in a battle which prompted a member of Congress and the Secretary to bring order out of the prevailing chaos.

CHAPTER 6

The Organizational Issue Resolved

PRESIDENT HARRY S. TRUMAN'S unexpected victory in 1948 insured that the Department of Agriculture would be directed by men who were not disposed to enhance the power of the Extension Service–American Farm Bureau Federation axis at the expense of the Soil Conservation Service (SCS). The effect of continued Democratic control of the Department of Agriculture upon relations between the Production and Marketing Administration (PMA) and SCS, however, was far from clear. Although Secretary Clinton P. Anderson had revealed the move to consolidate SCS and PMA in 1947, it is not certain whether the proposal was actually initiated by the Secretary or by persons in one of the two agencies. During 1948 and early 1949, it is possible that PMA leaders acted on their own initiative to gain control of the two soil conservation programs at the expense of SCS. The scattered available evidence, however, suggests that the Secretary must have known about the interagency competition and silently permitted it until he was publicly embarrassed by the ill-concealed fratricide in his Department. Certainly, Representative Jamie L. Whitten of Mississippi was aware of their competition, and moved in 1949 to end the struggle. His efforts were not wholly successful, however, until 1951 when Secretary of Agriculture Charles F. Brannan imposed an additional measure of discipline among the agencies administering the conservation programs.

PMA RIVALS SCS

In 1948 and 1949, PMA made a number of procedural changes in its Agricultural Conservation Program (ACP), which apparently were intended to limit the role of SCS and the soil conservation districts as agricultural conservation agencies. For example, relations between PMA and the soil conservation districts in Oklahoma were regulated in part by a formal agreement made in 1945 on the state level. It permitted the farmer-elected county committees which administered the programs of PMA and its predecessor, the Agricultural Adjustment Agency (AAA), to make agreements with the governing boards of soil conservation districts for using the districts as "vendors" of services. A procedure which PMA called "CMS" (conservation materials and services) allowed the county committees to purchase goods and services directly from vendors for the conservation program. The chief purpose of the PMA-district agreements in Oklahoma was to use equipment owned by conservation districts to construct ponds, terraces, stock tanks, and other works requiring heavy equipment not owned by most farmers. The Oklahoma PMA committees, however, reserved the right to approve other vendors, including those "furnished by a committeeman or alternate community committeeman and (given) approval when justified."[1] These agreements also stipulated the conditions under which the PMA county committees would use SCS technical assistance. Generally, they provided that committees would use SCS where arrangements could be worked out, but that they would not be bound by SCS determinations of a farmer's needs for practices nor by SCS certifications that practices had been completed to meet the PMA's specifications. This arrangement was limited to practices installed by farmers who were conservation district "cooperators"; it did not apply generally to all farmers who participated in the PMA's conservation program—and those participating in this PMA program amounted to the majority of Oklahoma farmers.

In March 1948, the "working agreement between SCS and AAA," according to Oklahoma's State Conservationist, was abruptly abrogated by the state PMA Committee "without consulting any representatives of our Service." This cancellation ended the use of SCS technicians to service AAA engineering practices on the farms of district cooperators, and thereby permitted the PMA committees to allow their own men or selected contractors to do this work. It meant that if a farmer were a cooperator with a conservation district, but wanted to use PMA conservation payments to install a practice called for in his SCS farm plan, he had to allow the work to be laid out and supervised by personnel chosen by PMA. This change meant that SCS would be able to work only with those few farmers who did not use PMA conservation payments to help pay for the installation of practices. The practical result would be that SCS could not expand its work through conservation districts to aid farmers who might want their

---

[1] This quotation, and others which follow dealing with the situation in Oklahoma, are from *National Land Policy Act,* Hearings on H.R. 6054, House Committee on Agriculture, 80 Cong. 2 sess. (1948), pp. 284–91, *passim.*

services. Since SCS was already under heavy fire in Congress from the American Farm Bureau Federation, it was evident to anyone familiar with the situation that SCS would be in serious trouble if the Oklahoma plan were put into effect nationally.

The Oklahoma action was aimed, also, at the districts. The state PMA Committee told its county committees that the 1945 agreement had hampered work in some counties so that, in "light of the policy to decentralize authority and to place more responsibility in the hands of county committees, the State committee decided to discontinue the memorandum of agreement with the Soil Conservation Service." County committees were free to enter into agreements with soil conservation district boards for keeping records and making reports, but if "other points are to be covered by written agreement they should first be cleared with the State committee in order that uniform policies will exist throughout the State." This directive was transparent on its face. The state PMA Committee ruled further that in counties where the soil conservation district boards had been approved as vendors of services for PMA, such agreements should be canceled if "any of the personnel assisting the districts in the preparation of purchase orders and making delivery of materials or services are federal employees or receive any portion of their salary from Federal Funds." This order was intended to cripple any district's equipment services to farmers, if it permitted SCS personnel to give clerical assistance to the governing board. Most districts relied on the PMA work to earn enough money to cover the costs of an equipment program. And the availability of the proper equipment at low rates had been one of the very attractive features of the SCS program through the districts in most areas where construction of conservation practices is necessary. Since SCS had been driving hard in Congress since 1945 to secure funds for the purchase of more equipment for use by the districts, the impact of this move by the state PMA Committee, again, was evident to all but the uninitiated. Legally speaking, such action could not eliminate soil conservation districts. But it is difficult to see how the districts could serve any useful function, since they could no longer claim to offer to farmers the benefits of materials, equipment, and services which had been available through SCS when the Civilian Conservation Corps (CCC) camps and the Works Progress Administration (WPA) were functioning. At most, the district supervisors could continue only to serve as advocates of conservation unless they turned to state and local governments for much larger amounts of financial assistance than they had been receiving. Few farmers could be expected to reject the PMA subsidy for installing conservation practices simply because they preferred to have SCS technical assistance and SCD equipment to do the job.

The PMA break with SCS and the districts in Oklahoma was made in an area where SCS and the districts were strong and supported in the state legislature. If this move succeeded in Oklahoma, it seemed that a pattern could be set to undercut SCS everywhere. Apparently, the inability of Congress to restrict the Department by taking any steps to terminate the rivalry between SCS and PMA during 1948 encouraged officials to continue the struggle after the election in November 1948. In that month, PMA issued new instructions to its field offices for using CMS procedures, and in December 1948 officials discussed CMS functions in future operations at PMA state conferences. An apparently slight change in administrative procedures was made to achieve results as significant as any which might be gained at that time through formal reorganization. The revised instructions to the field included the following proviso:

> It is not permissible to give priority to any class of participants under the program for materials or services to be furnished by reason of their affilia-tion with any organizations or because of their participation in the pro-gram of any other agency.[2]

Spelled out for the uninitiated, this order meant that farmers who were cooperators with soil conservation districts had no special claim on PMA's conservation payments. It showed that PMA was ready to intensify its drive to secure ascendancy over SCS and become the Department's conservation agency.

Within two months, the PMA drive against SCS and the districts had reached such proportions in the Northern Plains states that Gladwin E. Young, the Secretary's special representative in the Missouri Basin, at that time feared the Department might become the "loser." The bad part of the conflict, he said, lay in the fact that "PMA at its state meetings in this region is coming out in open opposition to the program of SCS." He noted firsthand accounts from Minnesota, North Dakota, South Dakota, and Montana. According to Young's informants, PMA state officials in conversations with county committeemen were saying SCS technical assistance was unnecessary for a soil conservation program and that, in addition, it was costly and inefficient. There was

---

[2] "Procedure for Furnishing Conservation Materials and Services," Production and Marketing Administration, U.S. Department of Agriculture, ACP-231, mimeo-graphed, November 1948, p. 2.

an "unconfirmed" report that the county committeemen were told that no Agricultural Conservation Program payments would be made for any work planned by SCS. These moves were interpreted in the region as evidence that "PMA has apparently determined to come out in the open and state the ground on which they intend to wage a battle for gaining the leadership within the Department of Agriculture for carrying on the soil conservation program."[3]

The North Dakota state PMA Committee instructed its county committees to employ personnel to "do *all* of the necessary field jobs in connection with conservation practices." The North Dakota state PMA Committee also directed its county committeemen to use SCS or extension personnel to lay out practices, but to have PMA field workers inspect the layouts before approving them for need and payment. In May 1949, Senator William Langer, Republican from North Dakota, forwarded a copy of these instructions, together with a letter of protest from the President of the state's Association of Soil Conservation Districts to Secretary Brannan and threatened to "take this matter up" on the floor of the Senate. He argued that "duplication" seemed to be scarcely justified in view of the fact that seventy-three counties in the state were in soil conservation districts and only five were not. Secretary Brannan hastened to chastise the North Dakota State Committee for this action and to reassure Senator Langer that it was not in line with official PMA policy. County committees were authorized to hire their own personnel for this work only to the extent that "assistance offered by the Soil Conservation Service, Extension Service, or other Federal or State agencies is inadequate."[4] Thus, the state and county committees were left free either to determine the "inadequacy" of SCS services and to hire their own men or to play off the state extension services against SCS wherever the county agents could find the time to do this sort of work.

Despite Brannan's assurances to Senator Langer that the North Dakota order was out of line with national policy, there were signs that

[3] National Archives, Record Group 16 (Reports 3–3), Gladwin E. Young to Ralph R. Will, Feb. 18, 1949. Prior to his appointment as special representative of the Secretary in the Missouri Basin, Young had served in the Bureau of Agricultural Economics and, apparently, was a neutral observer of the PMA-SCS fight. He later became Deputy Administrator and then Associate Administrator of SCS. (National Archives records cited hereinafter usually will be identified by the abbreviation NARG.)

[4] NARG 16 (Soil—1), Charles F. Brannan to William Langer, May 26, 1949, and attached file including North Dakota PMA memorandum No. ACP 49–5. The Secretary answered a similar complaint from the President of the South Dakota Association of Soil Conservation Districts, Frank Feser, March 25. 1949.

the struggle had spread and was probably intensifying. At its annual meeting in November 1949, the National Reclamation Association took the unusual step of adopting a resolution recommending the consolidation of the PMA and SCS conservation programs. This opinion was voiced because of reports throughout the West of "a lack of cooperation . . . that in some places . . . was a state of open warfare." Agricultural representatives of the Western railroads were particularly critical of the situation.[5]

It would be erroneous to create the impression that SCS and the district boards were completely innocent of any maneuvers to counter PMA's activities. Although the point was never made explicit, SCS drew tactical advantage from the claim that the district supervisors formulated a "district program" and determined how and when it would be carried out, with SCS merely providing "assistance." Insofar as it could be claimed that the district boards wanted to concentrate "their" assistance solely on the farms of district "cooperators," SCS could refuse to service farmers who were not cooperators and used ACP funds. The SCS was serving a preferred clientele—those farmers who had signed the conservation pledge in the form of the farmer-district agreement.

Kent Leavitt, President of the National Association of Soil Conservation Districts (NASCD), opposed a transfer of funds from PMA to SCS in 1949 for precisely this reason. "I am frank to admit that I cannot see how it can actually be effective." If the PMA committee in Dutchess County, N.Y., offered 10 per cent of its ACP funds to the district supervisors to be given to SCS for technicians, "it would completely upset our program. How can we allocate funds for a few months in order to carry out jobs at certain farms when we are trying to carry out an overall program." He wanted to know whether the transfer was mandatory, because, if it were, particularly on a national basis, the result for "local programs" would be "very embarrassing."[6]

The imputed independence of the district boards to direct the technicians' work became a stratagem in the efforts of SCS to become the exclusive soil and water conservation instrumentality of the federal government. Representative Clifford R. Hope, Republican of Kansas, and other friends of SCS would have insisted that farm plans made by conservation districts be required before federal aid to farmers would

[5] NARG 16 (Soil—1), memorandum, Ralph R. Will to the Secretary of Agriculture, Nov. 25, 1949.
[6] Department of Agriculture Appropriation Bill for 1950, Senate, 81 Cong. 1 sess. (1949), pp. 792–93.

be given were it not for the fact that such a small percentage of farmers had them in 1948.

Members of the Subcommittee on Agriculture of the House Committee on Appropriations were fully aware of the chaotic struggle in the field. As the PMA drive gathered headway in 1949, Representative Whitten, Chairman of the Agricultural Appropriations Subcommittee, took the initiative in discussing the situation with members of the House Committee on Agriculture and spokesmen for the Department. Whitten later explained that he had spent his time preaching and fighting for conservation and continued: "When I first came on this committee, we actually came close to losing both services because one of the big farmer organizations was in here trying to cut the Soil Conservation Service, and then the friends of the Soil Conservation Service brought their friends in next, and they set out to cut the PMA program, and they both almost succeeded one year, and it took a long time for them to recognize that they had to pull together, the friends of each."[7]

Whitten said he also had field reports that PMA would not certify construction on farms for payment if it were supervised by an SCS technician, even when it had been laid out according to PMA standards. Instead PMA insisted that it had to be rechecked by someone chosen by the county committees. There were numerous charges that the two agencies could not agree over the proper standards and specifications for practices.[8] At bottom lay the fact that PMA did not necessarily use SCS specifications for practices, but often relied, instead, on those recommended by the Extension Service or simply used specifications set by its own personnel. If PMA had used SCS employees for this purpose where they were available, it would have had no need to hire its own personnel—and to provide more jobs in each rural community so affected. Whitten was certainly no opponent of PMA—nor of SCS, but he said he thought that the PMA system of committees responded easily to pressures from farmers to approve practices which they "will do anyway as against practices which will eventually lead toward meeting this soil conservation need."[9]

In order to overcome the claims of PMA supporters that SCS failed to provide enough technical service to assist all the farmers who used PMA conservation payments, Whitten in June 1949 secured authori-

[7] *Department of Agriculture Appropriation Bill for 1951,* House, 81 Cong. 2 sess. (1950), Pt. 5, p. 1442; Pt. 4, pp. 1188–89.

[8] *Ibid.,* Pt. 5, pp. 1439–40.

[9] *Ibid.*

zation in the Department's appropriation bill for 1950 to permit PMA to transfer 5 per cent of its conservation funds to SCS to pay for additional technical assistance. The allocation of PMA funds was to be effected on a county basis, so that SCS would be able to hire more technicians and aid to service farmers in more counties than previously. The 5 per cent transfer was to be negotiated by the PMA county committee and the local technician of the SCS. This procedure also was intended to meet the criticism of farmers partial to SCS that PMA did not provide sound technical direction for its conservation program so that much work of poor quality was done. Whitten's original proposal had been for a 10 per cent transfer of funds, but this was opposed by the limestone interests and the National Association of Soil Conservation Districts, and was cut to 5 per cent by the Senate. Despite this cut, the congressional action was a warning to the parties that they should end their feuding.

Representative Whitten, during discussion of the 1951 appropriation for the Department, claimed that the "suggestion about the percentage of transfer came from me, but I found that each service was a little suspicious of it because knowing that they didn't offer it, they were afraid that the other had." He had hoped that the transfer "would at least get the two services acquainted at a working level and that it might have some beneficial effects. I understand that the notice didn't go out from the PMA group until the latter part of November [1949]. So I don't believe you have had too much time to find out how it will work." Whitten complained that he had "considerable difficulty and various conferences with certain people connected with each of the services. They were afraid that it was a case of letting the other get just a little bit over the fence. And I had further trouble with the Soil Conservation Service. They thought this was a case of them being able to get this transfer to do their own work." J. C. Dykes of SCS denied this allegation, but Whitten insisted that it properly characterized some of the SCS leaders who, he said, argued that the transfer funds could be used "to do just what you wished to do for the ACP group. The ACP group, of course, didn't do much about it anyway."[10]

Whitten reminded the PMA Administrator that many people in SCS were suspicious because they "thought it originated with PMA. They thought you were going to move in on their bailiwick, and I found folks in PMA were suspicious because they did not offer it, and they were afraid the Soil Conservation Service had offered the provision and was

---

[10] *Ibid.*, Pt. 4, pp. 1190–91

trying to get in on PMA." Officials of PMA assured Whitten, however, that they had acted promptly to put the new procedure into effect. They had issued a policy statement to their state committees in July and secured agreements with SCS in 6 states and 15 counties; and during 1950, negotiations were under way in 26 states and 287 counties, but county committees in other areas reported that there was no need to transfer funds, since "SCS and other agencies are already furnishing sufficient technical assistance." Officials indicated PMA was willing to continue the transfer procedure for the 1951 program year on a trial basis.[11]

Both PMA and SCS dragged their feet for another year, however, before Secretary Brannan decided that "consolidation" was a dead issue and took a positive step in the direction of ending the PMA-SCS squabble. During 1951, he held discussions with members of the Subcommittee on Agriculture of the House Appropriations Committee and the House Committee on Agriculture. Special investigators were sent to 13 states and 48 counties where they found the Department's conservation programs "permeated with duplication, overlap, conflict, and lack of coordination, and what has been aptly described as a state of 'civil war' . . . in many areas, involving not only PMA, SCS and Extension, but also the Farmers Home Administration which had hired agricultural engineers on its regional staffs."[12] When the evidence of maladministration long known in the field became public knowledge, the Secretary finally was constrained to action.

There was one overriding reason why Secretary Brannan had wanted to solve the issue of PMA-SCS relations without consolidating the two agencies in a formal reorganization. This accounts for his apparent indecisiveness and the peculiar character of his eventual solution to the problem. The Department's soil conservation programs were—and still are—secondary features of its over-all programs. The most important characteristic of federal agricultural policy, since 1933, has been intervention by the national government in the price structure. Cash subsidies and technical assistance for conservation are income supplements which either increase, or reduce losses in, productive capacity. They, therefore, are necessarily means of governmental intervention in the price system. In 1949, Brannan revealed a new plan, calling for a two-price system for certain commodities, including wheat.

Details of this plan are not pertinent here, but it is relevant to

[11] *Ibid.*, Pt. 5, pp. 1439–40.
[12] *Department of Agriculture Appropriation Bill for 1952,* House, 82 Cong. 1 sess. (1951), Pt. 1, pp. 545–58.

observe that when Brannan came to the Department he was familiar with views held by the National Farmers Union, whose support was greatly needed for the new price plan. The Farmers Union is numerically strong in much of the Wheat Belt, especially in the Plains where farmers have liked the conservation services of both PMA and SCS and where both agencies have emphasized water conservation. There was no area where SCS was stronger politically than in most of the Plains states and, yet, it was generally in the West that PMA had launched its drive to absorb SCS. In addition, the Farmers Union supported SCS in its conflict with the Extension Service, and there is some evidence that SCS and its supporters occasionally reciprocated by helping the Farmers Union break the legal tie between the state extension services and state farm bureaus. This latter issue was the subject of active agitation during this period under review.[13]

Under these circumstances, Secretary Brannan needed to end the rivalry over conservation because it was hurting the Department politically; but in 1952 the Democrats could not afford the loss of support which would probably result from a formal reorganization of PMA and SCS. This lesson, incidentally, was one which Brannan's successor, Ezra T. Benson, was to learn as he tried his hand at reorganizing the Soil Conservation Service in 1953.

BRANNAN'S MEMORANDUM 1278

On February 15, 1951, Secretary Brannan issued Memorandum 1278 to temper further criticism by Congress.[14] This memorandum proved to be the single most important action taken to solve the organizational problem which had plagued the Department since the beginning of the federal soil conservation programs. Brannan's order gave PMA, SCS, the Extension Service, and the forestry agencies a share of responsibility for soil conservation, while allowing each to retain its historic identity. It did, however, retain the PMA state and county committees as the Department's agents responsible for program formulation and execution.

Memorandum 1278 incorporated two important principles. First, the

[13] William J. Block, *The Separation of the Farm Bureau and the Extension Service* (Urbana: University of Illinois Press, 1960), pp. 229–30 and 247.
[14] Secretary of Agriculture Charles F. Brannan, Memorandum No. 1278, Feb. 15, 1951. See also *Reorganization of the Department of Agriculture,* Hearings before the House Committee on Agriculture, 82 Cong. 2 sess. (1952), Serial Q.

special competence and responsibility of each of the Department's agencies for soil conservation was spelled out, although the role of the state extension services was left somewhat ambiguous. Second, authority to fix technical recommendations for the PMA program was placed in a new committee in each state. The new committees consisted of representatives of PMA, the U.S. Forest Service, SCS, and the state extension service. Thus, for the first time, SCS, PMA, and the Extension Service were required to meet face-to-face in order to recommend practices and specifications for state and county conservation programs.

This change was almost certain to undercut the power of the SCS regional offices to ignore the technical recommendations of the state extension services. The loss was balanced, however, by the prospect that SCS could promote districts in areas where they were not organized. However, SCS could no longer confine its services to a preferred clientele—that is, to cooperators with the districts.

The immediate upshot of the memorandum was anger on the part of some leaders in both the Extension Service and SCS, the latter because they still thought that the Secretary of Agriculture intended for PMA to "swallow up" SCS and that this directive was a very serious first step in that direction. Oddly enough, many of the participants in this interagency struggle were absolutely convinced that Memorandum 1278 was a clever move to consolidate PMA and SCS at the expense of the Extension Service in advance of the presidential election of 1952 because it directed all agricultural agencies to operate their offices at a single location in each county. These reverberations and others will be dealt with after consideration of some of the features of Memorandum 1278.

The role of PMA under Memorandum 1278 was fixed simply. It retained general responsibility for formulating an annual national conservation program and it kept control of the money to pay farmers for applying practices authorized in state and county programs. The Memorandum also specifically directed each state PMA committee to initiate negotiations among the three agencies which were to have technical responsibility in the future—the Extension Service, SCS and the U.S. Forest Service. Each SCS state conservationist was to be "responsible for all technical phases of the permanent type of soil-conservation work, except forestry," undertaken by PMA and SCS (but not extension work) within a state under the direction of the Chief of the SCS. The Forest Service was directed to assume similar responsibility for forestry practices paid by PMA. Also, SCS and PMA officials

were directed expressly to "jointly encourage the creation and development of soil-conservation districts."

Memorandum 1278 was studiously vague in fixing the exact responsibilities of the state agricultural colleges and extension services. They
were not assigned responsibility for encouraging the formation of soil
conservation districts despite their long-standing claim to be the "educational" arm of the Department. Secretary Brannan also failed to
specify which agency would recommend the specifications for the recurring practices such as seeding, liming, fertilizing, and a host of others which had long been worked out between the PMA and the various
state extension services as a part of the accommodation between these
two agencies. Brannan apparently had no intention of bearding the interested lobbies and their allies at the colleges and in Congress. Thus,
Memorandum 1278 did not specify that PMA committees should accept SCS recommendations for these "agronomic practices" over which
the latter frequently differed with state extension services.

The PMA county committees and the local SCS technicians were
directed to formulate jointly their local programs within the terms of
their state program, with the county PMA chairman assuming the
initiative. The county agent (extension) and the supervisor of the
Farmers Home Administration were to be "invited to participate" in the
deliberations. Where soil conservation districts had not been organized,
the county committees were ordered to "arrange" for the participation
of an SCS technician.

Although conflict between SCS and the U.S. Forest Service over the
former's role in "farm forestry" theoretically had been put to rest in
1945, there was still agitation to give SCS some responsibility for
certain types of this work. In 1950, the Cooperative Forest Management Act placed farm forestry under the direction of the Forest
Service.[15] In Memorandum 1278, Brannan vested responsibility for all
of the Department's forestry activities in the Forest Service. He
authorized SCS to recommend "as part of its farm conservation planning, the land to remain in trees, the existing wooded areas to be
converted to other uses, and where new woodland areas should be
established by planting." This directive left the Forest Service to assign
further responsibilities to the state forestry agencies in the PMA
program. The state foresters could thereby be authorized to give some
technical direction to forest management practices subsidized by PMA.
In states where SCS had provided this kind of service as the result of

[15] Charles H. Stoddard, *The Small Private Forest in the United States* (Washington: Resources for the Future, Inc., 1961), p. 63.

successful negotiations with PMA, this directive could have the effect of displacing SCS. There were some states, however, in which the forestry agencies lacked the personnel to service all PMA requests, and the Forest Service might be forced to use SCS to do the job. In time, this happened.

The response of the National Association of Soil Conservation Districts to this part of Memorandum 1278 gave a clue to the underlying issues. W. F. Hall, NASCD's vice president for the Southeast, from Sparta, Georgia, and a friend of Senator Richard B. Russell, went to talk with Brannan. The outcome apparently was not wholly satisfactory, as the NASCD Executive Committee on April 24 adopted a resolution opposing the "complete" transfer of forestry programs to forestry agencies exclusively.[16] Hall explained that the directors of NASCD disliked the 1950 report of the Chief of the U.S. Forest Service which recommended compulsory forest management by the states or the federal government. "Compulsion or regulation of any kind is against the principles and beliefs of our Soil Conservation districts." Instead, the NASCD favored a continued "education program to private landowners."[17] If this official view of the attitude of district supervisors were noised about the piney woods of the Southeast, there could be little doubt as to the outcome of a struggle between SCS and the professional forestry agencies to win friends among thousands of small woodlot owners. Apparently the NASCD directors preferred to forget the *raison d'être* of the standard districts law—if they knew it.

Because SCS was not heavily staffed in ten states having few districts, it was not well prepared to assume responsibility for servicing the PMA practices during 1951. The need to spread personnel thin in these areas reduced service within existing districts, but had the effect of adding more than 36,000 new district cooperators as a direct result of the requirements in Memorandum 1278. Less progress was made, according to the new Chief of the SCS, Robert M. Salter, in developing the state and county PMA conservation programs. In fact, he reported that "little" had been done toward joint formulation, and it was not until the 1954 program year that procedures were fully developed to iron out differences. As a temporary measure, SCS and PMA negotiated agreements in several states for SCS to give technical supervision to the work of PMA's employees in applying the permanent practices for which SCS was responsible under Memorandum 1278. Even in such

[16] National Association of Soil Conservation Districts, "Minutes of the Meeting of the Executive Committee," mimeographed, April 24–25, 1951.

[17] NARG (Soil—1), W. F. Hall to Charles Brannan, and reply, May 11, 1951.

states, however, the agreements did not necessarily cover all counties which were not within soil conservation districts.[18]

However, PMA continued to push its conservation program, and in 1952 it featured a special one. One county in each state was selected for an experimental "administrative" program. Community committeemen were scheduled to visit each farmer in their areas and discuss the farmer's needs for practices, using the SCS farm plans or "others" in the farmer's possession. Where no such plans existed, the community committeeman and the farmer decided on the needed practices. These recommendations were then made to the county committee for review and approval of the amount of assistance the farmer was to get for subsidized practices. It was also proposed in the budget estimates for the 1953 program year that an additional $8 million be authorized to pay community committees in order to put this system on a national basis.

These funds were authorized, to be used whenever the 1953 program year started—usually on September 1; in this case during the calendar year 1952. This increase in the activity of community committeemen did not clarify in any way the role of the soil conservation district supervisors in promoting soil conservation, especially since the direct contacts with the farmers were by PMA community committeemen who were paid to sponsor this work. Their technical qualifications could be expected to be no different, better or worse, from those of the soil conservation district supervisors. Representative Hope asked whether this contact by the community committeeman might not be the occasion for some PMA committees to discourage farmers from applying for practices requiring SCS technical assistance. He wondered whether the PMA representative might talk a farmer out of requesting work to be laid out by SCS.[19]

One result of the added SCS responsibility for supervising the PMA permanent practices was an important change in the SCS method of making conservation plans for farmers. Under procedures prevailing since the beginning of SCS, a farmer was not signed as a cooperator with the district until he had been contacted and convinced that he wanted to join, had his farm surveyed, had a complete plan made for it, and had recorded his decisions stipulating the manner in which he would carry out the plan. In 1951, however, a new name was given to

[18] *Department of Agriculture Appropriation Bill for 1953*, House, 82 Cong. 2 sess. (1952), Pt. 1, p. 513; Pt. 2, pp. 1272–83.

[19] *Reorganization of the Department of Agriculture*, House Hearings, *op. cit.*, pp. 88–90 and 134.

the farm planning process "progressive planning." This change expanded the process to three steps: the "initial conservation stage," the "advanced conservation stage," and, finally, the "basic conservation plan." In the first stage, the farmer merely signed an agreement with the district, stating the desire to use his land within its capabilities and to treat it in accord with its needs. He was asked to express his intention to accept a basic conservation plan. Once a farmer entered the initial conservation stage he became a cooperator with the district and was eligible to receive assistance from the district; that is, the SCS technician could aid him in installing any practice for which he was ready. The farmer did not advance to the advanced stage until he had allowed SCS personnel to examine his land and make a "conservation survey" upon which a map showing land capability was prepared, together with other information and guides necessary to develop a basic conservation plan. The third stage was reached when, as in the past procedure, the farmer agreed to accept the complete farm plan. It is at this point in the process that the farmer definitely agreed to the terms of his plan and stated his intention to carry it out with the availability of assistance and the passage of time.

In principle, there was nothing new about "progressive planning," yet it enabled SCS to increase its good will among key congressmen who admired the thoroughness of the SCS work, although they were critical of its failure to reach farmers in large numbers. Since SCS was required to service PMA permanent practices under Memorandum 1278, it would assist many farmers who were not cooperators with districts, but it was expected that many would be converted by SCS technicians upon contact. The potential was limited, however, because SCS was responsible for only permanent practices paid by PMA, although it offered farmers more than sixty separate conservation measures in addition to such familiar ones as terraces, ponds, and contouring. From its start, SCS had taken the position that it was the sole agency of the Department which offered the farmer "complete" conservation planning and assistance. Memorandum 1278 did not recognize this claim to primacy and did not even hint that SCS had any special technical competence except in agricultural engineering. Progressive planning, therefore, enabled SCS to make the best of the situation by increasing the number of cooperators and farm plans. These data were always prominently featured in its justifications of budget estimates to Congress, along with reports of the number of districts, the land areas included within them, and the number of acres treated with various practices.

The second, or advanced, stage allowed the SCS conservation surveys to catch up with the technicians who were getting to farmers to make initial agreements faster than the mapping crews and farm planners could do their work. This change diluted the agreements entered into at the initial conservation stage, but allowed SCS to increase the number of agreements while spreading the available personnel. No doubt this move was necessary in view of the fact that about 150 new districts were being formed each year and SCS had added responsibilities for PMA work.

The switch to progressive planning helped SCS to survive the attacks of its rivals in PMA and the Extension Service. Certainly, however, there is reason for judging cautiously the meaning of SCS reports of the number of farm plans in effect, unless the stage of each plan and the date on which it was made are reported.

SCS REORGANIZED IN 1953

When Congress initiated the transfer of funds from PMA to SCS on the county level and Secretary Brannan issued Memorandum 1278, the outer limits of departmental reorganization were approximately established. Among the warmest partisans of "decentralization" in the Extension Service, however, the battle had not been won. The presidential election of 1952 seemed again to afford some hope for the undaunted, although the 1948 upset had not been forgotten. Memorandum 1278 angered some leaders of the agricultural colleges very much. And Milton S. Eisenhower, then President of Pennsylvania State College, vigorously rubbed all the old wounds in a speech at the meeting of the Association of Land-Grant Colleges and Universities at Houston, Texas, in November 1951. He made his remarks almost fifteen years to the day after the contentious session between Henry A. Wallace and the leaders of the Extension-Farm Bureau axis in the same city in 1936.

Milton Eisenhower said the Department of Agriculture and the state extension services had reached the point where it was generally conceded that the "action" programs were properly to be administered by the Department, although this had been a sore point for a decade. The one truly vexing issue was the continuation and expansion of the Soil Conservation Service which had, in the judgment of extension leaders, developed its activities to the point where there were two extension services, duplicating and conflicting with each other at many points. By Eisenhower's definition, SCS was no longer an action agency, since it no longer provided federal "subventions"—grants of materials, funds, labor, and equipment to farmers. It had become merely another "educational" arm of the Department. Eisenhower thought it was important for the general public to realize that "Extension agents and SCS specialists are today, in most counties, working with the same people, on the same farms, and with the same basic elements on those farms." Nevertheless, in "most" states relations were cooperative, with extension helping to organize districts and the same men serving as "Extension leaders and district supervisors." In others, however, "there is discord, particularly at the local level, with one group of farmers working with Extension, another group with the Soil Conservation Service . . . technical advice given to farmers in the same community is often at variance, even competitively so. One set of specialists may advise farmers to use uniform-width stripcropping in order that standard farm machinery may be employed efficiently in the farm operations; the other technicians may advise farmers to modify the width of the strips in strict accordance with the contours of the field. Extension agents may advise farmers to eradicate multiflora roses, while SCS technicians may advise that this plant be used in a good many circumstances."

He conceded that Hugh Bennett "more than any other individual, deserves credit for awakening our nation to the dangers of soil and fertility losses, and for bringing about the conviction that the problem can be met and must be solved." He developed the demonstrations of complete erosion control measures, but when he did so it was expected, Eisenhower said, that it would be the task of the extension services to "*extend* the lessons learned in these concentrated areas to all the farms and ranches of the nation." Instead, by 1951, the SCS was working in all but "600 agricultural counties of the nation." He conceded that before Bennett started the movement "experiment stations and Extension services neglected this important area, at least so far as many of the States are concerned." Now that the Extension Service demanded that it be given the work of SCS, it had to face the charge that "SCS has succeeded in doing what you fellows failed to do . . . [and] you want to take over."

Eisenhower explained that there were originally three reasons for operating through soil conservation districts: the need for a local sponsor, the expectation that conservation work would be organized on watersheds, and the belief that land use regulations would be needed to bring about observance of sound practices. These reasons were no longer valid in Eisenhower's view; therefore, SCS should be consolidated with the Extension Service. The state extension services "cooperate with a local organization in each county; two separate local organizations would not be needed, if the two agencies were combined." The watershed "has not materialized. Most district boundaries are coterminous with county boundaries." To the best of his knowledge, he said, "land-use regulations have not been used in a single district of the nation. It has been found that results can best be achieved through educational methods, that coercion is not needed." He concluded that the only reason for farmers to organize a soil conservation district "today" is that "Congress has provided that most of the SCS funds should be used for help within districts, and thus farmers who have desired to obtain SCS technical help have first had to organize the districts. In 600 counties farmers did not sufficiently desire SCS assistance as to cause them to form local districts. Here only county agents have helped farmers install conservation practices."[20] Eisenhower had opposed districts in 1936; he preferred soil conservation associations.[21] Fifteen years of experience in the Department and as president of two land-grant colleges had not changed his mind.

The disturbing part of the situation, so far as Eisenhower and other extension leaders were concerned, was that Secretary Brannan had directed SCS to "take over, in these counties, much of the conservation help heretofore given to farmers by the County agents." Eisenhower deplored this lack of confidence in the extension services, but cautioned his audience that the Extension Service and SCS would have to cooperate. Extension leaders should continue to advocate consolidation of the two agencies by open public discussion, but he warned them that, if they were to spend some time in congressional offices, they would not find "as many as fifty members of Congress [who] today would vote favorably for it." Memorandum 1278 had ended any hope on the part of PMA or the Extension Service that the Secretary would reorganize the conservation agencies either by letting Extension take over the entire range of technical service to farmers while PMA paid part of the bill with federal subsidies, or having PMA provide the

[20] NARG 16 (Soil Conservation Service, Organization 1), Milton S. Eisenhower to Ezra T. Benson, April 2, 1953, and the enclosed typescript copy of Eisenhower's speech of November 15, 1951.

[21] NARG 16 (Soil Conservation Districts and New Agricultural Act), Milton S. Eisenhower to Paul H. Appleby, Feb. 27, 1936. In an interview on Sept. 12, 1962, Dillon S. Myer said Milton Eisenhower served virtually as chairman of the Secretary's interbureau Committee on Land Conservation of 1935. Myer said that Eisenhower was not anti-SCS in 1935.

money and the "engineering" services and allowing Extension to continue to recommend practices and specifications for "soil management" as it had done since 1936.

Despite these warnings, this speech by Milton Eisenhower was regarded in some quarters as a reorganization plan to be put into effect at an appropriate time. It put to rest speculation about the future of soil conservation districts, if SCS were consolidated with Extension and other leaders acted on his views. The districts would simply atrophy. He said plainly that there was no need for two local sponsors of soil conservation—districts and extension sponsors. It was easy to see how the power of the extension services and their sponsors, whether county commissioners or farm bureaus, would be increased by their capacity to extend or deny technical service to any farmers. Few soil conservation district supervisors required much explanation to understand that if the Extension Service took over SCS, in many, if not all, states they would soon be out of a job. Soil conservation districts would wither like summer weeds.

It requires little imagination to understand the dismay with which some supporters of SCS beheld Milton Eisenhower's brother as the Republican nominee for President of the United States in 1952. It was apparently carefully guarded dismay, however, except in Oklahoma and Texas, where state Republican leaders charged that SCS personnel distributed copies of Milton Eisenhower's speech to district supervisors and claimed it was a Republican blueprint for reorganizing SCS if Dwight D. Eisenhower became President.

Anxiety over the future of SCS and the districts was intensified after Eisenhower became President and the new Administration was granted powers to reorganize the executive agencies.[22] When Secretary of Agriculture Benson announced his intention to make changes in the Department, tension was heightened, for he brought into the Department known opponents of SCS (especially Assistant Secretary J. Earl Coke) who had been with the California Extension Service, and delegated to them the task of reorganizing the Department.[23] Plans were very carefully guarded and inquiries about them, even from influential congressmen, were given noncommittal answers. Rumors spread, perhaps with encouragement.

[22] These comments on the reorganization of the SCS in 1953 are based on an examination of the very extensive files on the subject in NARG 16 (SCS, Organization 1, 1953–55).

[23] *Congressional Record*, Vol. 99, Pts. 3, 4, 5, 83 Cong. 1 sess. (1953), pp. 2856, 4954, 5414, 5582, 5637, 5672, 6000, and 6015. See also *Hearings* on S. 100, Senate Committee on Governmental Operations, 83 Cong. 1 sess. (1953).

Clarence Cannon, Democrat from Missouri, and ranking minority member on the House Committee on Appropriations in 1953, told True D. Morse, the Under Secretary of Agriculture who was also from Missouri, he had heard that both SCS and PMA were to be "turned over to" the Extension Service. Cannon, who had sometimes cross-examined SCS witnesses critically at budget hearings, had sponsored the rider to the Department's appropriation bill for 1945 to require SCS to recognize the power of state soil conservation committees to formulate state programs, and was generally considered a friend of Extension in this controversy. Despite this reputation, however, he warned Morse: "After long study of these activities and many years' experience, both as a farmer in Missouri, and as a member of Congress, I am apprehensive over the results of the change of these activities to the Extension Service, am certain the Department will give earnest consideration and exhaustive study to such a proposition before taking action and trust it will have your personal attention."[24]

Representative Hope, the Republican Chairman of the House Agriculture Committee, sent a telegram to Benson just before the Department's plan to reorganize SCS was to be announced. Hope said that he was "terribly concerned about the Soil Conservation reorganization program. If my information is correct it would seem that full information with regard to changes about to be put into effect was not presented to the Committee on Agriculture when we met with you and Assistant Secretary Coke on October 10." Hope said he had information that many of the services previously provided by SCS were to be eliminated, that approximately 800 technicians were to be dropped from employment and that as many as 3,000 would be shifted about as senior employees "bumped" their juniors. "If this is true I am afraid it will have a very demoralizing effect upon the entire service and will be particularly harmful to [the] watershed program which is now getting under way." Rumors about the character and effects of the reorganization were causing "great anxiety among some of our very best and most influential farmers and if report[s] are correct I feel this anxiety may develop into outright hostility." Hope urged Benson to defer the reorganization, "including elimination of regional offices," until Congress reconvened and further discussions had been held.[25]

[24] NARG 16 (SCS, Organization 1, 1953–55), Clarence Cannon to True D. Morse, Sept. 19, 1953.
[25] NARG 16 (SCS, Organization 1, 1953), telegram, Clifford R. Hope to Ezra T. Benson, Oct. 23, 1953.

Representative W. R. Poage, Democrat of Texas, also accused Benson and his assistants, Morse and Coke, of lacking candor, when they discussed their plan to reorganize SCS with members of the House Committee on Agriculture on October 10. According to Poage, Benson assured Representatives Hope and Carl Albert, Democrat of Oklahoma, that the Department would not transfer any "functions or personnel" of SCS to the state extension services. Yet, less than two weeks after this meeting and Benson's public announcement of the SCS reorganization on October 13, Poage heard from sources at Texas A & M that SCS state offices were in the future to depend entirely on specialists at the colleges for technical recommendations for agronomy, range management, biology, and forestry, and that SCS would not be allowed to have specialists in these subjects on their state staffs. They would no longer have them on regional staffs, either, since the principal feature of the reorganization which Benson publicly announced was that the SCS regional offices were to be discontinued at once. Poage said that Benson's plan was gradually to reduce the Department's budget estimates for SCS and to increase the funds appropriated to the state extension services under the Smith-Lever Act of 1914. Poage had heard also that effective at once SCS was to reimburse the agricultural colleges from its own funds to pay for the services of the subject-matter specialists whose services SCS would be directed to use.[26]

Other rumors circulated and several of them were quite plausible, especially one to the effect that farmers in the future would have to pay for soil surveys on their farms. The important rumor, however, was the one which Representative Poage reported. It is likely that it was completely correct. Certainly it was plausible in the judgment of people familiar with the history of relations between SCS and the extension services. If Benson and his associates really intended to effect a reorganization of the SCS soil conservation program, this method appeared to promise the best chance of success. Most important of all, for the first time since 1934, it would have recognized the claim that federal funds for soil conservation should be allocated, not to a specialized national conservation agency, but to the state extension services so that they might increase, or at least use federal funds to pay for, subject-matter specialists whose work was relevant to soil conservation.

There were divisions within SCS and among other interested parties

[26] NARG 16 (SCS, Organization 1), W. R. Poage to J. Earl Coke, Oct. 26, 1953. In his reply, Nov. 6, 1953, Coke did not allay any of the rumors.

over the regional offices. A retired regional chief of woodland management with seventeen years of service in SCS charged that his organization had deliberately duplicated services offered by some state extension services, particularly by recommending different practices and specifications. He also pointed out that SCS personnel in the regional offices frequently held only a Bachelor of Science degree, or at most a Master of Science, and yet they differed with college specialists who were Doctors of Philosophy and renowned scholars.[27] The specialists and administrators at the land-grant colleges were very sensitive about this matter and it is possible to play the cynic and attribute their feelings entirely to status or pocketbook deprivation. A more balanced view is that the issue of technical qualification is a proper matter to be weighed in judging the claims and counterclaims of the two antagonists.

Regardless of the merits of either position, a great many interested persons voiced their sentiments, including representatives of the National Association of Soil Conservation Districts. The NASCD President, Waters Davis of Texas, violently denounced Benson, apparently on the assumption that he intended to carry out the kind of reorganization demanded by militants in the extension services. Several of the state associations of districts, however, opposed Davis and urged Benson to eliminate the regional offices. The split among farm interests over this issue was great and, whatever Benson may have intended eventually to do with SCS, he settled for eliminating the regional offices without making other suggested changes of any consequence.

There were many reasons why Benson probably hesitated to decentralize SCS through any more drastic formal reorganization than the one effected in 1953. First, it is apparent that Milton Eisenhower correctly gauged the temper of Congress in saying that not more than fifty members would support complete decentralization of the sort recommended by the American Farm Bureau Federation in 1948. Also,

[27] NARG 16 (SCS, Organization 1), George W. Hood to J. Earl Coke, Feb. 9, 1953. This file contains numerous letters from SCS personnel, including work unit conservationists, sent either directly to Secretary Benson or through contacts, in which field situations highly critical of the SCS administration were reported. There were also letters and telegrams from several state farm bureaus and state associations of soil conservation urging Benson either to eliminate or retain the regional offices. Apparently, several SCS state conservationists thought their own careers would be advanced, if the regional offices were distributed to the states. Some state associations of districts were convinced that this change would improve technical services to farmers. Consequently, they, too, supported Secretary Benson. Thus, neither federal agencies nor their constituency groups are necessarily united upon proposals for administrative reorganization.

rumors and the campaign of denunciation which Waters Davis waged during 1953 hurt the Republican party in some areas, according to reports from the field. In addition, Brannan had pointed the way to a partial, but peaceful, solution to the differences between the Extension Service and SCS in Memorandum 1278. When he provided that SCS would participate on the state and county level in recommending practices and specifications for permanent practices, Brannan virtually eliminated the reason for SCS regional offices. As reorganized, SCS shifted its subject-matter specialists primarily to its state and national staffs. Procedures have been established, since 1953, for consultation among the parties to iron out differences in their recommendations and specifications, although these efforts have not been uniformly successful in eliminating all the old antagonisms.[28] Both parties settled for half a loaf.

Another factor probably induced some caution in Benson and his associates in 1953, although they did recommend drastic cuts in the appropriations for the Department's action agencies and increases for the Extension Service. In 1953, influential members of the House, such as Whitten and H. Carl Andersen on the Appropriations Committee and Hope and Poage on the Committee on Agriculture gave their support to legislation which launched the Department of Agriculture on a program for small watersheds seventeen years after Henry A. Wallace tried to tie the erosion control program administered by the Soil Conservation Service to upstream flood control.[29]

[28] The Soil Conservation Service recently has been reorganized to include four "regional technical service centers" to provide "essential services for our State program staffs." This change appears to restore the technical divisions to their status prior to the reorganization of 1953, although it has not been publicized. See *Department of Agriculture Appropriation Bill for 1966*, House, 89 Cong. 1 sess. (1965), Pt. 3, p. 362.

[29] The battle of the budget was waged again in 1955. Representative Jamie L. Whitten carefully scrutinized the Extension Service's budget for 1956. His questioning of Clarence M. Ferguson, then the Administrator of the Federal Extension Service, was trenchant and incisive. The budget estimates provided for 1,000 new county agents for fiscal 1956 and a reduction in the budgets of the action agencies including SCS flood control items. No additional SCS personnel were to be provided for an estimated 58 new soil conservation districts. Whitten also revived the reports of interagency differences based on the reports of the special investigators for the House Subcommittee on Agricultural Appropriations. He forced Ferguson to admit that the Department has little control over the programs and activities of county agents. See *Department of Agriculture Appropriation Bill for 1956*, House, 84 Cong. 1 sess. (1955), Pt. 2, pp. 919–43. For a discussion of the general problem illustrated here, see Avery Leiserson, "Political Limitations on Executive Reorganization," *American Political Science Review*, Vol. 41 (February 1947), pp. 68–84.

RECENT DIVISIONS OF RESPONSIBILITY

During the past decade, Congress has vested new responsibilities for conservation programs in the Secretary of Agriculture, and successive secretaries have placed them upon both SCS and the Agricultural Stabilization and Conservation Service (ASCS, successor to PMA) without following a clear pattern. In 1956, Congress enacted the Great Plains Conservation Program which authorized the Secretary to enter into long-term contracts with farm and ranch operators in designated counties of the Plains states. The idea of using contracts to achieve proper land use is not new, of course, since the Soil Erosion Service required farmers to sign binding 5-year agreements as a condition for receiving technical assistance, labor, and materials. M. L. Wilson explained in 1935 that one device which the Department anticipated that districts might use was a contract binding upon all future occupiers of land under agreement. Wilson's experience in the Plains states convinced him of the need for governmental regulation of land use to prevent practices harmful to the public.

This Great Plains Conservation legislation permits the Department to enter into 10-year contracts with individual farmers who agree to make land use adjustments in accordance with a complete farm plan. The contract is binding and enforceable, and the plan is made with the Department's technical assistance, and cash payments are provided for following the practices agreed upon in the plan. This program differs from the regular Agricultural Conservation Program and from the work performed through soil conservation districts. The farmer or rancher who agrees to accept a contract must put his entire acreage under agreement and he is not free to install some recommended practices and to ignore others.

When the legislation was under consideration in the House Agriculture Committee, Representative Karl C. King, Republican of Pennsylvania, asked why the Department did not propose to purchase abused lands in the Plains states, restore them, and then rent them to private operators under federal management. As an alternative, he wondered aloud whether the Department should not regulate land use, relying on "penalties rather than subsidies." Speaking for the Department, Assistant Secretary E. L. Peterson said that regulation was a matter to be decided by state and local governments. He did not elaborate for King the fact that the Department had originally required states to adopt legislation substantially in accord with the standard soil conservation districts law, to include the power to enact and enforce land use regulations as a condition for receiving assistance from the United States. Peterson did not mention the difficulties involved in enforcing such regulations in areas of eastern Colorado, where regulations were invalidated after World War II.[30]

At this same hearing, Peterson was asked whether the legislation would be administered by the county committees of the Agricultural Stabilization and Conservation Service. He answered that they would, but when he was asked whether soil conservation districts would have administrative responsibilities for the act, he answered: "That has not been determined fully, sir. My own view is that the Soil Conservation Service and the Agricultural Conservation Program Service [ASCS] would both heavily be involved." He indicated that technical planning would be assigned to the Soil Conservation Service.[31] Later, however, the Secretary assigned full program responsibility to the Soil Conservation Service, but none to districts in the sense that the contracts which are the key device to secure compliance with program objectives are made between the Department and the individual farmer without the agency of districts in any way. District boards have been asked to promote this program, however.

This assignment of responsbility was interpreted in some quarters as a revival of the old conflict between SCS and ASCS's predecessors (AAA and PMA). An organization, Farmers Association for Resource Management (FARM), was created—initially in South Dakota. In that state, some individuals active in the organization allegedly took the lead in initiating legal procedures for disestablishing several soil conservation districts, although only one was acutally dissolved. Before Congress, this Farmers Association joined with the Limestone Institute in raising questions about the efficiency of the Soil Conservation Service.[32] It opposed the Great Plains Conservation Program as a "forerunner of compulsory 10-year plans for all farms and ranches." Extolling "administration of farm programs by democratically farmer-elected county and community committeemen," FARM argued that "farmers in a district should have the right to vote out a district" just as they have the right to create one.[33]

---

[30] *Great Plains Conservation Program,* Hearings on H.R. 11831 and 11833, House Committee on Agriculture, House, 84 Cong. 2 sess. (1956), pp. 16–17. This legislation was enacted as Public Law 1021, 84 Cong. 70 *Stat.* 115.

[31] *Ibid.,* p. 28.

[32] *Department of Agriculture Appropriation Bill for 1960,* House, 86 Cong. 1 sess. (1959), Pt. 2, pp. 1167–81.

[33] *Department of Agriculture Appropriation Bill for 1961,* House, 86 Cong. 2 sess. (1960), Pt. 4, pp. 21–26; *Bill for 1960,* House, *op. cit.,* pp. 987–88.

This outbreak of hostilities was viewed with some alarm after FARM organized nationally and was identified with several former county and state ASCS committeemen. It was assumed that some persons within ASCS looked upon the Great Plains Conservation Program as a serious threat to their organization so long as it was administered by the Soil Conservation Service. The attack on districts was construed in the same way. When Congress reviewed early administrative experience under the program, it revealed that the Secretary had imposed a restriction that no single contract could exceed $25,000 during the maximum of ten years. The Secretary took this step after he discovered that one contract was executed for $106,500 on a single unit covering 17,500 acres, and another was for $45,400 on a unit of 22,700 acres, mainly for irrigation practices. With the stakes this high, it is understandable that friction should develop between partisans of the agencies and their personnel. However, ASCS could still serve these farmers, since operators were eligible to receive some Agricultural Conservation Program payments above and beyond the amounts they received through the Great Plains contracts.[34] After Secretary Benson denounced FARM, its attacks slackened and died out, but not without raising charges that some districts in South Dakota were charging private contractors "kickbacks." The General Accounting Office investigated, but made no public report.

The Food and Agriculture Act of 1962 authorized the Secretary to start a Resource Conservation and Development Program which was intended to adjust land use from crop production to other economic uses, such as outdoor recreation and industrial development in rural areas where farm income is relatively low and is likely to remain so. This legislation is another effort to concentrate agricultural production on the most suitable lands and encourage owners to convert unsuitable lands to nonagricultural uses. It is a broader authority than many in the past, such as the Soil Bank which used short-term contracts to retire land but did not insure that the lands removed from production would be marginal or poorer. The Secretary has assigned program "leadership" to the Soil Conservation Service. Projects are to be sponsored by "local organizations, including soil conservation districts, with authority to plan, administer, operate and maintain project installations."[35] It remains to be seen how many districts will be in a position to assume

---

[34] *Department of Agriculture Appropriation Bill for 1961*, House, *op. cit.*, Pt. 1, pp. 686–87.

[35] *Department of Agriculture Appropriation Bill for 1964*, House, 88 Cong. 1 sess. (1963), Pt. 2, pp. 1011–12.

such broad responsibility for a program that involves more than merely the interests of farmers.

The 1962 legislation also provided for a Cropland Conversion Program in which the Secretary is authorized to enter into 10-year contracts with farmers to convert their lands to improved uses. The principle is the same as that of the Great Plains Program administered by SCS; but Secretary Orville L. Freeman assigned responsibility for the new cropland conversion effort to the Agricultural Stabilization and Conservation Service.[36]

SUMMARY AND COMMENT

The major issues which had plagued the organization of the Department of Agriculture's soil conservation programs for two decades came to a head in the late forties and were largely settled by 1953. The Production and Marketing Administration was allowed to continue to provide conservation subsidies for farmers, but it lost the authority to build a technical staff of its own to rival either the Soil Conservation Service or the extension services. Thereafter, PMA had to use the technical services of SCS and the Forest Service for program recommendations and supervision within their designated areas of competence. The state extension services were incorporated with these two agencies in a group made responsible at the state and county level for the Department's Agricultural Conservation Program. Other state agencies were invited to join in annual program discussions. Memorandum 1278 issued by Secretary Brannan eliminated much of what investigators in 1950 called "a state of 'civil war'."

Although this settlement satisfied none of the parties to the old interagency conflicts, it apparently appeared to Secretary Brannan to be the only kind of settlement which would not seriously disturb the distribution of political power in agriculture. He could not realistically expect the gains of an all-out reorganization to exceed the losses with a presidential election only a year in the offing. It is doubtful, moreover, that he could have secured congressional approval of any reorganization plan that cost the Soil Conservation Service its identity, since it would be assumed widely that he intended also for soil conservation districts to atrophy.

Brannan, therefore, continued the Department's commitment to

[36] *Ibid.*, Pt. 3, pp. 1736–37.

"assist districts." He directed PMA county committees and departmental personnel to encourage farmers to organize districts where they did not exist (in about 600 counties). On the other hand, he directed SCS to supervise the "permanent" practices subsidized by PMA on all farms, whether their operators were cooperators with districts or not. This arrangement left every farmer free to get technical assistance from any source he desired (including private ones), if he did not use ACP subsidies. If the farmer did use them, he had to let SCS supervise "permanent" ones, such as terraces, and a forestry agency supervise one or two forestry practices. If he wanted an "agronomic" practice such as pasture seeding, he had to follow the requirements set down by PMA, which was left free to get its advice for such practices from whatever source it chose to follow.

The final step in this process of reorganization was taken in 1953, when the Republicans staffed the Department of Agriculture with some former extension personnel. Secretary Benson finally divested the Soil Conservation Service of its regional offices which had incensed extension personnel since 1935. This change forced SCS to organize state program staffs and has undoubtedly influenced its specialists to meet and accommodate themselves to some degree with subject-matter counterparts in the colleges, since both now participate in planning the Department's annual conservation program.

When Secretary Benson and Secretary Freeman assigned responsibilities for the Great Plains, Resource Development, and Cropland Conversion programs, they followed no clear rationale to differentiate the functions of the agencies except insofar as SCS continues to provide technical supervision of permanent practices and small works of improvement.

CHAPTER 7

From Programs to Projects

THE EXPECTATION THAT the Department of Agriculture would develop a program of erosion and flood control on the upstream tributaries of major rivers to complement the works of the Army Corps of Engineers downstream was largely unfulfilled as late as 1951. Many persons were still sincerely convinced that whole river basins should be comprehensively planned and developed for multiple purposes, but the tie between soil conservation on the nation's farms and flood control had not materialized. There were many reasons for this situation. The job was so vast that the scale and cost of operations staggered the imagination. World War II erupted only a short time after the ambitious plans of the thirties were proposed. So far as the Department of Agriculture's share of this work was concerned, there were other factors. It is doubtful that either Secretary of Agriculture Henry A. Wallace or Hugh H. Bennett ever placed a very high priority on the upstream work. In 1938, the Department reportedly turned down a request that it assume responsibility for comprehensive development of the Washita River basin in Oklahoma.[1] On the other hand, in the same year the Department received from Congress an appropriation of $4 million in advance of any specific approval of particular projects. Active work, until the war curbed activities, was confined chiefly to the area around Los Angeles, where the Forest Service and the Soil Conservation Service (SCS) collaborated on a difficult project.

From the start, the Corps of Engineers seemed reluctant to express confidence in, or support of, the Department's upstream program. Arizona's Senator Carl Hayden originally took the responsibility for giving authority to the Department of Agriculture under the Flood Control Act of 1936 because the Corps of Engineers delayed comment and approval of SCS plans for small dams on the upper reaches of the Gila and San Pedro rivers in Arizona. From that time forward, spokesmen for the Corps repeated essentially the same argument: Since SCS upstream plans rested on very limited experimental data and prelimi-

[1] Interview, Don McBride, former Administrative Assistant to Senator Robert S. Kerr of Oklahoma, Feb. 2, 1962.

171

nary surveys, the effects of proposed works on streamflow could not be predicted reliably. Sometimes Corps officials stated more positively that the work done by the Department was not flood control in the accepted sense.[2] Until 1944, the Department of Agriculture continued to make surveys under authority granted by the Flood Control Act of 1936, as amended, but it changed the name of this work from "upstream engineering" to "upstream or watershed protection." Spokesmen for the Department admitted that one serious obstacle to success was the relative obscurity of its work.[3]

One clue to relations between the Corps and the Department of Agriculture is to be found in the fact that the flood control committees of Congress authorized surveys for all projects, and that the public works subcommittees of the appropriations committees passed on the estimates for all work under the flood control acts. Funds were appropriated to the War Department and later transferred to the Department of Agriculture. It is not farfetched to say that the Department until 1953 was in a position *vis-à-vis* the Corps not unlike that of the Extension Service in relation to SCS. Both agencies and their supporters in Congress and on the hustings were competing for limited federal appropriations for programs that had the same name, but rarely overlapped so far as specific projects were concerned. The Corps had first call on flood control money because it was supported by congressmen from constituencies having major rivers which could threaten valuable urban or rural property (as along the Mississippi flood plain). They simply were not interested in diverting funds to the Department of Agriculture for upstream projects. The Department of Agriculture had little bargaining power in this situation.

In 1944, however, when Congress authorized the partially competitive programs of the Corps and the Bureau of Reclamation in the Missouri Basin, it also authorized the Department of Agriculture to undertake land treatment and works of improvement on eleven small rivers—or watersheds, as they have been known. A partial explanation for this action is that it was part of the Roosevelt Administration's plan for postwar reconversion. Hugh Bennett testified for the Department in support of only one of the eleven projects, that for the Trinity River in Texas. Urban interests in Dallas and Fort Worth had given the Department valuable assistance in its struggle with the Texas Extension Service over the standard districts law in 1937. The interesting

[2] *Department of Agriculture Appropriation Bill for 1943*, House, 77 Cong. 2 sess. (1942), Pt. 1, pp. 955–57.
[3] *Ibid.*, p. 958.

thing about Bennett's 1944 testimony is that he confined himself to claims that erosion control measures on agricultural lands were effective deterrents to floods. He made no mention of structural improvements such as detention dams and gully stabilizers, although the Flood Control Act of 1944 authorized such works.[4] Funds, however, were not made available for these projects until 1947.

THE MISSOURI BASIN PROGRAM

Very soon after hostilities ended in Europe, the Secretary of Agriculture abolished the Office of Land Use Coordinator and reorganized his staff, placing responsibility for land and water use programs in an Assistant Secretary. He also brought in some new aides to form a small staff to serve the Assistant Secretary. One of the projects to which this staff gave particular attention and emphasis was the development of the Department's agricultural plan for the Missouri Basin to supplement the Pick-Sloan Plan, prepared by the Corps of Engineers and the Bureau of Reclamation. This work was placed under Charles F. Brannan, at that time Assistant Secretary, who dispatched Gladwin E. Young of the Bureau of Agricultural Economics (BAE) to Lincoln, Nebraska, to serve as the Secretary's personal representative in the region.

According to reports from the field in 1946, the Governor of Missouri opposed the high-dam philosophy of the Corps of Engineers and blocked increased appropriations for the Pick-Sloan Plan. The Missouri River States Committee, consisting of governors of the Missouri Basin states, respected his opposition and withheld its endorsement of increased appropriations. The Missouri Governor wanted small dams to preserve agricultural flood plains, and he was supported by the state's Extension Service which advocated an acceleration of "soil improvement."[5] Planning moved slowly so that in the following year, 1947, the governors of the Missouri Basin states "emphatically" recommended that Congress appropriate sufficient funds for the Department of Agriculture to carry out its authorized flood control functions and complained that the Department had not publicly shown much interest

[4] *Flood Control Plans and New Projects,* House Committee on Flood Control, 78 Cong. 2 sess. (1944), pp. 459–75.

[5] National Archives, Record Group 16 (Reports), J. W. Burch to Gladwin E. Young, Nov. 8, 1946; and Young to Charles F. Brannan, Nov. 19, 1946. (National Archives records cited hereinafter usually will be identified by the abbreviation NARG.)

in this work. The Corps of Engineers, it was reported, never hesitated to express enthusiasm for flood control. Receipt of this intelligence prompted Brannan to prepare policy statements for his field representative in the Basin to be "released to strategic areas of the press" so that the Department's active interest could be made known.[6]

One factor slowing the rate of planning was the lack of what the Department considered to be proper authority under the flood control acts for work on basin programs. In order to develop a complete and "integrated" program in the Missouri Basin, the Department wanted to utilize all of its facilities. Put another way, Secretary Clinton P. Anderson and Assistant Secretary Brannan did not contemplate that the Department's program would consist exclusively of physical treatment of the land under the operating responsibility of one agency alone. Discussions with the directors of extension and experiment stations in the states of the Missouri Basin revealed that they, too, were keenly interested in an "accelerated" agricultural program. R. E. Buchanan of the Iowa Experiment Station expressed the views of a committee of representatives from the land-grant colleges in complaining that "hundreds of millions of dollars" were being spent on "action" programs "with a pitiful driblet of relatively uncoordinated agricultural supporting research under way." In the Tennessee Valley area, he argued, "meticulous insistence upon working with and through the agricultural and extension agencies" by use of the grant-in-aid had resulted in "remarkable coordination." It was his opinion that it would be "highly desirable" for the land-grant colleges to develop the same type of "cooperative" program of research and education for the Missouri Basin, even if there were no development of a "Missouri Valley Authority."[7]

Young presented these views to Brannan because of their importance in placing the Department "in a position to cooperate with the Bureau of Reclamation and the Army Engineers on the agricultural phases of project proposals." The Department could not afford to rest merely on the negative contention that it and the colleges alone were authorized to carry out agricultural programs. Since the colleges were familiar with the TVA relationship, they found it easy to enter into agreements with any federal agency. Their support of the Department's basin program could be assured only if the Department recognized their common interest by working for congressional authorization and additional appropriations. Brannan made it known that the Depart-

[6] NARG 16 (Reports), Brannan to Young, July 22, 1947, and Sept. 22, 1947.
[7] *Ibid.*, Young to Brannan, enclosure, March 11, 1948.

ment was aware of the need for such a "partnership" and said that
authorizations under the earlier flood control acts should have enabled
the Extension Service to provide additional educational assistance in
support of the land treatment measures and works of improvement
installed by the "action" agencies.[8] One of Brannan's specific worries
was that the Bureau of Reclamation would secure soil and water
conservation funds, enabling it to enter into agreements with the
colleges for research not only in the Missouri Basin, but in others as
well. Also, he was interested in legislation which would authorize
agricultural programs in basins in addition to the Missouri—
particularly the Colorado and the Columbia.

Brannan soon discovered that the Department's commitment to a
partnership with the land-grant colleges could raise new problems.
When the extension directors in the Missouri Basin specified the
amounts of funds they wanted and the purposes for which they would
be spent, it was found that the Missouri group proposed annual
expenditures for soil conservation greater than the combined total of all
the other states. The Missouri Extension Service wanted to provide two
assistant county agents in every county to serve as farm planners. The
extension leaders from the other states were reported to be disturbed
because they regarded this work as a function of the Production and
Marketing Administration (PMA) and SCS, but went along with it
because they did not feel they had the authority to reject the demands
of one of their peers.[9]

These demands from the administrators of the state extension
services and experiment stations for a share of the federal funds in the
Missouri Basin throw light on the meaning of the term "comprehensive," when applied to river basin planning. Brannan knew perfectly
well that, if he failed to make provision for the agricultural colleges to
participate in the "agricultural phases" of his program, he would violate
the rule that a national or regional program financed by the federal
government ought to have something for everyone. Brannan, who
became Secretary of Agriculture in June 1948, knew that there were
many ways in which the colleges could oppose him.

The Department needed authority, which it lacked, to use flood
control appropriations, and not agricultural appropriations, for this
work. It is easy to understand why the Soil Conservation Service was
unenthusiastic about Brannan's Missouri Basin program. Missouri's
Extension Service had managed to win most of the battles in its war

[8] *Ibid.*, Brannan to Young, March 17, 1948.
[9] *Ibid.*, Young to Ralph Will, Feb. 18, 1949.

with SCS to keep soil conservation districts from being organized. Its price for supporting the Basin program was total victory in this struggle over soil conservation districts. There was no more reason why SCS should be enthusiastic about a comprehensive basin program than for it to applaud an authority for the Missouri Valley modeled on the Tennessee Valley Authority (TVA) such as Senator James E. Murray of Montana proposed.

There was another strand in these developments. In 1948, there were heavy floods in Kansas. Governor Frank Carlson urged Secretary Brannan to clarify the intentions of the Department of Agriculture toward flood control, although the Missouri Basin program was still under study. Pressures were building up in this region to substitute the Department's "watershed management" for large flood control works installed by the Corps of Engineers. Brannan was unwilling to make any such sweeping claim for the work of the Department. He supported the position that the programs were "complementary" and not "competitive."[10] It can be argued that Brannan and his representatives were perfectly sincere in saying this. It was also apparent to those who followed these matters closely that the Department was in no position to make any other claim, since the Corps had the support of members of the public works committees and appropriations subcommittees of Congress. The Department maintained this official position throughout the years of bickering which followed, although certain interests— situated chiefly in the area from Nebraska southward to Texas—forced a separation of the Corps and the Department of Agriculture, allowing the latter to operate on small watersheds.

One interesting result of the 1948 floods in the Southern Plains was that President Harry S. Truman ordered an acceleration of flood control operations throughout the Missouri Basin. Congressmen from Oklahoma and Texas importuned their colleagues to provide for more survey work in their states and interest in the Department of Agriculture's upstream work picked up. Truman's order to speed work in the Missouri Basin stirred Representative Jamie L. Whitten, Democrat of Mississippi, to charge that SCS had diverted funds from the Yazoo and Little Tallahatchie projects in his congressional district to the Washita and Little Sioux watersheds in Oklahoma and Iowa. The Little Sioux was in Republican Representative Ben F. Jensen's district. Both of these congressmen had been consistent supporters of SCS. The fact that Jensen was a Republican up for election along with President Truman

[10] *Ibid.*, Young to Brannan, March 8, 1948.

in 1948, and that Jensen was senior to Whitten on the Appropriations Committee, complicated the situation for SCS.

Conditions in his congressional district gave Whitten reason to be particularly interested in SCS operations, and they account in some measure for his long record of support for the agency. The Corps of Engineers had constructed reservoirs in the Yazoo Valley, but they had not been entirely satisfactory, since silt was deposited in accelerated amounts in the small tributaries to the mainstream. This silt destroyed croplands on the flood plains of these streams. In addition, the U.S. Forest Service planned to purchase submarginal land in his district, reforest it, and add it to national forests under federal management and regulation. The plans of the Corps and the Forest Service, Whitten claimed, would make the federal government the owner of 48 per cent of the land in his district. Instead, SCS recommended that the poor lands be seeded and used for grazing, but under private ownership and management. This proposal had considerable appeal, in part because it was expected to slow or halt depopulation of the land. In order to stop the Corps and the Forest Service, the Mississippi legislature forbade the United States to purchase lands without the consent of the governing board of the county in which the lands were located. Whitten backed a law to the same effect with a rider to the Department's appropriation bill.[11]

Whitten was not convinced, however, that SCS was operating as effectively as it could in either his district or others. He was familiar with the mode of operations in the eleven watersheds authorized by the Flood Control Act of 1944. He liked the results in his own district because he thought that SCS personnel achieved more in a year or so than was accomplished in "10 years" through regular operations in soil conservation districts. He said that Congress was "probably responsible," but in districts the work was "a case of drifting along and having over-all plans and hopes that more farms will come under, but not having a definite plan and definite objective" unlike the "sub-watersheds" of the eleven authorized projects. In the latter, he said, "you have a plan; you have a definite objective. They start on this little tributary today and they know what they are going to do. They make a drive on the people that live right there, and they live through with it."

J. C. Dykes replied that SCS had advocated the "neighborhood group approach," but that "decision rests with the soil conservation district governing body." If SCS advocated the sub-watershed ap-

[11] *Department of Agriculture Appropriation Bill for 1948*, House, 80 Cong. 1 sess. (1947), Pt. 1, p. 960.

proach, many boards would resist just as they had the neighborhood approach. Most boards insisted that farmers be served in the order in which their applications were made. Representative H. Carl Andersen remarked that the voluntary character of the SCS program was its greatest asset. Whitten persisted and took the matter up with Bennett, whose response was that "our normal soil-conservation work has stopped floods pretty much in some places, but that is in just one type of watershed." In others, violent floods made detention structures necessary. Bennett explained further that SCS had not used the sub-watershed approach because, when work was first started on the Washita, it was thought that 100 per cent of the land should be treated before any detention dams were built. On the Little Sioux, it was decided that 80 per cent of farm land treated would be sufficient. But "we are beginning to feel that a reasonably good job of soil conservation should be completed when we go in with the flood control program." Bennett insisted that "flood control is a program to supplement . . . normal soil conservation work, not in any sense a substitute."[12]

As the time approached for Secretary Brannan to submit his program for the Missouri Basin, Whitten grilled one of Brannan's assistants, Ralph Will, to find out why he was appearing for the first time to justify the Department's estimates for flood control, although he had been in charge of planning for three years. Whitten said he had sympathy for the "complexities" of planning in the Missouri Basin, but he was "determined that the work that is being done in some areas not be retarded on the basis of putting another coordinator at the top of the heap."[13] Whitten questioned Sherman E. Johnson, Assistant Chief of the Bureau of Agricultural Economics, to find out just how much review of flood control projects his agency did at the "request" of the action agencies, and how much was done at the "direction" of Secretary Brannan. These questions grew out of Whitten's conviction that Brannan's land and water resources staff was "diverting" SCS flood control funds to other agencies of the Department in the name of "comprehensive planning." He did not like the economic analyses which BAE made of some SCS flood control projects and tried to show that new review procedures were being used to put SCS cost-benefit analyses in an unfavorable light.[14]

[12] Department of Agriculture Appropriation Bill for 1950, House, 80 Cong. 1 sess. (1949), Pt. 2, pp. 25–30, 62–64.

[13] Ibid., p. 4.

[14] Ibid., pp. 20–22.

When Brannan submitted his program for the Missouri Basin in 1949, Congress did not seem disposed to authorize it at once. In fact, no hearings on the program were ever held. Brannan was also unable to get any congressmen in the Basin to introduce legislation authorizing the program. In March 1950, Governor Val Peterson of Nebraska urged Representative Jensen of Iowa to introduce the legislation. But Jensen attacked the Missouri Basin program as a "piece of political encroachment" on the people of the area and said he would introduce a bill to authorize a small watershed program instead. James Lawrence, editor of the Lincoln, Nebraska, *Star,* and later chairman of the Missouri Basin Survey Commission, revealed an important source of the opposition to Brannan's program by characterizing Jensen as "a friend of the 'friendly' private electric companies," who did not want any basin program or valley authority.[15]

THE SMALL WATERSHED PROGRAM

The Department was not able to secure authorization for its basin-wide program, and was not willing publicly to claim that its flood prevention program was a substitute for large dams built by the Corps of Engineers. This situation stimulated the rapid growth of watershed associations reflecting diverse interests, including many in urban places. As usual in such cases, geographic location and the specific effect of works on the well-being of individuals appeared to determine the attitudes of members of these groups. For example, the Kansas Valley Association and the Kansas Chamber of Commerce were at odds over the issue of little dams *versus* big dams. Gladwin Young reported to Ralph Will that, while no one from the Department of Agriculture was promoting the watershed associations, it should not discourage them, since the associations developed public consciousness of conservation problems. Young thought it was probably significant that the Salt-Wahoo and Big Blue associations in Nebraska "are not sponsored by soil conservation districts." He was not sure what their development would mean to the Department because "these particular watersheds may or may not be the highest priority for initial effort" in a basin-wide program. He did think, however, that these associations would not

[15] NARG 16 (Reports), Charles F. Brannan to Gladwin E. Young, March 27, 1950; Young to Ralph Will, April 11, 1950; editorial in Lincoln (Nebraska) *Star,* April 10, 1950.

"pull our proposed basin program into bits or pieces."[16] Within three years, he was proved wrong as a prophet.

Broadly speaking, the Department of Agriculture proposed to attack the problems of land use on a broad front in the Missouri Basin. It planned to control erosion and flooding through the sort of physical treatment which had been developed in the eleven watersheds authorized by the Flood Control Act of 1944. One way was through agricultural land treatment, such as contouring, terraces, grassed waterways, and a variety of other practices installed on individual farms by operators on a voluntary basis. These practices were part of the regular erosion control program administered through soil conservation districts on private land. The benefits accrued directly to owners and operators and only indirectly to the public. Other measures of flood control installed by the Department were physical structures such as "wet" or "dry" dams—normally quite small—on land used for agricultural production, and it also planned and built a variety of other structures to stabilize gulleys and small stream beds. Frequently the plans required relocation or reconstruction of bridges on county roads where flooding caused frequent losses. In the Little Sioux Watershed in Iowa, plans normally were made for "sub-watersheds" varying in size from 1,000 to 10,000 acres. Stress was laid on the orderly development of plans so that necessary land treatment preceded the construction of flood and gully control structures. Construction of such structures, however, was a supplementary and not a primary measure, as Bennett pointed out in 1949.

In 1950, practices and procedures were still being developed to execute these projects, but it was recognized that the governing boards of soil conservation districts would have to concentrate technical and fiscal assistance on areas critically in need of treatment. Priorities had to be established for watersheds, districts had to make agreements with farmers to install specific practices within a time schedule, easements were needed to permit works to be installed, provisions for maintaining

[16] NARG 16 (Reports), Gladwin E. Young to Ralph Will, Nov. 18, 1950. On March 20, 1951, Young also reported to Will that a bill had been introduced in the Nebraska legislature to authorize the creation of watershed districts and Governor Val Peterson had submitted this bill to Young for comments. Young also reported that Ray McConnell, editor of the Lincoln (Nebraska) *Journal*, was talking of seeking federal appropriations of $25 million for small watersheds to be offered as a substitute for the "grandiose" Missouri Basin program. An Iowa watershed group under the leadership of Barr Keashler, an intimate of Representative Ben F. Jensen and a power behind the scenes in the Iowa legislature, was also working to sponsor a watershed program.

structures had to be worked out and funds for doing so found, since the Department did not want to be responsible for maintenance costs. The Secretary of Agriculture did not use his power of eminent domain to purchase lands for these projects. Thus, the Department stressed that it would not administer works of improvement once they were installed.

In the late forties and early fifties, there were a number of major and costly floods on various rivers in the Missouri Basin. These floods made the small watershed phase of the basin-wide program very attractive to a variety of interests. Many vociferously argued that the Department of Agriculture's little dams and watershed treatment were an alternative to big dams constructed by the Corps of Engineers. For example, farmers whose land was taken by federal purchase to permit the Corps to construct large dams and impound large areas of water to provide flood control for major urban centers downstream frequently looked upon the Department of Agriculture's program as a favorable alternative. Others, such as businessmen in small towns who feared the economic consequences of a continued loss of farm population, tended to support the agricultural program. For obvious reasons, private electric utility companies also supported it. Other groups were interested in securing flood control and more water for the smaller urban areas where the Corps had not developed projects, or where its proposed works were not entirely satisfactory to local interests for various reasons. Among these groups, there was a feeling that the Corps of Engineers was not very interested in the relatively small projects important to the smaller urban centers and, certainly, the Corps had no experience which qualified it to deal with the agricultural phases of flood control.

The leaders of several groups in Kansas and Nebraska, irked by the Department's diffidence and troubled with serious floods, conferred with their congressmen. They finally decided that, if the time were not right for legislative authorization, it was at least ripe for an appropriation to start a number of "pilot" watershed projects to demonstrate the effectiveness of complete and integrated watershed management. Reportedly on the suggestion of Senator Kenneth S. Wherry, Republican from Nebraska, they tried to secure such an appropriation by having the Senate Committee on Appropriations add it to the Department's budget in 1951, but this approach failed.

After this initial setback, the watershed associations in Kansas and Nebraska stepped up the tempo of their demands that the Department immediately initiate the flood prevention phases of its program. The

Nebraska group insisted that such action was necessary so that the plans of the Corps of Engineers and the Department of Agriculture could be integrated to produce maximum results for Lincoln, Nebraska. Their threats and maneuvers to get appropriations for this project without the assent of the congressional public works committees aroused opposition from supporters of the Corps of Engineers— opposition despite the fact that the appropriation for the eleven watersheds was made part of the Department of Agriculture appropriations bill in 1950 and charged, therefore, as an agricultural program, not flood control. In November 1951, the Secretary of Agriculture's representative in the Missouri Basin was asked at a meeting in Kansas to state unequivocally his position on the flood control controversy. He said again that the Department's work was not competitive with that of the Corps, but that it had suffered for years from a lack of sufficient appropriations. It was his judgment, moreover, that if there were a question of choosing between the two agencies' programs, he would prefer that the agricultural phase be increased even at the cost of delaying for some years the construction of large dams by the Corps of Engineers.

These events brought the smoldering controversy between the Corps and the Department of Agriculture to the point where Congress intervened. In March 1952, a subcommittee of the House Public Works Committee, under Representative Robert E. Jones of Alabama, held public hearings on upstream and downstream programs of flood control of the Department of Agriculture and the Corps. Representative W. R. Poage, who was convinced that the Department's work was necessary in areas such as his own district in South Central Texas, tried to outflank the Public Works Committee. In May, he introduced a bill (not the first of this sort) which would have authorized the Secretary of Agriculture to "cooperate" with state and local governments to plan and execute works of "flood prevention." The House Committee on Agriculture held hearings on this bill and reported it favorably. The House Rules Committee, however, refused to give the bill a special rule, and the matter rested until December, when the Jones subcommittee reported. It castigated SCS for allegedly blocking a cooperative plan with the Corps to provide flood control upstream from the city of Lincoln, Nebraska, and emphatically rejected the Department's plans for watershed treatment as flood control in the accepted sense. It recommended that all future flood control work in which the Department of Agriculture had an interest be placed under the Corps of Engineers. There the matter rested, with neither committee of the

House the victor, until the Republican Administration came to power in January 1953.[17]

Shortly after President Dwight D. Eisenhower assumed office, the watershed associations scattered around the country formed the National Informal Citizens Committee on Watershed Conservation. Its Chairman was Raymond McConnell, editor of the Lincoln *Journal* and a vigorous advocate of a new national policy which would add flood prevention under the Department of Agriculture to flood control by the Corps. At the same time, neither McConnell nor most of the other leaders of this group was primarily concerned with sub-watershed flood control as a supplement to agricultural erosion control. Rather, they wanted urban benefits, and to this end secured passage of legislation in Kansas and Nebraska to insure that rural interests would not control "local" organizations through which the Department of Agriculture might function. They discussed their program with President Eisenhower after two Kansans, Senator Frank Carlson and Representative Clifford R. Hope, once again chairman of the House Committee on Agriculture, had first had a talk with President Eisenhower. Carlson and Hope had sought the President's public support for new legislation which would free the Department of Agriculture from reliance on the public works committees to secure authorization and funds for projects.

President Eisenhower sent a very revealing note to Secretary of Agriculture Ezra Taft Benson, after Hope and Carlson had paid another visit to the White House. Eisenhower said that Hope and Carlson proposed the "further development of Soil Conservation Districts and Natural Drainage Areas." They argued that their proposition was "politically wise" during a period of depressed farm prices. The President added that Carlson and Hope had convinced him that "there is very great support for positive action both on the farm and in the city. They point out further that much could be done with very little increase in Federal expenditures; I understood them to say that it was even possible that any Federal expenditure that might be necessary could be acquired by transfer from another agricultural fund." Although the President was not explicit, it appears that Hope and Carlson must have suggested a cut in the PMA's Agricultural Conservation Program and a commensurate increase in appropriations for flood prevention under the technical direction of SCS. The President ex-

---

[17] Some of this account is drawn from my article, "Pressure Politics and Resources Administration," *Journal of Politics,* Vol. 18 (Feb. 1956), pp. 39–60.

plained other features of the Hope-Carlson proposal and closed by telling Secretary Benson: "I told them that I would pass the idea on to you immediately, together with a request that as soon as convenient, you contact them directly in order to gather further explanation of their idea. Above all, they are convinced that this would make a splendid, positive, and popular program for the Administration to foster. Therefore they urge that we seize the opportunity and get busy."[18]

Benson and Under Secretary True D. Morse, who already were planning to reorganize the Soil Conservation Service in order to satisfy the state extension services, were flabbergasted. They sent Eisenhower's note to Assistant Secretary J. Earl Coke the next day with a single, pencilled question: "Mr. Coke—How do you suggest we proceed?"

Benson delayed his reply to the President for about six weeks—until Representative Hope about the first of May made clear his determination to act. When Hope appeared before the House subcommittee on agricultural appropriations to request that Congress add $5 million to the estimates for SCS to start "pilot" watershed projects, Benson explained to Eisenhower that the Department had cut the estimates for SCS operations in districts for fiscal year 1954 by 2 per cent. The budget prepared by Secretary Brannan for President Truman had included an item of $6.5 million to initiate operations on seven new watersheds for which Congress had not yet approved survey reports made by the Department under the flood control laws. This proposal, Benson explained, would have been a "new approach" to appropriating such funds, contrary to procedures used under flood control legislation. Therefore, Benson advised the President, it was "not wise for the Administration to sponsor the proposal at this time for several reasons." A committee was considering overlapping and conflicts in flood control work; the House Committee on Public Works would be likely to "resist a premature proposal to initiate this work outside existing legislative authorities"; and bills already introduced in Congress by several members ought to be allowed to take their course. The Secretary noted that Representative Hope had conferred with the staff members of the Department of Agriculture about his proposal to appropriate $5 million for pilot watershed projects, but "our plans . . . contemplate deferring such action until difficulties confronting us are resolved in both the executive and legislative branches."[19]

[18] NARG 16 (Soil 1), Memorandum for the Secretary of Agriculture, March 19, 1953, initialed "D.D.E." and signed "DE."

[19] NARG 16 (Soil 1), Ezra Taft Benson to President Eisenhower, May 4, 1953.

Despite Benson's opposition, the $5 million were appropriated to start pilot watershed projects to demonstrate the value of this work, although House members had to overcome the resistance of members of the Senate by restoring the item in conference. Meanwhile Representative Hope and his supporters got the President to send Congress a special message in July 1953, giving his support to new legislation. Hope also introduced a bill to authorize a program similar to the small watershed bill which Poage had advocated without success in 1952. Hearings were held on the new bill in 1953 by the House Agriculture Committee without any opposition voiced publicly. The bill passed without a murmur in the House from sources that might have been expected to delay or block the bill. At the President's request, Senator George D. Aiken, Republican of Vermont, introduced a companion bill which was sponsored by five other senators. Both bills authorized the Department to conduct surveys and construct flood control works of improvement, including dams with a total capacity of 5,000 acre-feet, without reference to existing flood control laws. These projects were to be undertaken by assisting local sponsors. The Senate Committee on Agriculture and Forestry held hearings on Senator Aiken's bill, which the Corps of Engineers opposed on the ground that the small watershed program had not been adequately tested in the pilot projects or by other research; the bill did not provide adequately for state participation; and local interests ought to pay 50 per cent of the project costs. The American Farm Bureau Federation advocated this last feature, also, and joined the U.S. Chamber of Commerce in proposing an amendment requiring local sponsors to assume actual construction costs. Initially, the National Reclamation Association strongly opposed Hope's bill.[20] A representative of the Farm Bureau explained that a further provision forbidding the Secretary of Agriculture to initiate a project after July 1, 1956, without a local sponsor was "a key feature of the original act." It was intended to make sure that real responsibility would lie with local sponsors rather than the Secretary of Agriculture, since the sponsor would let contracts and oversee construction. The Department did not operate under this restriction in the pilot projects and the eleven watersheds authorized in 1944. The Farm Bureau passed a resolution in 1954 at its annual convention, urging the states to enact legislation authorizing watershed districts to serve as local sponsors, but did not specify the function of soil conservation dis-

[20] *Amendments to Watershed Protection and Flood Prevention Act*, Hearings on H.R. 6146, Subcommittee on Conservation and Credit of the House Committee on Agriculture, 84 Cong. 1 sess. (1955), Serial CC, p. 7.

tricts.[21] The opposition evident at the Senate hearings was strong enough that a National Watershed Congress, consisting of a very wide range of conservation interests, met in May 1954 to show Congress the extent of popular support for the newly proposed program. The bill finally passed as Public Law 566 and was approved by the President on August 4, 1954.[22]

In its original form, this legislation permitted the Secretary of Agriculture to cooperate with states and their subdivisions to carry out programs requiring the construction of works of improvement for two purposes: "flood prevention" (including both structural and land treatment measures); and the "agricultural phases" of conserving, developing, using, and disposing of water. Within this narrow range of authority, the Secretary of Agriculture was authorized to assist local sponsoring organizations by: (1) conducting surveys necessary to prepare plans; (2) studying the physical and economic soundness of plans and determining whether benefits would exceed costs; (3) entering into agreements with sponsors and providing them with financial and other assistance; and (4) obtaining the "cooperation and assistance" of other agencies of the federal government. A "local organization" legally qualified to sponsor a project with federal assistance was defined to be a state or any of its political subdivisions. The latter would include soil or water conservation districts, flood prevention or control districts, or combinations of these, or any agency having power under state law to "carry out, maintain and operate works of improvement."

Certain conditions were established in the statute in order for a local organization to receive assistance from the Department of Agriculture. They were to provide, without cost to the United States, needed land, easements or rights of way. In addition, they were to assume a "proportionate" share of the costs of works of improvement installed with federal assistance, provided that the sponsor assumed all costs for purposes other than flood control. The Secretary of Agriculture, in this connection, was given discretion to determine the "equitable" share of local costs in consideration of anticipated benefits from the improvements. Local sponsors were also required to provide for operating and maintaining works of improvement, to give assurances that affected

[21] *Amendments to Watershed Protection and Flood Prevention Act,* Hearings on H.R. 6687, 8738, 8742, 8745, 8750, 8804 and 9192, Subcommittee on Conservation and Credit of the House Committee on Agriculture, 84 Cong. 2 sess. (1956), Serial JJ, p. 33.
[22] 68 *Stat.* 666, 16 *U.S.C.* §1001 *et seq.*

landowners have acquired water rights under state law, and to obtain agreements from landowners of 50 per cent of the lands situated in the drainage area above each detention structure installed with federal aid to carry out recommended soil conservation measures and farm plans.

There were some other significant restrictions written into the statute. The Department could not plan a watershed larger than 250,000 acres, and no single structure could be built with more than 5,000 acre-feet of detention capacity. Moreover, no appropriation of federal funds could be made for any plan calling for any structure with more than 2,500 acre-feet of capacity, unless approved by resolution of the public works or agriculture committees of both the House and the Senate.

These restrictions, together with the proviso that the local sponsoring organization would have to let construction contracts after July 1, 1956, left the Department with a narrow range of new authority to carry out a watershed treatment program. One provision of the 1954 law repealed virtually all of the Department's authority granted by the Flood Control Act of 1936, but continued its program in the eleven authorized watersheds. The net effect of the new law was to allow the Department to carry out nationally almost the same kind of programs it had developed in the eleven watersheds authorized by the Flood Control Act of 1944, although the requirement that local sponsors let contracts and be responsible for supervising the installation of structural works was new. The limit of 5,000 acre-feet of storage capacity for any single structure was new and did not apply in the eleven watersheds. The restriction on the total size of any one project limited basin-wide planning—and even projects as large as several of the eleven wa-tersheds. The provision for project authorization by four committees in Congress insured that a wide spectrum of interests would have a congressional forum in which to raise objections to watershed plans. The August 1954 legislation posed no serious threat to the Corps of Engineers or to the Bureau of Reclamation. The benefits it authorized were intended to be purely agricultural, and the projects were limited to 250,000 acres—or areas amounting to a little less than 400 square miles.

Donald A. Williams, SCS head, testified that the basic idea of Public Law 566 had been discussed for several years. When Hope's bill was presented to Congress, spokesmen for SCS "had in mind origi-nally . . . the flood control purpose as being almost the total concept of it." The bill passed the House with this objective, but in the Senate "after a great deal of consideration, there was an amendment added . . . which put in this agricultural phase of water."[23] Assistant Secretary of Agriculture E. L. Peterson testified that he and Secretary Benson understood that the "objective of this legislation was to bring under proper treatment and control the lands that needed further effort and could be benefited."[24] Representative Clifford G. McIntire, Republican of Maine, agreed with Peterson that Congress passed the law as "an incentive to develop more intensive application of these practices . . . [and] not a program to reimburse practices already done, but . . . to develop an accelerated program."[25] Representative Poage understood that Congress intended to authorize flood prevention, even in areas where there had already been intensive land treatment, as in his own district in Texas. He wanted a distinction between flood prevention, which the law authorized, and flood control, which in his judgment Congress did not intend to permit.[26]

There was complete agreement, however, that the law would permit the Department to expand to all areas of the country a limited program of agricultural flood prevention, and to include drainage or irrigation features so long as they did not require detention dams to store water for these two purposes. Supposedly, the work everywhere would be planned and executed in small watersheds, as Representative Whitten had urged upon SCS in 1949 as a means of speeding accomplishment. Each sub-watershed would have its own community plan and reliance would not be placed on achieving land treatment by waiting for each individual farm operator to make up his mind unaided to become a district cooperator and to practice conservation. Presumably, too, this new legislation reflected Bennett's contention that flood control is a supplement, but not a substitute, for agricultural land treatment. The new law, however, compromised this ideal. Bennett once said that when his organization first started land treatment, they thought that it was necessary to complete 100 per cent of the land treatment upstream from a control structure. Later this proportion was reduced to 80 per cent. Public Law 566 of 1954 merely required that 50 per cent of the land lying above a detention structure be under agreement and have a farm plan. There was, and is, no statutory requirement that any land actually be treated. Local sponsors have a supposed obligation to

[23] *Amendments to Watershed Protection and Flood Prevention Act*, Hearings of Subcommittee of the House Committee on Agriculture (1955), Serial CC, *op. cit.*, p. 6.

   [24] *Ibid.*, p. 45.
   [25] *Ibid.*, p. 46.
   [26] *Ibid.*, p. 56.

effectuate agreements for completing and maintaining these projects, but, since this responsibility does not have to be met, there is little ground for believing that it always will be met.

Although there was no dissent from the basic position that Public Law 566 was intended originally to provide a program limited in application to agricultural lands, one member of Congress indicated that some supporters of this program had a totally different objective in view. Representative Harlan Hagen, Democrat of California, remarked: "Of course, this whole program was set up in the mind of some people to defeat some of the larger hydro-electric producing projects and so forth."[27] He warned that if the Department were to seek broader authority under this law, that issue would be raised again.

AMENDMENTS TO PUBLIC LAW 566

In August 1954, the ink from President Eisenhower's pen was scarcely dry on Public Law 566 before a host of disappointed interests were importuning their congressmen to change it. Within the next two years at least twelve bills were introduced, and the results of three hearings were published for the record. The Watershed Protection and Flood Prevention Act of 1954 was considered by many interested parties to be entirely too narrow in its purposes and too restrictive in several other ways. Of course, the Secretary of Agriculture and some members of the Senate Committee on Agriculture and Forestry opposed both the original act and the key amendments which were considered in 1955 and 1956 hearings because they expected to reorganize the Soil Conservation Service eventually to suit the demands of the state extension services. The reorganization plan reported by Representative Poage in 1953 would have permitted SCS to continue limited responsibility for the engineering phases of a purely agricultural small watershed program and to increase the Extension Service's share in the land treatment program through budgetary controls. Major changes made in Public Law 566 in 1956, however, permanently ended any hope among extension leaders that they could absorb much of the SCS program.

The general problem created by Public Law 566 was highlighted by various members of Congress during the 1955 and 1956 hearings: the Law did not authorize a "national" program. Representative Iris F.

[27] *Ibid.*, p. 15.

Blitch, Democrat from Georgia, said that in her part of the country there were no watershed projects "of that type that can be used in flatlands where the flood-control problem, as we recognize flood control, is not a circumstance." Mrs. Blitch was distressed and puzzled: "I just do not understand the Department's attitude in being willing to go ahead and work under 566 without making provisions for all sections of the country to have the privilege of participating in that sort of program."[28] Senator A. S. Mike Monroney, Democrat of Oklahoma, claimed there was "positive proof" that floods could be reduced by "upstream prevention" and thought that the new program "ought to have more than the 2 or 3 pennies out of every flood-control dollar it was getting."[29] Senator Earle C. Clements, Democrat of Kentucky, said that P. L. 566 was enacted only to satisfy the need for retarding and utilizing water in the "small watersheds in the interests of agriculture. By so limiting this program . . . we are actually preventing an opportunity to provide solutions to many water problems not directly connected with agriculture."[30] Senator Wallace F. Bennett, Republican from Utah, argued that Public Law 566 was so useless in solving the problems of his state and region that it looked to him as if it had been passed "principally to aid the Midwest and only incidentally the mountain West."[31] Representative Lester R. Johnson, Democrat from a hilly district in Wisconsin which was drained by streams emptying directly into the Mississippi River, said that the "poor farmers" on the uplands heard that the "government employees are out trying to get them in watersheds" only to find that they would have to bear the cost of dams intended "to help the people down below."[32] Representative William H. Avery, Republican of Kansas, summed up the limitations of the legislation, saying that it "precluded any industrial or any municipal interest from coordinating their efforts with the Department of Agriculture . . . it has pretty well restricted the effort of the Department of Agriculture to strictly an on-the-farm program."[33]

Major amendments were enacted before dirt was moved in the first project authorized by the original Public Law 566. The several amend-

[28] Ibid., pp. 4–5.

[29] Amending the Watershed Protection and Flood Prevention Act, Hearings on H.R. 8750, Subcommittee of the Senate Committee on Public Works, 84 Cong. 2 sess. (1956), p. 16.

[30] Ibid., p. 22.

[31] Ibid., p. 26.

[32] Ibid., p. 27.

[33] Amendments to Watershed Protection and Flood Prevention Act, Hearings of Subcommittee of the House Committee on Agriculture (1956), Serial JJ, op. cit., p. 51.

ments made by Congress since 1954 have rather consistently added authority for the Secretary of Agriculture to extend planning, fiscal, and technical assistance to multiple-purpose projects within the area limitation of 250,000 acres. In doing so, Congress has with equal consistency developed a "national" program, but at the cost of shifting the focus of action away from the purely agricultural phases of soil and water conservation.

In 1955, amendments were proposed and considered by the Subcommittee on Conservation and Credit of the House Committee on Agriculture. Representative Poage, who worked closely with Representative Hope on the original bill, conducted the hearings, but the two bills under consideration were not reported out. In part, this inaction was due to prior interest in the Bureau of Reclamation's bill on small projects, which was the subject of rather intense legislative maneuvering.

Again, in 1956, the Subcommittee on Conservation and Credit of the House Agriculture Committee held hearings on seven bills, more or less identical. These would provide the Department of Agriculture with a more flexible legislative authorization to undertake projects, while operating under procedures essentially the same as those of the Corps of Engineers and Bureau of Reclamation, except for the area limitation of 250,000 acres. Representative Poage's bill providing this broader authority passed the House and was referred to the Senate Committee on Public Works, rather than the Committee on Agriculture and Forestry, which in 1954 had provided a sounding board for objections from the Corps of Engineers and had written the bill so as to limit the Department to a "purely agricultural" small watershed program. The Senate Public Works Committee turned the Poage bill over to its Subcommittee on Flood Control under the chairmanship of Senator Robert S. Kerr, Oklahoma Democrat, whose keen interest in developing the water resources of his home state was legendary in Washington. The Subcommittee gave a very sympathetic hearing to the proponents of Poage's bill and pushed through the Senate virtually all of the amendments favored by Poage, but not by the Department of Agriculture, without serious limitation.

One feature of the Poage bill aroused universal comment among supporters and opponents and was enacted into law. It was a requirement that the Secretary of Agriculture bear the entire costs of the flood control features of any structure installed under Public Law 566. Supporters contended that this feature was "fair" because it was the principle embraced in flood control legislation since 1936 and ought to be applied to all areas to the disadvantage of none. It was also argued that it was necessary to keep communities from outbidding each other for assistance from the Secretary of Agriculture to the disadvantage of those in economically depressed areas.[34] Representatives from Kentucky, for example, pointed to the limitations of the original law in the eastern part of that state.[35] The Department of Agriculture officially opposed this feature on the ground that it preferred to concentrate on the agricultural phases of the program.

A second major change made in 1956, together with certain later amendments, redefined authorized works of improvement to permit projects including structures to impound water for all kinds of water use and conservation, not merely the "agricultural phases" as in the original act. These purposes include irrigation, municipal and industrial water supply, recreation, wildlife, pollution control, and others. Under these amendments, local sponsoring organizations are still required to pay some of the structural costs of storing water for such features. They permit the Department of Agriculture to share up to 50 per cent of the costs of easements, rights of way and land for fish, wildlife, and recreational features; and also give it flexible authority to base the local sponsors' share of costs for the "agricultural phases" of a structure on "national needs and assistance authorized for similar purposes under other federal programs." Subsequent changes also made it easier to include municipal water supply as a structural feature. They authorized the Secretary to pay 30 per cent of the costs for municipal and industrial water supply, if he is assured of repayment within a time representing the effective span of life of the reservoir, but not to exceed fifty years after the reservoir is first used for water supply.

Although numerous other amendments have been made, those which clearly show the trend away from the purely agricultural flood prevention program of the early fifties are changes in the size of structures authorized in any one project. The 1954 legislation flatly limited construction to single structures with a maximum detention or storage capacity of 5,000 acre-feet for all authorized purposes. In 1956, when the Poage bill proposed to eliminate the ceiling on storage capacity so that any structures might serve multiple purposes, the Corps of Engineers again objected. At the suggestion of the Secretary

---

[34] *Ibid.*, p. 12.
[35] *Ibid.*, pp. 39–46.

of Agriculture, the Senate incorporated a maximum limit of 25,000 acre-feet of storage capacity for all purposes in any single structure.[36] It was argued with some force by the proponents of multiple-purpose dams that available sites, especially those on the smaller streams, ought not to be under-utilized.

The limit on 5,000 acre-feet on single structures has been maintained, although several bills have been introduced recently to raise this limit to 12,500 acre-feet. The Soil Conservation Service has estimated that about 20 per cent of the structures in the eastern two-thirds of the country would be over 5,000 acre-feet and that 35 per cent of them in the western third would be in excess of this size if the limit were raised in the interest of both feasibility and economy.[37]

Loan features permitting local sponsors to borrow from the Department of Agriculture have been added to the original law to facilitate projects in areas of low income. Even so, there are important limitations on their capacity to participate in the Department's broad range of projects, even where it is thought that new facilities would be a stimulus to a declining economy in rural areas. Donald A. Williams, Administrator of SCS, testified in 1964 that in some communities in the Appalachian area local sponsors are unable to meet the necessary costs for structures. Partly, this is a result of the depressed economy, but also it is partly because of the topography, a prominent feature of which is narrow valleys in which highways, railroads, and utility rights of way are crowded. Structures can be installed in such areas only at high costs for easements, land, and rights of way. Projects of this sort are part of President Lyndon B. Johnson's program to expand the type of multiple-purpose projects which the President characterized in 1964 as "small" but "of vital importance to rural areas." In fact, 40 per cent of the 535 small watershed projects approved for construction as of the beginning of 1964 had multiple purposes.[38] Amendments proposed in 1964, but not officially proposed by the Department of Agriculture, would permit local sponsors to use federal funds other than those appropriated under the authority of Public Law 566. This change would permit the use, for example, of funds made available by such

---

[36] *Amending the Watershed Protection and Flood Prevention Act,* Hearings on H.R. 8750, of Subcommittee of the Senate Committee on Public Works, *op. cit.,* p. 57.

[37] *Amending the Watershed and Flood Prevention Act,* Hearing on H.R. 9695 and 9938, Subcommittee on Conservation and Credit of the House Committee on Agriculture, 88 Cong. 2 sess. (1964), Serial 00, p. 9.

[38] *Ibid.,* pp. 3 and 18–26.

agencies as the Area Redevelopment Administration to local sponsors.[39]

CONSERVATION AND PUBLIC WORKS

The major changes made in Public Law 566 since its passage in 1954
mark a distinct trend away from agricultural flood prevention and
comprehensive integrated basin-wide planning. The Department of
Agriculture's Small Watershed Program now emphasizes multiple-
purpose projects to serve the needs of urban centers located on the
smaller streams.[40] The program is now national only in the sense that
the Secretary of Agriculture has authority broad enough to undertake
water projects which fit the special needs of small communities, within
the size limitations imposed by Congress. Localities once slighted by
the Corps of Engineers and not situated so as to be served by the
Bureau of Reclamation are now in a position to negotiate with the
Department of Agriculture for projects which include the whole
spectrum of multiple-purpose water development. The program's ratio
of total federal expenditures for public works is almost certain to be
more than the "2 or 3 pennies out of every flood-control dollar" which
Senator Monroney lamented before this legislation was passed. In some
areas, such as Appalachia, projects of this sort are almost certain to be
used as part of a national program to lift the economic level of the
region. In short, the program of erosion control which Hugh Bennett so
successfully started in 1933, partly to relieve unemployment, has slowly
evolved to the point where it is part of a federal program of public
works.

One difficulty in this trend of policy, however, is that the selection
of projects becomes the object of negotiations, and members of Congress are the brokers. The inescapable conclusion from the evidence is
that sooner or later public works programs become "national." That is,
they are planned, authorized, and supported on a project-by-project
basis, with distribution aimed at securing maximum congressional
support. Members are determined to get benefits for their home
districts and are not inclined to place much reliance on rational criteria
of "critical" needs outside their own constituencies.

This situation is not necessarily improved by some types of state

[39] *Ibid.*, pp. 35–38.
[40] 68 *Stat.* 666 as amended by 70 *Stat.* 1088; 72 *Stat.* 563; 72 *Stat.* 1605; 74 *Stat.*
131, 132, 254; 75 *Stat.* 408; 76 *Stat.* 605, 16 *U.S.C.* 1001–1016.

fiscal assistance. During the fiscal year 1963, states appropriated $6,845,001 for either planning or "works of improvement" on small watersheds. Total state appropriations in that year for soil conservation committees and "direct assistance" to districts were only $4,309,620. Appropriations for watersheds varied widely among the states. Connecticut, for example, accounted for 80 per cent of the funds for works of improvement ($3,857,500), while each of six other states (Delaware, Louisiana, Massachusetts, Oklahoma, Pennsylvania, and Washington) provided $100,000 or more for this purpose. Among 21 states providing funds for project investigations or planning, Oklahoma alone accounted for approximately 20 per cent ($342,980) of the total. California was second with $210,898; Illinois and Kansas followed with $137,500 and $120,000, respectively. Among the remaining 17 states, only Georgia provided as much as $100,000.[41] These state appropriations are not made to match federal funds on the principle of grants-in-aid, although there is some misunderstanding on this subject. In fact, the initiative has been left to the states to provide funds for either project planning or such assistance to local sponsors as the purchase of easements and rights of way to construct works of improvement. These states have an obvious advantage over those which do not appropriate funds, since projects are approved for construction in the order in which plans are approved by the Secretary of Agriculture.

The resulting wide disparity in the number and location of watershed projects among states has caused dismay in Congress for several years, but the issue was finally taken up in the Senate Subcommittee on Agricultural Appropriations in 1963. Since then, members of Congress have commented unfavorably on the disparity in both the distribution of projects among states and the proportion of costs borne by local sponsors and the federal government, respectively. At the hearings in 1963, Senator Roman L. Hruska, Republican from Nebraska, reported the enactment of legislation in his state to accelerate Public Law 566 watershed projects, of which 12 were under construction. He wanted Administrator Williams to assure him that SCS would not penalize Nebraska by assisting other states which provided more funds than Nebraska. Senator Spessard L. Holland, Democrat of Florida, clarified the issues when he said that there is need to make the program "more uniform" by "not permitting some states, simply by appropriating a little bit of additional planning money, to get much more than their share of projects and federal funds." He said he was hearing

[41] *Department of Agriculture Appropriation Bill for 1964*, Senate, 88 Cong. 1 sess. (1963), p. 193.

"more and more complaint" about the situation and did not like to use his own state for an example, but he knew the situation and thought the record ought to show that federal funds had been obligated in Florida for only eight projects for which the federal cost-share was 43.8 per cent. That was close to the average of 50 per cent which was the compromise proportion decided upon in 1954. "I handled that bill [Public Law 566], [and] I had considerable part . . . in getting the reluctant consent of the Corps of Engineers to approve this program." Senator Holland said he was reluctant to name other states, but he noticed that SCS was not following this "moderate approach" in some states where the federal share was as high as 80 per cent. At the time, there were two: Arizona (87.9 per cent), which had not passed enabling legislation authorizing local sponsors because of opposition by the irrigation interests; and Pennsylvania (80.4 per cent), where soil conservation districts had been formed with much difficulty. Senator Holland and Senator Allen J. Ellender, Louisiana Democrat who was Chairman of the Senate Agricultural Committee, told the Soil Conservation Service to find more multiple-purpose projects instead of projects for flood control only, for which the federal share of costs is 100 per cent. Senator Holland also advised SCS to see that in the future the program is "administered so that all States can get a uniform amount of money, that is, within a certain range."[42]

Within each state, watershed projects must be cleared and recommended by an agency designated by the governor. Procedures vary in detail among the states, but each state's soil conservation committee, commission, or board has been directed to recommend Public Law 566 projects for gubernatorial approval. According to one of its spokesmen, the SCS serves the state committees in making project decisions in "an advisory capacity . . . to sit with them in developing the criteria and guidelines that will be used in recommending priority of consideration."[43]

The conservation committee in most states is dominated by a majority of members who are supervisors of soil conservation districts. Some of these are appointed by their governors, as in Iowa, but most are chosen without any obvious consideration of partisan attachments.

---

[42] *Ibid.*, pp. 191–205. See also *Small Watershed Program*, Hearings, Subcommittee on Watershed Development of the House Committee on Public Works, 86 Cong. 1 sess. (1959), esp. pp. 104, 107, and 177–79.

[43] *Department of Agriculture Appropriation Bill for 1964*, Senate, *op. cit.*, p. 199.

Usually, members of the committees have worked their way up to positions of leadership in the state associations of soil conservation districts. When a state committee is politically independent of a governor, there is always the possibility that it is responsive to suggestions from agencies of the Department of Agriculture and members of Congress who are directly concerned with the selection of project sites and related matters. In contrast, when a committee is subject to some direct influence from the governor, it is likely that project decisions will reflect his wishes to a degree—although in states where the governor is forbidden to succeed himself the preponderance of influence would seem to lie with members of Congress. In any event, the Secretary of Agriculture cannot be insensitive to the wishes of congressmen who vote funds to the Department and oversee its responsibility for administering them.

Although the emphasis upon multiple-purpose projects usually is undoubtedly sound, it is not difficult to confuse a multiple-purpose project with a multiple-benefit program. So long as a program provides valuable benefits to individuals, the tendency seems irresistible for administrators and congressmen alike to distribute them as widely as the interests of their constituencies demand. The Agricultural Conservation Program and Small Watershed Program are both of this nature, despite the fact that single individuals are the direct beneficiaries of the first and communities of the second.

For example, in the fifties, when the Secretary of Agriculture attempted to reduce or eliminate ACP subsidies for some "conservation" practices, Senator Richard B. Russell of Georgia was incensed. He said Congress had been "appropriating money by the hundreds of millions of dollars to build just one irrigation dam in the reclamation areas of the country; to open up tens of thousands of acres of land for new production; and because we have a few little scattered projects around where they pipe out a little water from a fish pond for irrigation, now they want to do away with that and stop irrigation in the eastern areas that have been taxed to carry on this overall program of irrigation throughout the entire Nation for all these years." Senator Aiken of Vermont also was "disturbed over the proposal to eliminate to such an extent the program which has helped us through two wars, helped us diversify agriculture tremendously in the Eastern and Southern states." And at the hearing on the 1960 Agriculture appropriation, when Paul M. Koger, head of the Agricultural Conservation Program from 1955 to 1961, said that changes were proposed "just for this year," Senator

Ellender remarked that it "is just the beginning. . . . For 6 years you have been trying to cut it back."[44] Representative Whitten had voiced the same sentiments during hearings on the 1959 appropriation in reminding Assistant Secretary Peterson of the "disturbance that came about last year when it was understood that from Washington there would be a serious curtailment of some of the practices. . . . As you know there are a lot of practices of vital interest to the southeastern part of the United States . . . [and] certain others that are of vital interest in the New England area."[45] Two years later, the pulling and hauling had not abated, as indicated by Whitten's remarks during hearings on the 1961 appropriation that ACP is a "national program, and being national I realize that you have to let all your areas, whether they need it or not, be included." Continuing, he said, "there is no authority to take the money from my colleague's State and put it in any sections of this country where the erosion and the worn-out resources are terrible, as is true in many sections of this country."[46]

SUMMARY AND COMMENT

Since 1936, water conservation in the Department of Agriculture has retrogressed from programs to projects. Thirty years ago, the Department was granted authority by the Flood Control Act of 1936 to plan the agricultural phases of multiple-purpose comprehensive and integrated basin-wide programs. That authority now has been discarded and, instead, the Department plans and executes literally hundreds of unconnected "small watershed" projects scattered throughout the United States.

Certainly the Flood Control Act of 1936 was of little use to the Department of Agriculture for many reasons, most of them beyond its control. The Secretary tried, with known success in Texas, to use his authority as a carrot stimulating states to adopt legislation for standard soil conservation districts. With a single exception, the Department's flood control authority was used only to conduct surveys and investigations and to make reports.

The Flood Control Act of 1944 failed to give the Department

[44] *Department of Agriculture Appropriation Bill for 1960,* Senate, 86 Cong. 1 sess. (1959), pp. 371–77.

[45] *Department of Agriculture Appropriation Bill for 1959,* House, 85 Cong. 2 sess. (1958), Pt. 3, p. 1460.

[46] *Department of Agriculture Appropriation Bill for 1961,* House, 86 Cong. 2 sess. (1960), Pt. 1, pp. 668–69 and 679.

authority to carry out an agricultural program for the Missouri Basin which would complement plans authorized for the Corps of Engineers and the Bureau of Reclamation. This act did move, however, toward the watershed concept. It authorized the Department to undertake upstream flood control on eleven small rivers, but the work was of benefit to agricultural lands primarily.

When the Secretary of Agriculture planned a comprehensive Missouri Basin agricultural program in the late forties, he was stymied by the antagonism of some of the agricultural colleges of the region toward the Soil Conservation Service. The Secretary received no help from Congress. Instead, key members executed a series of legislative maneuvers which eventuated in the small watershed legislation—Public Law 566—in 1954. This original small watershed legislation was intended by some of its sponsors, at least, to authorize on a national basis a program of agricultural flood control of the sort authorized for the eleven watersheds in the Flood Control Act of 1944. But since this objective was too narrow for many members of Congress, they pushed amendments to broaden the purposes and to make it actually a national program rather than one restricted to agricultural flood control. It is not, however, national in the sense that it provides for comprehensive, orderly development of water resources to serve national needs. Members of Congress apparently are determined that federal funds shall be well distributed geographically. What is not at all clear, however, is the way in which the distribution of these community benefits is going to conserve both soil and water for agriculture where the need is most critical. In fact, there is no evidence that there are any criteria of need; at least of national need.

The Secretary of Agriculture has assigned chief responsibility to the Soil Conservation Service for administering the Watershed Protection and Flood Prevention Act of 1954, although he has assigned some duties to the Forest Service and directed the Agricultural Stabilization and Conservation Service to use funds of the Agricultural Conservation Program in aid of individual projects. Now, as so often since 1935, there is a tendency for some people to believe that the Soil Conservation Service, and not the Secretary of Agriculture, has been charged by Congress to plan and execute small watershed projects to conserve soil and water. It is appropriate, of course, for the Secretary to assign to one bureau primary responsibility for accomplishing a mission within the scope of his authority. Assignments of this sort should not, however, fragment the power of responsible executive officers to the point where they cannot utilize all the skills available within all their agencies.

Since soil conservation districts have been so intimately related to the Soil Conservation Service throughout their history, some people have thought that they ought to assume new responsibilities under Public Law 566. There is a fundamental difference, however, between the limited purpose for which districts were originally created in the thirties and the multiple-purpose objectives of Public Law 566 as it stands amended today. Districts were established to help the Department of Agriculture control erosion on agricultural lands as part of a broader program to adjust land use. In 1935, Henry A. Wallace, M. L. Wilson, and their associates considered and rejected a proposal to recommend conservation districts with powers broad enough to engage in "upstream engineering" to include flood control and municipal water supply. The head of the Soil Conservation Service, Hugh H. Bennett, said that districts were not "appropriate agencies" to carry out the Secretary's authority under flood control legislation. More than a decade later, he testified before Congress that flood control was only a supplement to his agency's soil conservation operations on agricultural lands.

From the point of view of future objectives, there are some questions as to where the responsibility under Public Law 566 should lie. The small watershed legislation is now clearly intended to provide relatively small communities with multiple-purpose projects attractive to a broad spectrum of urban and rural interests. Judged by the performance of soil conservation districts during the last quarter of a century and in view of future needs, are such districts appropriate units of government to undertake further responsibilities for achieving national objectives? The leaders of the districts apparently think so. For one thing, the name of their organization was broadened from National Association of Soil Conservation Districts at their February 1962 convention to National Association of Soil and Water Conservation Districts (NACD). They are frank to say, moreover, that the question at issue is whether rural interests—and their supporters whose identity is not always apparent—are to maintain a sufficient majority to affect the control of water policy in the future.

In 1962, the NACD President, William E. Richards, asked: "Who is to decide?" Alluding to an anticipated water shortage in which it would not be possible to provide "simultaneously for irrigation, navigation, industry and domestic requirements," he called for a federal policy which gives "local people" an effective voice in decisions.[47] In

[47] National Association of Soil and Water Conservation Districts, *Tuesday Newsletter*, April 10, 1962.

early 1964, Richards' successor, Marion Monk, asserted that soil and water conservation districts (as they were designated by then in many states) face a "series of new tests" which will determine their role in any future multiple-purpose conservation program. "Prompted by the widespread success of Districts in dealing with their original assignments—erosion control and soil and water conservation—State officials and private organizations are beginning to explore the possibility of broadening Soil and Water Conservation District responsibilities to cover a wider range of resource problems . . . Is there any better instrument than the District to develop and move forward with an orderly well-considered program of local resource development?" He reminded readers that Massachusetts had recently changed the name and authority of soil conservation districts to "conservation districts," and that proposals had been made in two other states to permit a single representative of urban interests to serve as district board members. Although these changes broaden the range of interests on soil conservation district boards, they leave effective control in the hands of a rural majority. Monk emphasized this point by suggesting that this concession to urban interests will insure that the "majority viewpoint will continue to originate in rural areas."[48]

The question whether the agencies of government used to accomplish future national objectives of resources policy ought to be dominated by a rural point of view deserves careful consideration. It may be useful first, however, to examine evidence of the extent to which soil conservation districts actually have had "widespread success in dealing with their original assignments," as claimed by their supporters.

[48] *Ibid.,* Jan. 14, 1964.

CHAPTER 8

Federal-State-District Relations

WHEN M. L. WILSON and his assistants from the Solicitor's office in the Department of Agriculture were drafting the standard conservation districts law in 1935, they were considerably vexed by the problem of fixing state-district relations. Their difficulty arose partly out of the need to recommend procedures for organizing districts in accordance with the requirements of state law. They also needed to solve certain tactical problems arising from their expectation that many of the state extension services would oppose districts intimately related to the Department through the Soil Conservation Service (SCS). To comply with prevailing doctrines of public law and provide a convenient and relatively simple means of establishing districts, the Department recommended that a new state administrative agency, called the state soil conservation committee, be established.[1] Once this decision was made, it was necessary to fix the committee's membership and powers. Perhaps even more important in achieving the Department's conservation objectives was the need to determine whether conservation operations were to be planned and administered through a system of federal-state-local relations or a direct federal-local relationship.

The first draft of the standard act which the Department submitted to the National Resources Committee in October 1935, under circumstances described in Chapter 2, provided that the state governor be authorized to appoint a single commissioner with the power to carry out procedures for organizing districts. This proposal was abandoned in favor of a state committee consisting of five members, including the directors of the state agricultural experiment station and of the state extension service. Once this change was made as a means of partially bringing the colleges into the committee, it was necessary to fix rather exactly the relations of the state committee with districts which were to be organized. M. L. Wilson said the Department wanted to avoid a situation in which farmers would feel that a district's program was "dictated" from the state capital.

[1] U.S. Department of Agriculture, A *Standard Soil Conservation Districts Law* (Washington: Government Printing Office, 1936), pp. 51–55.

In this situation, it is not surprising that the standard act recommended two kinds of powers for state committees. The first was to encourage and enable farmers to organize districts; the second, to offer assistance, information, and program coordination to farmers insofar as it might be done by "advice and consultation."[2] There was no specific provision for the state committee to have supervisory powers over district administration. The Department of Agriculture did not indicate how it expected district supervisors to learn and to assume their new duties. Neither did the standard act specify any power for the state committee to plan and execute a state program of soil conservation. No mention was made of the functions of the state agricultural colleges in relation to district programs. For all practical purposes, then, the Department wanted a weak state committee; that is, one with no effective powers over either the conservation operations or administrative activities of districts, once the districts were organized with the state committee's active aid.

In the model conservation districts law, therefore, the Department of Agriculture laid the foundation for direct federal-local relations to carry out its erosion control program. Between 1936 and 1940, the Department did make some effort to secure joint planning by SCS and the extension services for operations within districts and on projects outside them. However, when relations between the American Farm Bureau Federation and the extension services, on the one hand, and the Roosevelt Administration on the other, broke down almost completely during the presidential campaign of 1940, these limited attempts to work together were abandoned. From 1935, the state colleges had fought the regional offices of SCS on the ground that the technical competence of the college specialists was not properly recognized. In view of this hostility, it is hardly surprising that SCS did not welcome coordination of conservation district programs through state committees which had directors of extension and the experiment stations as members—frequently one or the other as chairman. Distrust became even greater during meetings between the Committee on Extension Organization and Policy of the Association of Land-Grant Colleges and Universities and the Soil Conservation Service in the spring of 1940. The college leaders made it clear that they were trying to sabotage the Department's efforts to get conservation district boards to make long-range programs and to hold farmers responsible for properly using the assistance they received from the federal government. Secretary

[2] *Ibid.*, Sec. 4(D).

Claude R. Wickard's memorandum of September 21, 1940 (described in Chapter 4), therefore, was a signal that the Department was terminating any effort to administer a program of soil conservation through a system of federal-state-district relations. Instead, SCS was charged with responsibility for developing and maintaining direct federal-local relations involving both conservation operations and administrative guidance.

Even as late as 1945, the hope and expectation that districts would receive state and, possibly county, appropriations as the Department had originally intended for the most part were not fulfilled. Districts still depended upon federal technical assistance and only a few had received grants from the Works Project Administration (WPA) before it was discontinued. Certainly, cooperative federal-state-local fiscal cooperation had not developed to a significant degree by 1945. And it was not likely to increase, since the Department was providing farmers with conservation subsidies through the Agricultural Conservation Program (ACP) administered through the Production and Marketing Administration (PMA) in 1945.

Each of these problems in federal-state-local relations has materially influenced the functioning of soil conservation districts and has determined the extent of their dependence upon the Soil Conservation Service. They are also factors which will have major significance in connection with the future role of the districts in agricultural conservation administration or resources conservation embracing a broader range of objectives. They, thus, are the subject not only of this chapter, but also of the three which follow it.

COMPOSITION OF
STATE SOIL CONSERVATION COMMITTEES

The primary purpose of the state soil conservation committees, so far as the Department of Agriculture was concerned in the thirties and forties, was to encourage and assist farmers to organize districts. Clearly they performed this function to a considerable degree—even after the SCS and Extension Service break-up in 1940—for 1,235 districts were organized by the end of 1944. Nevertheless, late in 1943, when the decision apparently was made to promote conservation districts in all agricultural counties, someone within the Department of Agriculture must have decided that changes would have to be made to dilute the influence of the agricultural colleges. Starting when the state legislatures met in 1945 in Oklahoma, Georgia, South Carolina, Michigan, and Arkansas, state laws were amended to provide for committees consisting of either elected or appointed members from the public. The representatives of state agencies were usually displaced. These changes marked the beginning of a trend which can be expected to continue until all of the state committees are composed of majorities of members having a special interest in the functioning of soil conservation districts. The speed with which these changes are made is often an indication of the degree of political skill and vigor of the leadership of a state association of soil conservation district supervisors.[3]

Although there are detailed and numerous variations in the composition of the state committees, they can be classified usefully according to the number of members who are district supervisors (or farmers) in contrast with representatives of state governments. These committees may be grouped in five classes.

First are six committees composed exclusively of members who are district supervisors. These supervisors are either elected or nominated by state association or soil conservation districts for appointment by the state governors.

Second are twenty-three committees consisting of a majority required by law to be either supervisors or farmers and a minority of members from state agencies who serve ex officio.

Third are fourteen committees which are variously composed, but serving technically within some department, board, commission or other agency of the state government.

In the fourth group, there are four committees with an exact balance between the supervisor and agency members.

Fifth are three committees having a majority of members from state agencies.

Texas provided a model for committees of the first class with the state committee which it created in 1939. Five members of the Texas Soil Conservation Board are elected at conventions of supervisors in areas designated by state law, although, from 1939 until 1953, the support of the politically potent county judges was sought by giving

---

[3] This analysis of the composition of the state committees rests on an examination of the relevant provisions of the 50 state soil conservation district laws cited in Appendix A. Naturally, the comments made about provisions of law affecting the districts rest also upon such an analysis. To avoid undue detail, citations are not given in connection with each of the summary comments made in the text. However, it should be noted that these comments usually are based upon laws in effect at the end of 1963, since that year was the last in which most state legislatures met before this study was completed.

them a voice in the selection process. Texas has never had officials from state agencies serve as committee members. In 1945, South Carolina and Georgia patterned their reorganized committees on the Texas model, but provided that the governor appoint district supervisors as committee members. The South Carolina law provides that these nominees shall be presented to the governor by the executive committee of the state's association of district supervisors. Georgia law provides only that the appointees shall be district supervisors, although in practice there is no difference between the procedures followed in the two states. Technically each also has a large advisory committee, consisting of the principal officers of several state agencies and the state conservationist of the SCS. Arkansas made a complete break with the past by setting up its Geological and Conservation Commission, composed of members appointed by the Governor with the consent of the state Senate. Idaho and Oklahoma followed the Texas model in having no advisory committee, but membership requirements reflect some common differences in detail. Members in Oklahoma are appointed by the Governor with the consent of the state Senate from five designated areas of the state. These members must be farmers who are active cooperators with districts and practice soil conservation on their lands in compliance with district requirements. In Idaho, which switched to a new type of committee in 1957, the law provides merely for appointment by the Governor, who is not legally restricted to choose supervisors or persons nominated by the district association in the state.

In the twenty-three states where supervisors or farmers make up the majority of the committee—either as originally constituted in Iowa, for example, or as more recently established in Louisiana—there are variations; not all members necessarily respond exclusively to a state's association of district supervisors. In 1945, for example, Michigan amended its district law by providing that the Dean of the College of Agriculture at Michigan State University, the Commissioner of Agriculture, and the Director of the Department of Conservation be joined by four "practical farmers" appointed by the Governor from among the directors of soil conservation districts. A more restrictive system of choice is imposed in several states which explicitly provide for the governors to receive nominations from the state associations of district supervisors. As an example, in South Dakota, the association of supervisors is required to submit a list of past or present supervisors from areas designated by law. Six supervisor members give a comfortable majority. The South Dakota Secretary of Agriculture is an ex officio member, and the Director of Extension, Director of the Experiment Station, together with the Commissioner of School and Public Lands, are by law directed to serve merely as "advisory" members. The South Dakota Soil Conservation Committee may also "invite" the U.S. Secretary of Agriculture to appoint the SCS State Conservationist as an advisory member.

Among the remaining states in this second classification, the statutes do not specify that district supervisors shall constitute a majority but they may be. At least, they constitute the largest single interest. For example: In Vermont, out of four members, two shall be district supervisors. Wisconsin merely requires that one of the appointed farmer members shall be a cooperator with a "county soil conservation district." Missouri requires only that the three members who do not represent state agencies be *bona fide* "farmers." In Iowa, Indiana, and Ohio, committee members normally are appointed by the governors only after consultation with the state associations of conservation district." Missouri requires only that the three members who do not mittee consists of five district supervisors, the Governor, two representatives of "agriculture," the Director of Extension, and the Director of the Experiment Station. The Governor of Virginia appoints one member who is his personal representative on the committee, which also has six members nominated by the State Association of Districts; this large committee includes, also, two representatives of the college of agriculture, the Commissioner of Agriculture, and the Director of the Department of Conservation and Development.

Nebraska has a large committee representing the complex interests of water users. Two members, one representing irrigation interests and the other the Nebraska Chamber of Commerce, are appointed by the Governor. One member of the Nebraska Irrigation Association is elected at its annual convention. A director of a watershed conservancy district, watershed district, or watershed planning board is elected at the annual convention of the State Association of Soil and Water Conservation Districts. The State Association also elects five supervisors, four from areas designated by law. Four ex officio members consist of two from the College of Agriculture of Nebraska University, the state's Director of Water Resources, and the Director of the Geological and Conservation Survey Division of the University of Nebraska. The Nebraska Committee may invite both the Governor and the U.S. Secretary of Agriculture to appoint an advisory member. Perhaps more than in any other state, the composition of the Nebraska Committee reflects the determination of the State Association of Soil and Water Conservation Districts to be the controlling influence in a committee which represents many interests having a common concern with the small watershed legislation, Public Law 566.

In the third group of fourteen states where the soil conservation committee, by law, is within some other agency of state government, the members of some serve technically only in an advisory capacity to an elected or appointed state official. The degree of independence of these committees, therefore, is a matter of practice, but in most of the states this is not a live issue. In some states, the arrangement reflects past power struggles, although sometimes the reflection is a strange one. New York provides that the State Committee is "in" the College of Agriculture at Cornell University. Like many of the other committees in the Northeast, its membership is overtly and formally tied to the farm organizations. The Governor appoints five "farmer" members, two from a list of nominees submitted by the New York State Grange, two from a list submitted by the New York State Farm Bureau, and one "at large." The two usual college representatives, together with the Commissioner of Agriculture, the Commissioner of Conservation, and the State Conservationist of SCS serve only as members of an advisory committee. The Massachusetts Committee is in the Department of Natural Resources; the Connecticut and Illinois committees are in the state departments of agriculture. In all three, the committee is "advisory" only. New Jersey's Committee was recently placed under the Commissioner of Agriculture, as it has been in Pennsylvania since 1937. Florida, Kentucky, California, Washington, and Alaska have boards within their respective departments of conservation. Rhode Island's is in the Department of Agriculture and Conservation, with three committee members who are representatives of the Rhode Island State Farm Bureau, plus the Commissioner of the Department of Agriculture and the usual two college representatives. Maryland provides that the State Board of Agriculture, which is also the Board of Regents of the University of Maryland, shall serve as the conservation committee with advisory members from state agencies. Arizona is in a class by itself. In 1945, it abolished a committee consisting of state agency representatives modeled on the standard act and, to satisfy the livestock grazing interests, placed all authority in the Arizona Land Commissioner. His assistant performs all the usual functions of a state committee and executive secretary.

In the matter of practice, however, a majority of these committees formed within state agencies are constituted much like the ones placed above in group two; that is, they draw a majority of members from among supervisors, or other farmers, whether with or without the usual ex officio members. In California, Pennsylvania, and New Jersey, at least, the choice of placing the committee in an existing department of state government was a deliberate stratagem to protect and facilitate the functioning of districts. In Alaska, California, Connecticut, Massachusetts, and Illinois, the committees are advisory only, but in California there appears to be little inclination on the part of the present Director of the Department of Natural Resources to intervene in committee decisions. The Division of Soil Conservation under his direction has the largest staff of any state and appropriations are among the largest in the country. In Pennsylvania, the Conservation Committee is in the Department of Agriculture and has members appointed by the Governor. Its work has flourished in recent years, despite a stormy earlier history. There is no doubt, however, that this kind of legal subordination of a state committee does make it possible for the departmental head or state board concerned (in Florida, for example, the Board of Conservation) to restrict the committee's activities.

The remaining seven committees are in classes four and five. Those characterized by a formal balance in the number of members of state agencies serving ex officio and the appointed public members are Alabama, Mississippi, North Carolina, and Wyoming. The group-five committees, where representatives from state agencies make up a majority of members are in New Hampshire, Utah, and West Virginia. Among these seven committees, the four with balanced membership provide specifically that the public members shall be officers of the state associations of district supervisors. In the others, Utah provides only that the President of the State Association of Soil Conservation Districts shall be an appointed member; West Virginia stipulates that three members be "representative" citizens, but only once in the past ten years have the recommendations of the State Association of Soil Conservation Districts been ignored by the Governor in making appointments; and New Hampshire, which provides for three farmers, in 1955 dropped its requirement that one of these be a district supervisor.

It should be re-emphasized that no single factor in a state, including the formal organization of the state committee, determines the operational environment for conservation districts and the various governmental agencies. But the organizational factor does reflect the power struggles with remarkable accuracy. The provision for representation from the state organizations of the Grange and Farm Bureau, the agricultural colleges, and usually the heads of the departments of agriculture and conservation in the Northeastern states serves as a particularly good example. In the East Central states, conservation district laws providing for public members on the state committees have been turned to the advantage of the districts, as the custom of gubernatorial consultation with the state conservation district associations has grown with the political strength of these groups. In Arizona, Utah, and Wyoming, the lingering suspicions of the livestock growers and some irrigation interests are reflected in a refusal to permit the conservation district associations to have a majority of members on their state conservation committees. The strength of the extension services in North Carolina, Alabama, and Mississippi has left relations in a stalemate which is reflected in their balanced committee membership. In neighboring Georgia, favorable contacts with the dominant wing of the Democratic party have resulted in a committee that is free of fettering ties, although it is heavily dependent upon its political connections. In the Plains states, from Texas to North Dakota and a few of their neighbors to the west, the associations of district supervisors have been strong enough to have state laws rewritten in such a way that the associations control the selection of a majority of all members of the state committees.

The selection of committee members is more clearly involved in the usual political processes in states where gubernatorial appointment is not restricted to the choice of nominees from the associations of district supervisors. Rarely is this factor openly alluded to in the statutes, although the Kentucky law provides that no more than five of the nine members of the Soil and Water Resources Commission be members of the same political party. Indiana's law stipulates that the Governor's appointments shall be made without regard to "partisan politics." Florida forbids the appointment of more than one member from any one congressional district. Among the states in which political clearance is possible, field interviews indicated this was practiced in Pennsylvania, Iowa, and Ohio. It is difficult to determine the extent of political clearance in those states in which the supervisor associations nominate candidates for gubernatorial selection, but the usual provision of law is such that the number of nominees shall be twice the number of vacancies to be filled. This arrangement provides for a kind of flexibility which is generally well understood. State committees, such as the one in Texas, where members are elected by supervisors at the annual meeting of the association are likely to be immune from selection on partisan grounds only.

SCS RELATIONS WITH STATE COMMITTEES

When the heads of state agencies serve ex officio on state committees, they have the opportunity to observe and oppose SCS policies. It is virtually impossible, however, to determine with any assurance how frequently they do so. Nevertheless enough incidents have been reported in the sample of states visited and from documentary sources to suggest that such opposition is sometimes troublesome. Members of the state committees who were not supervisors in both Idaho and Wyoming carried their differences with SCS to the Secretary of Agriculture in 1954. In both states, such committee members were critical of SCS for refusing to assist farmers not living in districts, except to serve farmers outside districts who were referrals from the Agricultural Conservation Program (ACP). Initially the Department of Agriculture agreed that the 1935 legislation did not require SCS to serve farmers only through districts. In 1954, after Secretary of Agriculture Ezra T. Benson reviewed the events leading to the House rider which terminated the SCS demonstration projects in 1945, this interpretation was changed. Donald A. Williams, the SCS Administrator, asserted that Congress had imposed this requirement for SCS to work through districts, and thus Congress would have to make any future changes.[4] Instead of waiting for U.S. congressional or departmental action, Idaho amended its districts law in 1957 to replace the former state committee (which had included the Director of Extension) with one consisting only of three "farmers." In 1961, the Wyoming legislation was finally amended so that four supervisor members were added; as a result, they and the President of the State Association of Soil Conservation Districts provided an even balance with state agencies.

In 1953, the Chairman of the Nebraska State Conservation Committee opposed leaders of the State Association of Soil Conservation Districts who agreed with SCS that state appropriations should be increased. The chairman, E. C. Condra, was also the Director of the Geological and Conservation Survey Division of the University of Nebraska. Condra opposed any increase, he said, because under his

---

[4] National Archives, Record Group 16 (Soil Conservation Service Organization 1), J. Elmer Brock to J. Earl Coke, March 26, 1954; Brock to Coke, May 4, 1954; Coke to Brock, May 27, 1954; C. C. Youngstrom to Coke, July 8, 1954; Youngstrom to C. M. Ferguson, Feb. 4, 1954; and D. A. Williams to Youngstrom, March 29, 1954. (National Archives records cited hereinafter usually will be identified by the abbreviation NARG.)

long tenure as chairman the districts had been "sold" to Nebraskans with the promise that they would not increase the costs of state and local government. His influence was enormous in the legislature because he had cultivated support throughout the state for his Division in the University and he had been the key figure on the State Committee in promoting districts. At first, Condra succeeded in blocking increased appropriations. In 1953, however, he made the mistake of writing to Secretary Benson, telling him that he had held the State Association in line at its 1953 annual meeting so that it did not openly oppose Benson's reorganization of SCS.[5] Whether carelessly, or for some other reason, Assistant Secretary James Earl Coke sent Condra's letter to SCS to prepare a reply. At the 1954 meeting of the State Association of Soil Conservation Districts, held jointly with the State Conservation Committee as an expenses-paid "short course" for supervisors, participants and witnesses recall that there was an uproar. The Association overrode Condra and decided to hold separate meetings at its own expense, to lobby for increased state appropriations and to enlarge the state committee by adding supervisor members.

Although supervisor-dominated state committees are favored in some quarters, their members are not always wholly compliant with either their state associations of soil conservation districts or the SCS. In Oklahoma, a bitter controversy developed in 1953. The SCS was accused of lobbying for the election of certain individuals as officers of the State Association, and of urging the Oklahoma Soil Conservation Board and the Association to oppose appropriations in the legislature for a special project proposed by the agricultural college.[6] More recently, in Georgia, differences between the State Association of Soil Conservation Districts and four members of the all-supervisor State Committee over the refusal of the latter to push for enlarged state appropriations resulted in the Governor's replacement of the four members when their terms were ended.[7] As another example, the Texas Soil Conservation Board has consistently refused to request increased

[5] NARG 16 (SCS, Organization 1), E. C. Condra to Ezra T. Benson, Dec. 11, 1953, supplemented by field interviews.

[6] NARG 16 (SCS, Organization 1), Floyd E. Carrier, Chairman, Republican State Committee, to Whitney Gilliland, Assistant to the Secretary, May 27, 1953, and enclosed report made for the Republican State Committee of Oklahoma.

[7] Field interviews in Georgia during November 1961. Significantly, however, the Chairman of the State Conservation Committee was reappointed. It is reported that this was because he is the son of a man who is commonly reported in Georgia to be the "kingmaker" of Georgia politics—Jim Gillis, Sr., Chairman of the Georgia Highway Commission.

state appropriations, much to the annoyance of the Soil Conservation Service.

Provisions for farmer members rather than district supervisors appointed by the governor do not necessarily shelter a state conservation committee from differences with the leadership of its state supervisors association, district boards, or SCS. In Iowa, for example, there have been differences in the past few years between incumbent members of the State Committee and leaders of the State Association of Soil Conservation Districts over some of the Governor's appointments after the death of a state Secretary of Agriculture who had normally cleared the appointments. The State Committee refused to seek legislative support for a policy allowing districts to spend their administrative budgets without approval of the State Committee, although there was considerable support for this arrangement in the State Association. The present Iowa Committee has heeded, however, the State Association's opposition to an increase in the field staff of the Committee as a means of permitting closer direct supervision of district activities. Similarly, the Iowa Association was reported to have opposed a proposed comparative study of some Iowa districts by staff members of Iowa State University, although State Committee members were said to be divided. Close observers claimed that SCS quietly and "unofficially" sided with the Association in these matters and that changes in the membership of the State Committee followed.[8]

It is impossible, of course, to determine accurately the influence of various groups on the selection of members and the results in policy decisions. There has been a clear and steady trend, however, for the past fifteen years in the creation of a system of selecting district supervisors and members of state committees which is as isolated as possible from the normal political processes provided for the selection of public officers. The decision-making apparatus of soil conservation has been placed largely in the hands of men who control a closed system to which the general public has very little access. This system is insulated from the urban sector of society, largely self-perpetuating, and focused almost exclusively upon farming operations. It may be true that in great measure the public is indifferent, except for occasional sentimental outbursts of enthusiasm for "conservation" from urbanites who love to look upon orderly and well-tended farm lands, or to hunt and fish in well-stocked fields, woods and waters. There is only a little evidence, except in some Northeastern states, that the state conserva-

[8] Field Interviews in Iowa during July 1961.

tion comittees and districts are governed by persons who represent the broad spectrum of interests which ought to characterize governmental units designed to provide the nation with leadership in both soil and water conservation.

RECRUITMENT AND SELECTION OF SUPERVISORS

The membership of the state committees and of district governing boards has a close relationship in most states. For one thing, the supervisors of the conservation districts are the principal sources of members of state committees. In addition, one of the functions which many of the state committees perform is to appoint some of the district supervisors and conduct elections for others. It thus becomes important in evaluating the soil conservation effort to know something about the individuals who serve on the committees and district boards. What are their backgrounds; what are their bread-and-butter interests; what are their attitudes generally toward the policies and administration of conservation programs; and with what organizations are they, or have they been, connected? Not all of these questions can be answered and it is not possible to arrive at a definitive description of a typical board or committee member. However, it is reasonable to assume that most members are rural organization men—leading farmers and farm organization leaders. In order to find out as much as possible on this point, RFF sent a questionnaire to roughly 700 representative districts which included questions about the experience as group leaders of district supervisors during the past decade.[9] Responses from the 278 districts which replied are summarized in Table 1.

The responses to the questionnaire are weighted in favor of the Central and Plains states. Even so, it is evident that district supervisors hold positions of leadership in a very wide range of farm organizations and governmental positions in addition to the district board. On the whole, this is an impressive show of cross-leadership, but certain facts stand out. Most striking is the extent to which the district supervisor is an officer of his state farm bureau, and serves as a member of his local organization sponsoring extension work. The record of participation in the other two general farm organizations—the Grange and Farmers Union—is much less impressive, although it is probable that these

[9] See Appendix B for an explanation of the questionnaire used to secure information on the background of members of soil conservation districts, and the geographical breakdown showing responses to the questionnaire.

TABLE 1. PARTICIPATION OF SUPERVISORS OF SOIL CONSERVATION DISTRICT BOARDS IN OTHER ORGANIZED FARM ACTIVITIES, 1961–62 AND DURING PAST 10 YEARS[a]

Past or present affiliation	Number of affiliated supervisors	Number of districts reporting a supervisor with affiliation
County or state:		
Agricultural Stabilization and Conservation Service committee (member)	177	125
Farm Bureau Federation (officer or director)	288	166
Grange (officer or director)	44	32
Farmers Union (officer or director)	43	31
Livestock growers association (officer or director)	157	111
Commodity growers association (officer or director)	49	34
District grazing advisory board (member)	27	17
County extension or agricultural board (member)	246	128
County governing body (member)	69	44

[a] Based on responses to a special 1961–62 questionnaire for this study sent to soil conservation districts, which is explained in Appendix B.

organizations would have made a better showing if the proportion of responses to the questionnaire had been higher in the areas where their major numerical strength is concentrated. Quite as impressive is the proportion of supervisors who have served on the county or state committees of the Agricultural Stabilization and Conservation Service (ASCS). The proportion of supervisors who have served in positions of leadership in local government and the various commodity and livestock organizations is also high.

This information throws considerable light on the failure of extension leaders and the Department of Agriculture under the leadership of Secretary Benson to decentralize the administration of the Department's technical assistance so long provided by the Soil Conservation Service. Representative Clifford R. Hope of Kansas did not exaggerate when he warned the Secretary that rumors of the Department's proposed reorganization of the SCS in 1953 had greatly disturbed the "best and most influential" farmers.[10] It is also quite possible that the very extensive cross-leadership between conservation districts and the county ASCS committees has helped to harmonize relations between ASCS and SCS. Moreover, there is no question that an outstanding array of leading farmers is committed to soil conservation by their service on district boards. Certainly, any changes in future policy will

[10] See Chapter 6, note 25.

have to take this into account. Undoubtedly, much can be done with the support of these farm leaders, and little without it, where agricultural soil and water conservation is concerned.

Impressive though this record of leadership may be, some pertinent questions remain—especially if districts are considered as potential local sponsors of a wider range of conservation programs. How are the district supervisors selected, for example, and how are they recruited? How responsive are they to the general public?

The standard districts act provided that each state soil conservation committee was to appoint two supervisors and three other supervisors were to be elected by the "land occupiers" of each district following nomination by petition of twenty-five "land occupiers."[11] The standard act made no provision for using existing election machinery, probably because it was assumed that county boundaries would not determine the location of district boundaries. Secretary Henry A. Wallace and his associates in the Department of Agriculture were, of course, also eager to avoid close ties with the local sponsors of the extension work, whether the local Farm Bureau or county governing board.

The general farm organizations and county governing bodies are authorized by law to participate in the selection of district supervisors to one degree or another in some states—chiefly in the Northeastern states but also in a few others. Sometimes curious mixtures of selection systems allow for varying degrees of influence. Both Vermont and Maine have three elected supervisors and two appointed by state committees dominated by the Farm Bureau and the Grange. In Rhode Island, all three supervisors are appointed by the same type of state committee, selected from nominees suggested by the Grange and Farm Bureau. New York provides that two district supervisors shall be members of the county board of supervisors which appoints three other district supervisors—one each representing the Grange and Farm Bureau and the third "at large"—that is, the choice goes to the stronger of the farm organizations, if anyone cares to fight about the matter. As Kent Leavitt, President of the National Association of Soil Conservation Districts (NASCD) remarked in 1947: "In districts, our directors [supervisors] are appointed politically rather than by being elected by the farmers themselves. The result is it is difficult to keep politics out of the work."[12] In New Jersey, the three supervisors of each district are appointed by the State Conservation Committee, and in the past, at

---

[11] *A Standard Soil Conservation Districts Law, op. cit.,* pp. 7–15.

[12] National Association of Soil Conservation Districts, "Minutes of the Meetings of Feb. 25–26, 1947," mimeographed, p. 21.

least, a provision that candidates be nominated by the county agricultural boards meant that the local Farm Bureaus and county agents had a dominant voice in the selection process. Pennsylvania law, since 1945, has provided that each county governing board shall appoint five district supervisors; four from nominations made by "county-wide" farm organizations and one who is a member of the county governing body.

There are some similiarities in the selection procedures in states relying on the farm organizations and county governing boards, but these are superficial in some instances. New York's system insures that district supervisors will not offend either the formal political power structure of the county or the two major farm organizations (the Farm Bureau and the Grange) functioning in the state. No one who is opposed by any of these forces becomes a district supervisor. No other local agency is in a position to sponsor the district supervisors in opposition to the wishes of the county board and the local leadership of the two chief farm organizations. In New Jersey, there is no formal overt tie with the county governing body, since nomination of supervisors is by the county agricultural board, and this board consists of the leadership, or its representatives, of the principal farm organizations. In Pennsylvania, "clearance" with the county board is insured by having one member—normally one interested in the rural interests in his county—serve as a district supervisor. This supervisor often, if not always, has the dominant voice in determining the appointment of four additional supervisors from the list of nominees provided by the farm organizations. The crucial point in this case is that the Pennsylvania Commissioner of Agriculture has authority to determine which "county-wide" farm organizations are qualified to nominate candidates. The list is large, so that county boards are not limited to the choice of the nominees of the politically dominant Pennsylvania Grange. A key feature of the system is the location of this authority outside the Pennsylvania State University and Grange axis of power. Delaware, on the other hand, provides that the county agent shall be the secretary of the district board, ex officio, joined by the chairman of the county levy board and four elected members.

The three New England states—New Hampshire, Connecticut, and Massachusetts—which delayed adoption of districts legislation until 1945 do not provide for formal involvement by either local governments or the farm organizations in the selection of supervisors. In New Hampshire all supervisors are appointed by the State Conservation Committee, which includes the Commissioner of Agriculture, a representative from the agricultural college, and two farmers. Nominations are received from the incumbent district boards and, if there is more than one nomination for each vacancy, the vote is taken at the annual district meeting. Connecticut vests authority completely in its Commissioner of Agriculture, who has passed his authority on to eight areas called "districts," coinciding with former county boundaries. Although the supervisors of these districts are elected at annual meetings, this procedure is based solely on the administrative discretion of the Commissioner of Agriculture, who also has an advisory committee reflecting a variety of interests, including the farm organizations. Massachusetts provides that its State Conservation Committee make initial appointments of supervisors, but that the successors of the appointed supervisors be elected. Amendments to the Massachusetts legislation enacted in 1963 not only avoid ties with local farm organizations in selecting supervisors, but also omit any requirements that supervisors be farmers. They need only to be land occupiers. The names of the districts have been changed from "soil conservation" to "conservation" and given broadened powers. The former State Soil Conservation Committee has been replaced by one reflecting a broader range of interests in the Department of Natural Resources.

With some variations in law and practice, Maryland, West Virginia, and Virginia, like most states of the Southeast, follow the standard act providing for two supervisors to be appointed by the state committee and three to be elected locally. Maryland's system is much like those of the Northeastern states insofar as its State Conservation Committee is composed of the State Board of Agriculture, which is also the Regents of the Maryland College of Agriculture. In West Virginia, the number of elected supervisors is variable, depending principally upon the number of counties embraced within a district. In Virginia, the State Conservation Committee customarily appoints one county agent from within each district to be a supervisor. South Carolina, Georgia, Mississippi, Louisiana, Oklahoma, and Tennessee follow the standard act, but the Alabama State Committee appoints all supervisors. In the Southeast, only Florida elects all supervisors. Overt association with farm organizations in the selection process normally is avoided, although in both Alabama and Mississippi the strength of the agricultural colleges is reflected in the state soil conservation committee membership. North Carolina's system is so complicated and variable, it virtually defies description. Essentially, however, the law provides for the members of county governing boards to serve ex officio, together with three elected members of a soil conservation committee in each county within a district. When a district includes fewer than four counties, the State Conservation Committee appoints two additional members. In districts containing more than four counties, the county committee may appoint one member. Under this system, involvement in local politics appears to be as probable in North Carolina as in New York and Pennsylvania.

Most of the Central and Western states, from Ohio to California, provide for the popular election of all supervisors. There are exceptions, however. In Wisconsin, the agricultural committee of the county board of supervisors serves as the district governing body. District affairs obviously are tied to extension and courthouse politics in this state.[13] Missouri and South Dakota provide that the county agent shall serve ex officio, with four elected members. Indiana, Kansas, Colorado, Utah, Idaho, Washington, and Hawaii follow the standard act. Alaska, much like Connecticut, provides for five appointed supervisors.

In thirty states, some or all of the supervisors either are appointed by or serve ex officio. In the remaining twenty states, all supervisors are popularly elected. Interestingly, however, the total number of appointed supervisors is almost exactly 3,000, while the number elected is approximately 10,700. This apparent anomaly is explained by the fact that many of the states which appoint some supervisors have districts organized on a multiple-county basis, while most of the states which provide for elected supervisors either organize each county as a district or permit more than one district in a county. This information shows why the supporters of the Agricultural Conservation Program administered by the farmer-elected county committees of the ASCS place great stress on their "democratic" character by way of implied contrast with districts in some states.

In the states which provide for the appointment of two or more supervisors by the state soil conservation committee, it is difficult to determine the details with any real confidence of how appointees are

[13] This Wisconsin arrangement has disturbed officers of the National Association of Soil Conservation Districts who have found that such "districts" have virtually no separate identity in public opinion; that supervisors of "districts" hardly view their functions as separate from their primary and other public capacities; and that many of the supervisors are not farmers, and there is relatively little support for the sort of state association of districts which functions, for example, in North Dakota and Oklahoma. State quotas of dues from the districts to NASCD have not been met, and the Wisconsin State Association (like some of its neighbors in the Central states) has not had a notable record of activity or success in lobbying for state legislation and appropriations. At the invitation of the SCS Conservationist an officer of the NASCD spent a few days in Wisconsin inspecting the situation. See National Association of Soil Conservation Districts, "Minutes of the Meetings of March 19–22, 1956," mimeographed, pp. 5–6.

chosen—that is, with whom the members of these committees actually discuss appointments, and why. Representatives of the state committees were asked in the special questionnaire for this study to report the sources consulted before these appointments are made.[14] All respondents, except Vermont and Mississippi, reported that the existing district boards were the chief sources of recommendations. Half added that they relied heavily also on the recommendations of the local personnel of agricultural agencies, especially the county agents and SCS work unit conservationists. In Mississippi and Vermont, the agricultural agencies were named as the chief source.

In West Virginia, there is an unusual effort to seek out and prepare future supervisors, both for appointment and election. Field agents of the State Conservation Committee get in touch with representatives of all the agencies cooperating with districts to obtain their recommendations. These are discussed with the members of district boards, who, in turn, send their recommendations to the state committee. Normally, it is reported, there is complete agreement on the candidates finally recommended by this process. Particular efforts are made to use conservation awards paid by the State Committee out of public funds to identify farmers who have demonstrated in practice their enthusiasm for the conservation program. These men are then groomed, tested, and, if they display the desired qualities, appointed.[15] Alabama's State Committee requires that all prospective appointees fill out and return a qualification questionnaire.[16] In Maryland, a mimeographed policy statement sets forth desirable qualifications for district supervisors. Aside from stating that appointees shall have exhibited "leadership and management in soil and water conservation activities" and related matters, this statement points to categories of persons to be avoided in making appointments. These include those doing business with a district, such as selling supplies, renting property, or contracting for equipment or construction work. Before appointments are made to the Maryland districts, prospective appointees are interviewed to make certain that they understand their duties.[17] No other state reported such a positive effort to make clear the possibilities of conflict of interest on the part of supervisors and, in fact, to avoid it. Of course, in many states the districts have little or no equipment and supply few if any

[14] For an explanation of the special questionnaire for this study sent to state conservation committees, see Appendix B.

[15] Response to special questionnaire sent to state conservation committees: from West Virginia.

[16] *Ibid.*, response from Alabama.

[17] *Ibid.*, response from Maryland, and enclosed policy statement.

materials, although even in those, there is some petty patronage to dispense in the form of appointments of clerks, aides, equipment operators, or similar employees.

The state soil conservation committees are responsible under state law for conducting elections of supervisors which usually are held at different times from those of other regular primary and general elections in the states. Only in Virginia are the supervisors chosen at regular elections under the complete supervision of regular local election officers who report their tallies to the State Board of Elections which publicly reports the results. One result of this system is that the total number of votes cast for supervisors in Virginia apparently greatly exceeds the totals in other states.[18] This pattern of voting behavior is singular in the United States, however, since the limited data available point to low voter participation in supervisor elections in those states which keep any official record on the state level.[19] This pattern of limited participation in part is probably because most elections of supervisors are not held in connection with regular elections of other state and local officers. It is not unusual for an extremely small proportion of the eligible voters to participate in elections for officials of local government. In several states, not even the state soil conservation committee has any record of the votes cast, since the district boards conduct the elections of supervisors in the majority of states, and a locally responsible officer is required merely to certify to the state committee the names of the persons elected. The only official or public supervision of the elections which are conducted by the incumbent boards is in the form of instructions from the state soil conservation committees. The typical practices seem to be like those in the state of Washington where the State Conservation Committee "furnishes instructions on procedure, notice of election for official publication, oath of officer for polling officials and official ballot. The State Committee receives completed tally sheets, Certificate of Due Notice of Election and oath of office. The district handles its own publicity."[20] In short, the incumbent district board members conduct the elections in which

---

[18] *Ibid.*, response from Virginia, which reported that the range in number of votes cast for district supervisors was from 988 to 17,689, depending upon the population of the district.

[19] *Ibid.*, response from the 45 states reporting. Since only the Virginia State Committee provided a tabulation of all supervisor elections, it is impossible to generalize accurately. A few scattered reports from state committees showed a typical range from approximately 50 to 250 votes cast. The higher figure was reported from some of the districts in the Central states.

[20] *Ibid.*, response from Washington State.

either they are returned to office or their successors are chosen.

In Louisiana, one of the states where there are a few variations in election practices, the State Committee examines nominating petitions before the elections to insure compliance with law and administrative practice, and pays district supervisors expenses of $4 a day for sitting as judges of the elections—often their own re-election. The South Carolina State Committee appoints and pays a local election judge. The Mississippi elections are conducted by the Executive Secretary of the State Committee and the appropriate county agent—a procedure which suggests that the Extension Service has influence in the state. South Dakota varies this procedure somewhat by having the district boards and the county agents conduct the elections. A half dozen or more states hold elections of supervisors at annual district meetings to which the public is invited. Among these are Minnesota and North Dakota, in which the district and township boards hold the election. In Georgia, Vermont, and Nevada, other regular local officials perform this function, and inform the state committees of the names of persons elected. California provides for election by the district supervisors and the county clerk; but if no one runs for the office, or there is only one candidate for a vacancy, the election may be dispensed with; or, if a proper nominating petition is not presented at the required time, the county board of supervisors may appoint the single nominee, or lacking a nominee, may appoint any qualified person.

Nearly all the state soil conservation committees reported that they attempt to encourage qualified men to run for the office of district supervisor, although most are careful to disclaim any intention to meddle or to dictate to the district boards. A theme often repeated in reports and interviews was stated in the response from the South Dakota State Committee: "Soil and Water Conservation Districts are local subdivisions of State Government. They are organized by local people and operated by local people. They should make their own choice of supervisor."[21] Nevertheless, only a very few state committees claimed flatly that they do not give any encouragement to incumbent boards to cultivate qualified successors to the supervisors. The usual practice is for the executive secretary or a field representative of the state committee to discuss this matter and to suggest ways of finding willing candidates. One method used by a dozen or so state committees is the appointment of either an advisory committee or assistant supervisors without vote. Another is for the state committee or districts

[21] *Ibid.*, response from South Dakota.

association to hold occasional training sessions in which the matter is discussed as part of a formal agenda. Perhaps the commonest device, however, is similar to that used by West Virginia—to make some type of annual conservation award in a district to some farmer with an outstanding record of practices applied. It helps, of course, if he is also active in local farm organizations and is prominent enough to have served on the county governing board, the school board, county ASC committee, irrigation district board, or in similar positions. Use of these awards ranges from subtle to direct as a means of cultivating both supporters and future supervisors of districts.[22]

In many states, it is sometimes necessary actively to seek out men who are willing to serve and who understand their duties and responsibilities. One executive secretary of a state committee in the West remarked that not more than 10 supervisors out of nearly 300 in his state understood and properly performed their duties. Another executive secretary in a different part of the country said privately that he opposed the reorganization of multiple-county districts into single-county districts because it was hard enough already to find good supervisors. Some of the more thoughtful leaders among supervisors have privately indicated their desire to attract more businessmen with agricultural interests to the ranks of supervisors, since the urban population supplies much more sophisticated leadership. Equally in private, the personnel of some of the agencies have expressed their dismay at the differences in the quality of urban and rural leadership. This matter is now coming to the forefront because of the expanded coverage of the Small Watershed Program.

It is difficult to say to what extent, if at all, the personnel of various governmental agencies influence the choice of district supervisors. Certainly, so far as the Soil Conservation Service is concerned, official instructions are clear and unequivocal: "SCS employees are not in any way to be involved in or concern themselves with the election or selection of the governing bodies of soil conservation districts."[23] Despite this directive, however, personnel of SCS have some influence on the selection of district board members. It is impossible to say

---

[22] This issue is the subject of considerable discussion among the executive secretaries of the state soil conservation committees at their annual meetings, which are held with the convention of the National Association. At the 1956 and 1961 NASCD sessions, for example, this matter was discussed. The West Virginia system was described in detail at the 1956 meeting, and remarks were mimeographed for interested parties.

[23] U.S. Soil Conservation Service, Districts Memorandum SCS–2, Dec. 22, 1954, p. 2.

exactly how widespread the influence is, but evidence was found during field interviews in six out of thirteen districts located in Pennsylvania, Virginia, Georgia, Louisiana, Texas, Ohio, Iowa, Nebraska, and Arizona. Particularly well-informed observers in three Southeastern states said that this SCS influence on selection of supervisors was common practice. Most SCS state conservationists denied it; two admitted that they would like to get rid of lazy or indifferent supervisors, but that the risks were worse than the disease. The area and work unit conservationists, however, especially the former, are responsible for maintaining satisfactory working relations with district boards. Under normal conditions, they have literally dozens of opportunities annually to discuss the selection of a new supervisor with the members who tend in most districts to lean on SCS personnel for advice concerning district affairs.

SCS AND ADMINISTRATION OF THE DISTRICTS

In 1952, W. R. Parks completed his study of soil conservation districts during their first fifteen years of operation and concluded that they had not demonstrated their "value in future agricultural administration." Mindful of the bitter feelings among agency personnel, Parks gave districts the benefit of his doubt, judging that they had "promising potentialities" for integrating lay and expert opinion in conservation programs. "District administration" had not yet become to a sufficient degree "supervisor administration," in his judgment. "Even the stronger district boards, in their management activities, depend heavily upon the professional workers."[24]

At about this time, Hugh H. Bennett's long, colorful, and controversial career as head of the Soil Conservation Service was ended. Although Bennett reached retirement age in April 1951, his service as SCS chief was extended until November, when his illness prompted Secretary Charles F. Brannan to appoint Robert M. Salter to Bennett's old position. This appointment amazed the officers of the National Association of Soil Conservation Districts; they were not consulted and did not know Salter, who had served for many years in the Bureau of Plant Industry.[25]

[24] W. Robert Parks, *Soil Conservation Districts in Action* (Ames: The Iowa State College Press, 1952), pp. 105 and 224.
[25] National Association of Soil Conservation Districts, "Minutes of the Meetings of Oct. 29–31, 1951," mimeographed.

Salter quickly learned that Parks' analysis of district administration was sound and that relations between SCS and the districts required several adjustments. One of the most important of these was the role of SCS in the internal administration of district boards. The need for district operating funds also gave rise to vexing problems, and there were other difficult, but important, issues.

Brannan and Salter were disturbed by reports of unsatisfactory relations between SCS and districts, especially since a number reached the Secretary through members of Congress. Senator Tom Connally, Democrat of Texas, forwarded a letter from the McLennan County District; the District Board complained that it could not get along with the local PMA County Committee and staff because of interference by the SCS Area Conservationist. The Board said its relations with PMA could be greatly improved if the position of area conservationist were eliminated and work unit conservationists were permitted to function freely with district boards.[26]

Far more damaging and specific was a long and angry letter addressed to Salter and the Secretary from the Chairman of the Calcasieu Soil Conservation District in Louisiana. He detailed a series of events considered by the members of his District Board to be intolerable acts of interference by SCS personnel in the Board's proper activities. The controversy had become so bitter as to produce a state of open hostility between the two parties. As a result of this report, Salter sent one of his assistants, Donald A. Williams, accompanied by Fred Ritchie of the PMA-ACP Branch, to Louisiana to conduct an investigation. They were joined by Nolen Fauqua and Herbert Eagon, both officers of the NASCD. At a meeting with the parties concerned, agreements were worked out whereby SCS transferred "without prejudice" one of the offending work unit conservationists and an SCS aide who had once served as an employee of the Calcasieu District. The District Board agreed to force the resignation of its "district manager," and to schedule all future technical services by SCS and the District on the basis of area priorities. Salter said, although there was "inconclusive" evidence that SCS personnel had solicited candidates for the office of district supervisor, "it is not to be done in the future." The District had threatened to cancel its memorandum of understanding with SCS if conditions were not corrected to its satisfaction. Salter countered with a promise to do the same thing if the District Board did not meet the Department's conditions. The terms of the treaty were

[26] NARG 16 (SCS, Organization 1), Dave Simons to Senator Tom Connally, Jan. 26, 1952.

agreed to by the parties and "peace" was declared at the next meeting of the Board on May 15, 1952.[27]

Senator Edwin C. Johnson, Democrat of Colorado, forwarded a complaint from a conservation contracting firm, charging that SCS work units in Colorado were directing the use of district equipment in competition with private business. Details were given of instances in three districts in which personnel of SCS supposedly had persuaded farmers to use district equipment, rather than deal with the private contractor, because the districts' rates were lower. On one occasion, the firm had lost $350 because an SCS technician had persuaded a farmer to stop in the middle of a job and bargain for private rates in keeping with the lower district rates. On other occasions, it was alleged, district equipment was used on farms situated outside of an organized district. In one instance, an SCS employee operated a district "cat," it was charged. The Secretary's reply was that an investigation had been made and revealed no violations of SCS policy. The SCS personnel was permitted to quote rates, mention the availability of contractors, give other information, and were often, of necessity, the first persons to discuss such matters with farmers. These actions did not, in the Department's view, constitute SCS "management" of district equipment, although its employees "are intimately associated with the preparation of plans and designs for particular jobs to be done by the district and people generally associate Soil Conservation Service personnel with the district very closely."[28]

In July 1952, the Acting Chairman of the Grady County District in Oklahoma demanded that the SCS State Conservationist be removed for investigating the records of the District without the District Board's permission and after its Clerk had been sworn not to reveal the event. The Board's spokesman said that the State Conservationist had charged the District Board with "corruption in the handling of flood control grass seed. His charge is absolutely false." The State Conservationists had "turned loose a barrage of loose talk which smears and blackens the name of" the Board and its employees.[29]

[27] NARG 16 (SCS, Organization 1), Robert D. Shaefer to Robert M. Salter, Feb. 22, 1952, with copy to Secretary Charles F. Brannan; Salter to Shaefer, May 12, 1952. Calcasieu Soil Conservation District, Minutes of the Meeting of Feb. 20, April 16, April 25, and May 15, 1952, mimeographed (the District held no meeting in March 1952).

[28] NARG 16 (SCS, Organization 1), Ted D. Smith and W. E. Lucas to Senator Edwin C. Johnson, March 7, 1952, forwarded with covering note to Secretary Brannan, March 11, 1952; reply, Knox T. Hutchinson, Assistant Secretary of Agriculture, to Senator Johnson, April 10, 1952.

[29] NARG 16 (SCS, Organization 1), H. U. Golty to Secretary Charles F. Brannan, July 29, 1952.

Other instances of conditions in the field were reported. A former work unit conservationist charged that the State Conservationist in South Dakota had pressured the Brookings County District Board into paying dues to the State Association of Soil Conservation districts instead of putting the money into a sinking fund to buy replacements for worn-out equipment. The machines belonged to SCS, he reported, but were operated by the District, which built up its own funds from charges to farmers for its use. He claimed that the District Board strongly objected to this diversion of its funds and that when he had protested the State Conservationist's action, he had been transferred over the protests of the Board's members. He asked whether an employee of a "line agency" must close his eyes to "questionable" practices, or was there proper redress? He said that he left SCS in 1950.[30]

Another work unit conservationist wrote to Ezra Taft Benson outside of "channels," calling soil conservation districts ["SCD's"] an SCS "propaganda device" and a "front." He said, "Very few Districts do much of themselves without the SCS people there suggesting, needling, and even doing most of the work for them. Most of the reports of the SCD's are written by SCS men. In my years with the SCS I have never seen a SCD that is independent or self governing." He claimed that he had recently attended an SCS state meeting and that the "point of the conference was: The SCS is being accused of using the SCD's as a front for their activities. Deny it. But use the SCD's for a front, nevertheless."[31]

It must be emphasized that these allegations do not constitute legal evidence. However, Salter apparently found sufficient need for changes of policy to effect an immediate reorganization of SCS and to report his intentions to the NASCD. He told the NASCD Board of Directors in mid-1952 the "time had come when there should be a re-statement of policy concerning working relations" between the two groups. He suggested that SCS prepare such a statement in collaboration with the NASCD and announced that he had already taken steps to reorganize the Soil Conservation Service.[32]

Salter remained as chief of SCS only until November 1953. During his short tenure, however, he initiated a revised supplemental memorandum of understanding with the districts. The new SCS Adminis-

[30] NARG 16 (SCS, Organization 1), J. T. Paulson to Ezra T. Benson, Dec. 1, 1952.

[31] NARG 16 (SCS, Organization 1), letter [name withheld] to E. T. Benson, Jan. 7, 1953.

[32] National Association of Soil Conservation Districts, "Minutes of the Meetings of May 27–29, 1952."

trator, Donald A. Williams, had been Acting Chief of ACP in the ASCS on detail from his position of Chief of Operations of the Soil Conservation Service. (He had previously served as Assistant Regional Conservator in the Pacific Northwest, and later was in the Office of the Secretary of Agriculture.) During his temporary tenure in ACP, Williams had been responsible for effecting some major changes which were intended to increase the emphasis on "enduring" conservation practices, commencing with the 1954 program year. Shortly after assuming his new position as Administrator of SCS he, like Salter, took up the problem of relations with the districts. Within a few months after assuming his duties, Williams issued the first of a series of policy directives which reflected his continuing concern to improve relations.

Williams' first move was explicitly to direct SCS personnel to refrain from handling districts' funds.[33] By December 1954, there had been sufficient time to review relations and to prepare and issue a more extensive statement of relations, although it appears that, in part, this directive was prompted by the Secretary of Agriculture's Memorandum 1368 dealing with the relations of departmental employees with farm organizations. Williams emphasized that: SCS assistance was to be made available in accordance with an annual schedule, after a joint review of district needs and SCS resources; SCS would work primarily, but not exclusively, with districts; SCS responsibilities for ACP activities were to be discharged regardless of whether a farmer or rancher concerned had a cooperator's agreement signed with a district; SCS personnel was not to secure signatures on farmer-district agreements, become involved in supervisor elections, or deal in any way with the administrative problems of districts. The Administrator's memorandum made clear that SCS policy was to "continue . . . to encourage the maximum of self-reliance on the part of district governing bodies."[34]

In September 1955, Administrator Williams again raised the issue of SCS-district relations by making it the theme for the annual planning conference. This matter had been pushed to the forefront because SCS had just assumed responsibility for the Small Watershed Program, Public Law 566, to the distress of some NASCD leaders who had already voiced the fear that "watersheds" might soon become the "tail that wags the dog."[35] A joint committee of SCS and NASCD members

---

[33] U.S. Soil Conservation Service, Districts Memorandum SCS-1, June 16, 1954.

[34] Districts Memorandum SCS-2, Dec. 22, 1954, *op. cit.*

[35] At the meeting of the Board of Directors of NASCD in May 1954, the prospective Small Watershed Program, later enacted as Public Law 566, was discussed and fears were voiced that districts might be by-passed (NASCD,

was formed at this time to study the problem. The Committee, during 1956, produced a report which was reviewed and approved by both parties at the top level. Concurrently, Williams issued a series of memoranda in which he told personnel that the SCS would work with districts in the future, but not at the price of subordinating or abandoning its new program responsibilities. He said in substance that SCS had moved away from its former policy of providing technical service only for "district cooperators" and sometimes giving service reluctantly to farmers wanting single practices subsidized by the conservation program of the ASCS.

The NASCD-SCS joint committee barely had time to act before SCS was assigned complete responsibility for administering the Great Plains Conservation Program in 1956. This assignment alarmed some district leaders because SCS administers the long-term contracts between the Department of Agriculture and individual farmers, leaving the district boards with no administrative functions in connection with this program.[36] Also, SCS was given responsibility for the technical supervision of some practices authorized under the conservation reserve portion of the Soil Bank Program in 1957. This move annoyed those district leaders who thought it would reduce further the amount of time the SCS technical staff had available for the "district programs." Many felt that this program—like the assignment of technical responsibility to SCS for some practices under Secretary Brannan's Memorandum 1278 in 1951—would dilute the quantity of service available exclusively to "district cooperators."

In order to counter the "fears that additional assignments to the SCS will divert attention from the primary job," Administrator Williams in his policy statement outlined a number of functions in which SCS state conservationists should consult with their respective state soil conservation committees and state associations of districts. These were: (1) Supervisor "handbooks" should be developed to provide up-to-date information to include the duties, powers, and responsibilities of

---

"Minutes of the Meetings of May 3–8, 1954," p. 7). Administrator D. A. Williams attempted to set SCS personnel and the NASCD at ease by saying that districts would "be used to take advantage of the newly authorized approach to the watershed phase of soil and water conservation work." See U.S. Soil Conservation Service, Districts Memorandum SCS-3, Dec. 20, 1955.

[36] U.S. Soil Conservation Service, Districts Memorandum SCS-4, Dec. 12, 1956. Williams stressed "our continuing policy" of carrying out new responsibilities, other than the Great Plains Program, through districts. All of these activities "help to implement the district's program. . . . Furthermore, many potential district cooperators can learn to understand and appreciate the district's program through contacts made while providing services through the additional authorities."

district boards, procedures for elections, limits of the "supervisor's responsibilities in relation to SCS administration," relations with the Extension Service, and the responsibilities of ASC county committees. (2) Attention should be called to the need for state committees to "assist district governing bodies in becoming better acquainted with their duties and responsibilities." (3) The state committee and association should be informed about changes in the SCS program as they affect SCS operations with districts. (4) Attention should be directed to the experience of several states which "indicates some very strong points in favor of all States providing soil conservation district funds for district operations." Williams said such funds would not only stimulate "a more active and positive interest" among supervisors, but also permit an expansion of SCS technical assistance as districts "perform these clerical and managerial jobs." Boards should be urged also to review other sources of technical assistance available. Within the SCS, work unit conservationists should be directed to the "desirability of reviewing the duties of both organizations; representatives of districts should be invited to attend "appropriate" SCS meetings for "planning conservation operations." In addition, SCS area conservationists should be directed to maintain "active contact" with district boards by attending one or more of their monthly meetings each year for the purpose of supervising directly the relations between work unit conservationists and the boards. "Working relationships with districts should be part of the measured job of the work unit conservationist."[37]

SUMMARY AND COMMENT

Theoretically, the standard soil conservation districts law provided for a federal-state-district cooperative program of erosion control. Since the beginning of the relations of the Department of Agriculture with districts, they have received both technical assistance and administrative guidance almost exclusively from the Soil Conservation Service. Early efforts to provide joint SCS–Extension Service program planning for districts were given up in 1940. The state soil conservation committees established by law to organize districts, hold supervisor elections, and assist them with "advice" were not intended by the Department to provide administrative guidance for districts.

Between 1945 and 1952, the rivalry of the Soil Conservation Service,

[37] U.S. Soil Conservation Service, Districts Memorandum SCS-6, Dec. 13, 1956.

the Production and Marketing Administration and the state extension services in seeking control of soil conservation policy and administration was intensified. As a result, in order to undercut the influence of the agricultural colleges, state legislatures successively changed the composition of the state soil conservation committees so that the majority of committee members were district supervisors. These state conservation committees, like district boards, are now composed of leading farmers and farm leaders, but the system of recruiting and selecting committee members tends to isolate them from the general farm organizations—with a few notable exceptions. The system of selecting supervisors, as well as conservation committee members, minimizes the influence of political party organizations in the process of recruiting them as public officeholders.

The intimacy of the relationship between district boards and the Soil Conservation Service became the subject of increasing criticism, in many cases from supervisors themselves, during the 1952 presidential election year. Whether any of the allegations of SCS interference in district administration were prompted by the prospects of a change in the party controlling the national administration, or for other reasons, is not clear. Nevertheless, since 1953, when the Republicans took over the executive branch, the NACD and the SCS have held numerous joint meetings to develop policies which will permit SCS to withdraw its administrative assistance to districts. For the first time since they were created, the state soil conservation committees are being urged to provide districts with both improved administrative guidance and increased financial aid. In view of their 30-year history and the rapid decline in farm population, the extent, if any, by which state soil conservation committees will alter the Department's association with districts and make the relationship a meaningful federal-state-local one is still uncertain.

CHAPTER 9

State-District Relations

THE DEPARTMENT OF Agriculture officials who first conceived the idea of soil conservation districts believed that the districts would not be acceptable to state legislators and farmers if they were given the power to tax, borrow, or assess and collect benefits. These powers would make the districts as unpopular as prohibition, in M. L. Wilson's judgment.[1] His fears were confirmed in more than one state, but the situation in Kentucky can serve as an example.

When the Kentucky Farm Bureau voted in 1939 to oppose the standard districts law, its Executive Secretary explained that the opposition stemmed from a dislike of "coercion" and a preference for an "educational" program under the county agents, modeled after the arrangements supported by the Tennessee Valley Authority (TVA). "Another alarming feature" of the proposed law was that it would create a "new division of government," the spokesman continued, when "we have consistently fought for fewer units of local government" and lower property taxes. He conceded that while districts might continue to receive all of their financial support from the federal government, "they can very easily become pressure groups to demand local and state appropriations, with resulting higher local and state taxes." The Kentucky Farm Bureau was very much in favor, however, of the program of the Agricultural Adjustment Administration (AAA), for which over "$1,000,000,000 are being appropriated to conserve our soil. . . . Tremendous progress is being made under this program."[2]

Wilson was not wrong in 1935, when he said that for the "next two years, at least, it is likely that the operations of these . . . Districts must be financed chiefly with Federal funds."[3] However, neither he nor others in the Department accurately gauged the strength of the opposition to state and local financial assistance to districts. Even more

---

[1] National Archives Record Group 16 (Erosion File), M. L. Wilson to Charles W. Eliot, 2nd, Oct. 2, 1935. (National Archives records cited hereinafter usually will be identified by the abbreviation NARG.)

[2] NARG 16 (Soil—1), Ben Kilgore, Executive Secretary, Kentucky Farm Bureau, to H. K. Gayle, State Coordinator, Soil Conservation Service, Oct. 24, 1939.

[3] NARG 16, Wilson to Eliot, *op. cit.*, Oct. 2, 1935.

important, they did not anticipate that the AAA conservation payments which were then under discussion would be adopted and become a major feature of the Department's soil conservation program for the next thirty years or more.

Only in Colorado and California were districts authorized to levy taxes, and, in California, it is quite probable that the power was included in the districts law to discourage farmers from organizing them rather than to raise revenue for district operations. As late as 1953, there were sharp skirmishes between the Soil Conservation Service (SCS) and the California Extension Service which opposed activities to organize some new districts. County agents used the argument that districts would increase the burden of property taxes unnecessarily, since the California Extension Service was prepared to render comparable services without districts.[4]

In addition to federal assistance, other sources of income for districts have been earnings from operation of conservation equipment, contributions from various sources, appropriations from state governments, and regular or occasional appropriations from county governing boards. Of all these methods, the ones now preferred by the Soil Conservation Service are regular state or county appropriations. In order to understand the reasons for this preference, the extent to which it had been realized by 1964, the purposes for which districts use funds, and the relation of these matters to the Department of Agriculture's subsidies under the Agricultural Conservation Program (ACP) some details delineating the character of state-district relations are necessary.

FINANCING DISTRICTS:
CONTRIBUTIONS AND ASSESSMENTS

Since the Department of Agriculture frowned on taxes as a source of funds for soil conservation districts until 1948, it supplied conservation equipment for use by districts either as loans on generous terms or through outright grants to the governing boards. Earnings from operating equipment became a valuable source of income, and this probably was the chief source for most districts until state and county appropriations and contributions were made available to them. However, information about district finances is so very sketchy that any conclusion on this matter is speculative. Once Congress terminated appropri-

[4] NARG 16 (Soil Conservation Service, Organization—1), William H. Alison, Director of Merced County Extension, to J. Earl Coke, March 24, 1953.

ations to the Soil Conservation Service for purchasing equipment to lend or grant to districts, no earnings from equipment operations were available to districts organized after about 1948. Since the cutoff, SCS has urged districts not to operate conservation equipment in competition with private contractors. The exceptions to this policy are cases where it seems necessary to provide relatively scarce items or to bring down high charges by the locality's private contractors. This policy is the product of a complex set of factors which have led SCS to urge districts to secure their funds from county and state governments. Of them all, the most important factor has been the identification of SCS with district activities to such an extent that almost any practices by the district reflect on SCS.

Some districts which needed funds, but had no equipment program, formerly raised funds by adding a fee to the charges made by private contractors to farmers for conservation practices requiring earth-moving (such as ponds). Soon these districts were accused of demanding "kickbacks" from contractors who were installing practices with SCS technical assistance and ACP subsidies. The Soil Conservation Service was greatly embarrassed because it appeared that the fees added by districts were increasing the costs of ACP subsidies, financing district activities with ACP funds, making farmers provide funds for districts as a condition for receiving their ACP subsidies, and creating the impression that farmers had to pay for federal technical assistance—which, in fact, is available free of charge. When this practice was uncovered, SCS Administrator Donald A. Williams in 1957 directed his subordinates to review relations with supervisors and to caution them against the unfortunate consequences of following this policy.[5] Nevertheless, districts in Arizona and Colorado were making assessments as late as 1961.[6]

Some other district boards raised funds, not by making equipment available to farmers at a nominal rate, but by selling it a short time after it was acquired from the federal government with SCS assistance. The manufacturers of farm and construction equipment and dealers became alarmed; the dealers were especially annoyed, since sales of surplus equipment made by districts reduced the opportunities for

[5] U.S. Soil Conservation Service, Districts Memorandum SCS-7, June 5, 1957, and attachment.

[6] Gila Valley Conservation District, *Nineteenth Annual Report* (Safford, Ariz.: 1960), and Redington, Willcox, and San Simon Soil Conservation Districts, *1960 Annual Report* (Willcox, Ariz., jointly published). See also, Colorado State Soil Conservation Board, "Supervisors Guide," mimeographed, Denver, Feb. 1962, p. 13.

retail merchants to sell through normal trade channels at a profit. In addition, both SCS and the National Association of Soil Conservation Districts (NASCD) were concerned.

For several years, NASCD had urged conservation districts to enter its "Dealer-District Conservation Program" as a means of obtaining financial and promotional assistance from farm equipment dealers. Typical of these is the program in Texas where the State Association of Soil Conservation Districts, bankers, newspapers, and other interested parties have formed a State Steering Committee to help district boards to initiate local programs. It is frankly recognized that "implement dealers are salesmen and they can help sell soil and water conservation." The program is based on the argument that conservation "pays— not only to the farmer and rancher but to the town or city businessman whose products farmers and ranchers buy." The dealer who agrees to join the program can earn "conservation awards" in various ways. He may "discuss" conservation with "five good farmers" to find out from each whether he is a "district cooperator," how much "conservation farming" he has done, and what his reactions are to the soil conservation program. Or the dealer may join with the district board in advertising in district newsletters or local newspapers to promote soil conservation. He is required annually to undertake two additional promotional activities, such as demonstrations or tours. He may show farmers a film, "Partners in Profit," available through the district; maintain an appropriate display of pamphlets and other materials "available from manufacturers, related commercial sources, or government agencies," or participate annually in a variety of meetings held with a district board.[7] A dealer who enters this program receives an award in the form of a decal, preferably at a ceremony given suitable publicity; then he is normally invited by the district board to become an "affiliate" member of the district by making a contribution, which is often $25 dollars or more. A few affiliate memberships can be of great help to a district board which is attempting to raise its quota of dues to the state and national district associations.

The NASCD also has relied on the interested support of the large equipment manufacturers for contributions to its budget. It was with some dismay, therefore, that the NASCD Board of Directors received a report in 1959 that the major manufacturers of heavy equipment were working to discourage, or prevent, the sale of federal surplus equip-

[7] "Texas Dealer-District Conservation Program," Texas Association of Soil Conservation Districts, Temple, Texas, mimeographed.

ment on the market, even though districts were not special targets of this campaign.[8] The Soil Conservation Service also was placed in an awkward situation by the districts' fund-raising efforts with dealers. For example, its relations with the Caterpillar Tractor Company had changed materially since 1935 when two of the company's southeastern representatives urged that the soil conservation program be administered through the state extension services instead of the federal government. (See note 7, Chapter 2.) Since then, the Caterpillar Company has promoted the Small Watershed Program with advertisements in magazines of national circulation; it has produced a handsome color movie and an equally impressive brochure picturing the benefits of small watershed projects.

Both the NASCD and SCS reacted to this problem of relations with manufacturers and dealers by initiating measures to improve the administration of finances and equipment by the district boards. In December 1958, the Administrator issued a policy directive intended to curtail severely the sale of surplus federal equipment to soil conservation districts. He authorized state conservationists to continue to give SCS assistance in locating equipment for districts, but only under strictly defined limits. Purchases must be made at no cost to SCS, and only for districts which had demonstrated a special need for services which were not available through private contractors. "It is the policy of the Service to encourage cooperating farmers and ranchers to deal directly with independent contractors in applying conservation practices requiring heavy equipment." However, SCS does recognize that there are areas where circumstances "would necessitate contractor charges in excess of a fair rate for the community. In such cases districts have been able to acquire and operate their own equipment to the advantage of all." The Administrator said that under current policies a state conservationist could assist a district in locating surplus equipment, but only with the understanding that the equipment was not to be used in work of a "commercial type" or leased or rented to private contractors or other units of local government. No district could obtain such equipment until it had made a plan of operation and demonstrated its capacity for operating and maintaining the equipment without any direct assistance from SCS personnel; SCS employees were forbidden to manage, operate, handle in any manner, or route such equipment. If a district acquired surplus equipment, the equipment

[8] National Association of Soil Conservation Districts, "Minutes of the Meetings of Oct. 18–21, 1959," mimeographed.

must be retained for the duration of its economic life or until there is no further need of it. Districts were expressly forbidden to resell such equipment "to obtain operating funds." All sales made during the first year after the purchase, or grant, of equipment must have the written concurrence of the state conservationist, and sales after a year's ownership must be preceded by thirty days' notice in writing to him.[9]

The problem of district financing was by no means a new one for SCS. In 1957, for example, Administrator Williams had required state conservationists to estimate the total local, county, and state funds made available to districts. And he followed that request with a policy statement urging SCS personnel "to encourage Soil Conservation Districts to use assistance available from all agencies, groups and organizations that have any type of assistance to offer." He stressed that "SCS resources are used more efficiently and more effectively in areas where local, county and state funds are available to a Soil Conservation District to complement our technical assistance."[10] Again, in 1959, Williams reiterated SCS policy that there "shall be no fees or donations required for personal services and materials furnished by the Federal Government to cooperating farmers through soil conservation districts." He conceded that district administration of state, county, and private funds in support of their programs "is outside the responsibility of the Federal Government." He warned, nevertheless, that SCS had a continuing concern with district financial operations under the following conditions: (1) When arrangements require compensation for personal services and materials furnished free by the federal government; (2) when districts assess fees or charges at such "inequitably high rates" for "nonfederal" services and materials through districts as to increase the cost to the federal government of work accomplished through the ACP cost-sharing system; and (3) when a district utilizes employees of SCS to handle its business affairs. Under no circumstances, Williams emphasized, should districts make the payment of a fee or a "donation" to a district for "non-federal services" a condition either "directly or indirectly" for receiving federal assistance. Such practices, Williams said explicitly, are "improper" and a "misuse" of federal resources. He directed state conservationists to call the attention of district boards, state soil conservation committees, and associa-

---

[9] U.S. Soil Conservation Service, Administrative Services Memorandum SCS-16, revised, Dec. 8, 1958.

[10] U.S. Soil Conservation Service, Districts Memorandum SCS-8, Nov. 22, 1957.

tions of districts to the terms of their memorandum of understanding with SCS and to cancel the memorandum of understanding with any district which violated the terms.[11]

Criticism of district administration reached Congress, as might be expected. Following a request by the Senate Committee on Agriculture and Forestry for an investigation by the General Accounting Office, a report was completed in March 1961, but was not made public.[12] The NASCD Board of Directors recognized the seriousness of the situation and discussed district administrative problems at its meetings in October 1959. Although there was some inclination to lay the trouble at the door of personnel in the Production and Marketing Administration (PMA) and in reorganized agencies which came after PMA, and to view criticisms of the districts as being inspired by the smoldering conflict between the SCS and PMA adherents, it was agreed that the time had come to urge state soil conservation committees to improve district administration. A new NASCD committee—the Committee on District Operations—was created to work on the assignment. In addition, a committee which had been working on a statement of relations with SCS reported that it had prepared and would distribute a brochure, "Working Together." And in an effort to secure more complete and accurate information on district financing, the NASCD Board decided to send a questionnaire to districts to determine the proportion, but not the amounts in dollars, of funds received from various sources.[13]

FINANCING DISTRICTS:
APPROPRIATIONS AND EARNINGS

Soil Conservation districts do not receive federal appropriations. The NASCD consistently supports requests for increased federal appropriations to the Soil Conservation Service to expand its technical assistance to farmers "through" districts, but the national association has never requested direct national grants to districts. The most obvious reason for their failure to do so is that conditions would probably

[11] U.S. Soil Conservation Service, Districts Memorandum SCS-10, June 10, 1959.

[12] Comptroller General of the United States, "Operating Practices of Soil Conservation Districts," audit report to the Senate Committee on Agriculture and Forestry, No. 114833, March 9, 1961. See also, *Department of Agriculture Appropriation Bill for 1960*, Senate, 86 Cong. 1 sess. (1959), pp. 340–43.

[13] NASCD, "Minutes of the Meetings of Oct. 18–21, 1959," *op. cit.*

be attached to their use so as to maximize the availability of technical service to farmers. It is doubtful, however, that either the Department of Agriculture or the Soil Conservation Service wishes to see district boards assume responsibility for administering any federal funds which would strengthen the centrifugal forces that weaken departmental administration. It is also doubtful that the Soil Conservation Service wants districts to administer the Agricultural Conservation Program in place of the county committees of the Agricultural Stabilization and Conservation Service. The NASCD has expressly opposed occasional suggestions that districts assume this burden and the reasons are not hard to find.[14]

The official attitude of the NASCD is illustrated by the action of the Thomas Jefferson Soil Conservation District in Virginia in 1952–53. Under the chairmanship of E. L. Bradley, who was one of the organizers of the NASCD in 1946, the District Board issued a letter to its "friends and supporters" asking them to "help acquaint the public with what has been done and what is being done. It needs your assistance in telling your neighbors how they can improve their farms and their income by Farming the Conservation Way. We need some money, too, to promote educational activities and to help inform the public about the District's work." Bradley suggested that the recipients "make a contribution not to exceed $10.00." The reason why this District's Board was begging for contributions from "friends and supporters," instead of financing its operations with public funds derived from the tax base, was given in the letter:

> The District is set up under law as a local, self-governing unit. It receives no funds from any governmental agency, and is *responsible to no authority other than the people of the district*. . . . Your district is in need of assistance *if we are to continue operating independent of Federal funds and control*.[15]

Since soil conservation districts do not receive any funds from the federal government, nor administer any funds for it, the districts are

[14] NASCD, "Minutes of the Meetings of March 17–20, 1958," mimeographed.

[15] Thomas Jefferson Soil Conservation District, mimeographed letter, Nov. (no day), 1952, emphasis added. On January 23, 1953, the District's supervisors sent out another duplicated letter to the "Contributors" to the District, again saying that: "We supervisors hoped we could continue to operate independent of governmental Funds and control thereby continuing as a local self-governing unit. You agree with us and by your contributions make it possible to continue. . . . Knowing we have your backing will help us through these flat times when we seem to be making little if any progress."

thrown upon four major sources of income: (1) state appropriations, (2) county appropriations, (3) earnings from operating equipment, and (4) miscellaneous arrangements such as gifts, contributions, or assessments. Data available to the general public concerning district finances have consisted of generalized totals reported to Congress in support of the Soil Conservation Service's budget estimates. Therefore, the special questionnaire, described in Appendix B, requested detailed information on district financing from the state soil conservation committees. The information on district income which follows is based on the responses of state committees reporting annual income for 1960–62. There have been changes in detail, of course, since these reports were made, but it is not likely that the broad patterns have changed materially. The income totals given in this chapter are for regular soil conservation district operations only. They do not include appropriations for watershed projects under Public Law 566, which may be only incidentally related to soil conservation programs on farms. Neither do such totals include appropriations to state soil conservation committees to be used solely for their own internal administration.[16]

The total income for all districts in the United States is shown in Figure 1, which also shows state appropriations as the largest single source. A word of warning about county appropriations and district earnings is in order. The totals given by the state committees were often estimates; frequently, the amounts were either unknown or unreported if known. Efforts to gather information of this sort in some states have not been very successful, since there is a tendency to believe that it is the exclusive concern of the supervisors of individual districts.

The presentation of national totals of income by major sources obscures extreme variations among regions, states, and individual districts, as Figures 2 and 3 show. Although there are some distinct patterns of district financing among the regions of the country, there are wide differences among states within the regions. And there are further significant differentials in the sources and levels of income among districts which depend upon community support in the form of county appropriations or occasional grants, earnings, assessments, and contributions. For example, the total earnings reported from the 25 districts in Maryland was $219,624, but only 5 of these districts operated heavy equipment which brought in income. In South Caro-

---

[16] See Appendix B; for information on recent appropriations for watershed Projects, see Chapter 12.

lina, 27 of a total of 45 districts were reported to operate equipment which returned income of $37,946 in 1961. There were similar variations in most other states where districts operate equipment. Florida reported that only 8 districts of 59 in the state had income from equipment—a total of $37,243. Out of 87 districts in Nebraska, 40 reported earnings estimated to net $25,000 annually. South Dakota districts rely entirely on earnings or contributions, but only 6 of its 65

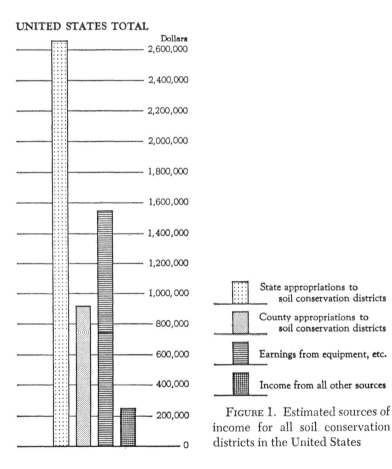

UNITED STATES TOTAL

State appropriations to soil conservation districts

County appropriations to soil conservation districts

Earnings from equipment, etc.

Income from all other sources

FIGURE 1. Estimated sources of income for all soil conservation districts in the United States

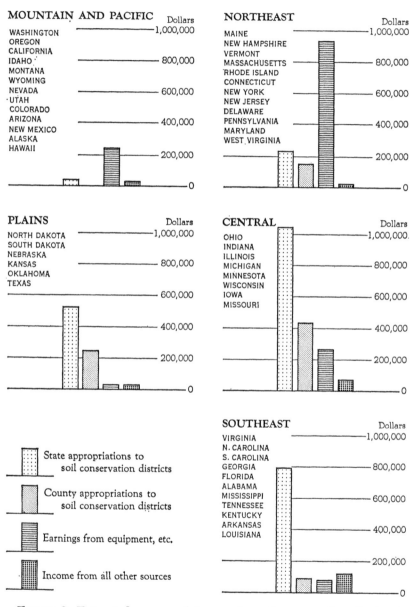

FIGURE 2. Estimated sources of income for soil conservation districts in five regions

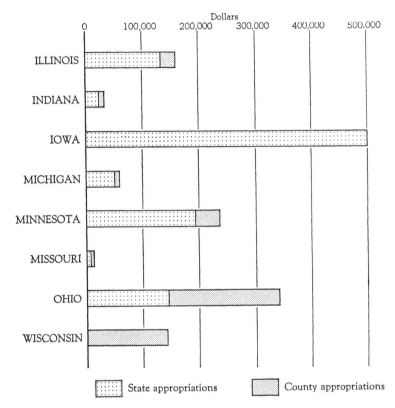

FIGURE 3. Estimated state and county appropriations for soil conservation districts in eight central states

districts were reported to operate income-yielding equipment programs which returned $65,520 in 1961.

Differences between neighboring districts are often startling. For example, the Gila Valley Conservation District in Arizona had a gross income in 1960 from equipment operations amounting to $218,266. Its major expenditures totaled $173,201, leaving a net income of approximately $40,000.[17] The Duncan Valley District, which is contiguous with

[17] Gila Valley Conservation District, *op. cit.*, p. 24. In contrast, the annual reports of the Redington, Willcox, and San Simon Districts (*op. cit.*), and Buckeye, Gila Bend, and Roosevelt Districts do not mention equipment programs or earnings. The Redington District reported that district cooperators make "voluntary contributions each year." They vary from "$5.00 to $10.00 depending on the size of the farm or ranch. San Simon SCD asks for a $5.00 contribution upon receipt of application for 40 acres or less, plus 5¢ per acre of cultivated land over 40 acres or 10¢ per section

the Gila Valley District and was organized earlier, operates no equipment program and is virtually defunct in all but legal contemplation, according to persons interviewed in the area.

In 1962, there were 32 states making appropriations for some type of conservation district expenditures (see Table 2). Iowa had the highest, $500,000; Louisiana was second with $375,000; and Oklahoma was third with $337,300. Alone, these three states accounted for approximately 50 per cent of all state appropriations to districts in 1962 ($1,212,300 out of $2,655,292). Only five other states appropriated more than $100,000: Arkansas, Illinois, Minnesota, Ohio, and West Virginia, provided approximately another 25 per cent of the national total. Texas and Nebraska appropriated $95,000 and $80,000, respectively, but the remaining 40 states appropriated at the level of $50,000, below it, or not at all. California appropriated $100,000 for special projects within districts, but not for regular district operations. Thus, 10 states provided 80 per cent of the national total of state appropriations in 1962. Eight of these (all except Louisiana and West Virginia) were among the 12 states in which there was the greatest demand by farmers for ACP practices which are serviced by the Soil Conservation Service. (See Chapter 11 for discussion of demand for ACP practices.)

Twenty-four states reported that some counties appropriated funds regularly or occasionally to districts in 1962 (see Table 3). Use of this total, however, requires caution. Apparently, about one-third of all districts receive funds from county governments, but such appropriations are spread very unevenly among states and even within them. In only 13 states did approximately half the districts receive county funds. In the 13 Mountain and Pacific states, only 27 districts out of 746 were reported or estimated to receive funds from counties. Apparently, only a little more than one-third of the districts in the region have equipment programs from which they derive earnings. Since only four states in the area appropriated funds for districts (a total of less than $25,000 for the four states), at least half the districts in the Western states relied on contributions, or assessments, or they had no income. Districts in 23 Northeastern and Southeastern states displayed the same uneven pattern, except that well over half of these states provided state funds for districts. District earnings in these states were scattered. Only in the eight Central states were county appropriations commonly made to

---

of range land. The Willcox SCD prospective cooperators make contributions to the Willcox SCD in the amounts of $10.00 per application plus 10¢ per agricultural acre above 40 acres and $3.00 per section on range land."

TABLE 2. MAJOR OBJECTS FOR WHICH SOIL CONSERVATION DISTRICTS SPEND STATE APPROPRIATIONS[a]

32 states with appropriations for conservation districts	Expenditures from state appropriations[b]			
	Conservation aides	District clerks	Supervisor per diem	Supervisor travel and expenses
Alabama			X	X
Alaska			X	X
Arkansas[c]				
Connecticut			X	X
Georgia			X	
Illinois		X		X
Indiana		X		
Iowa	X	X		X
Kentucky[c]				
Louisiana	X	X	X	X
Maine		X		X
Maryland	X	X	X	X
Michigan		X		X
Minnesota	X	X	X	X
Missouri		X		X
Nebraska	X	X		X
New Jersey		X	X	X
New Mexico			X	
North Carolina			X	X
North Dakota				X
Ohio	X	X		
Oklahoma	X	X	X	X
Oregon				X[d]
Pennsylvania				X
Rhode Island			X	X
South Carolina				X
Tennessee[c]				
Texas	X	X		X
Utah[c]				
Vermont			X	X
Virginia				X
West Virginia		X	X	X
Total reported	8	15	14	24

[a] Based on responses to a special 1961–62 questionnaire for this study sent to soil conservation districts, which is explained in Appendix B.
[b] Expenditures for office supplies and equipment are not included unless noted.
[c] There were no responses to the special questionnaire.
[d] Oregon reported $30 per year for each district for supplies and expenses of annual meetings.

supplement funds from the states. Even in this region, however, three states—Iowa, Michigan, and Missouri—had either no counties which appropriated funds to districts or had very few. More than three-quarters of the counties in Kansas appropriated $134,500; 90 per cent of

TABLE 3. MAJOR SOURCES OF LOCAL INCOME FOR SOIL CONSERVATION DISTRICTS AND OBJECTS OF EXPENDITURE[a]

States	Total districts fully organized July 1, 1960	Districts with income from:		Districts which use locally derived income for:				
		County appropriations	Equipment programs	Clerks	Conservation aides	Supervisor Per diem	Travel and expense	Dues to state and national associations
Alabama	52	25[b]	0					
Alaska	18	0						3
Arizona	47	0	18	4		12		
Arkansas[c]	75							
California	158		50	d			d	d
Colorado	95	4	20[b]			d	d	d
Connecticut	8	0	4	3			d	
Delaware[c]	3							
Florida	59	23	8	9			30	48
Georgia	27	2	6				18	27
Hawaii	16	0	5					d
Idaho	51	0	18	6		22	26	
Illinois	98	55	5	96				
Indiana	80	83[e]	4	d			d	d
Iowa	100	0	2					
Kansas	105	88	25[b]	85			50	95
Kentucky[c]	121							
Louisiana	26	f	26				26	26
Maine	15	5	13	15			15	15
Maryland	24	4	5	2				24
Massachusetts	15	4	3	d		3	3	
Michigan	76	15	0					
Minnesota	80	71	41					81
Mississippi	74		15[b]					40
Missouri	37	19	4[b]					
Montana	60	20[b]	15[b]	5			3	50
Nebraska	87	62	40	25			60	87
Nevada	35		25[b]					
New Hampshire	10	0	2				4	
New Jersey	14	0	5				14	14
New Mexico	57	0	20[b]	6			2	
New York	46	46	6	40	2	46	46	46
North Carolina	37	46[e]	0[b]	32	4	4	42[e]	76[e]
North Dakota	75	g	g					
Ohio	87	87	0	87	52			
Oklahoma	87	0	84				62	87
Oregon	57	3	33	3			d	d
Pennsylvania	48	48	0				d	d
Rhode Island	3	0	3					
South Carolina	45	21	30	d				10
South Dakota	68	0	65		10		68	
Tennessee[c]	95							
Texas	176	0	150					
Utah[c]	41							
Vermont	13	0	12[b]	d				
Virginia	29	10	9[b]	2				31
Washington	77	0	51				15	g
West Virginia	14	4	14					
Wisconsin	71	66	16	38	1	71	71	71
Wyoming	44	0	9	46				

[a] Based on responses to a special 1961–62 questionnaire for this study sent to state conservation committees, which is explained in Appendix B. Blank spaces indicate the response did not include information on the particular item. The report of district equipment programs may include some minor items made available to farmers but, generally, the figures show districts having heavy, or major, items of equipment to carry out construction practices.
[b] Estimated by the state soil conservation committee.
[c] No report from the state soil conservation committee.
[d] An unreported number of districts spend for this purpose.
[e] The number of counties exceeds the number of districts.
[f] County governments usually supply office space, free utilities, etc., for district office.
[g] North Dakota reported for county appropriations: "unknown"; and for equipment programs: "65 per cent." Washington reported "60 per cent" for dues.

Wisconsin's counties provided $145,900 as the major source of income for districts in both states. In Nebraska, two-thirds of the counties supplemented state appropriations with $110,000 coming from the local source. County appropriations well in excess of $100,000 annually in these three states reflect the fact that in 1962 they had very high ratings when judged by demand for ACP practices installed with technical assistance from the Soil Conservation Service in the region. Their ranks on this basis were Kansas, 1; Nebraska, 4; and Wisconsin, 10. Only Indiana, which stood eighth in this ranking, provided extremely low levels of state and county appropriations (a total of $35,000) as well as contributions.

DISTRICT EXPENDITURES

Two of the most important purposes for which district funds can be used are the employment of clerks to relieve technicians from some administrative duties and the employment of conservation aides who can assist technicians in laying out practices on farms. But relatively few districts have used their available funds for these purposes, as Tables 2 and 3 show. Districts in only 8 of the 32 states making appropriations in 1962 used these funds to pay aides and only 15 used them for clerks. State funds were used in 14 of the states to pay supervisors' per diem expenses, and in three-fourths of the states for travel to and expenses at meetings. Funds provided by local sources were reported to be used to hire aides in only five states, one of which also received state appropriations. Local funds were used in some districts in 23 states to hire clerks, but in 6 of these states the local funds supplemented state funds. In sum, districts hired aides in only 12 states and clerks in 17, according to the information available.

The principal uses of locally derived funds are to pay travel and expenses for supervisors attending meetings connected with their duties and to pay district dues to state and national associations. Some counties provide indirect support for districts by allowing them free office space and occasional assistance from county officers. The Soil Conservation Service frequently occupies premises provided by, or through, districts.

At various times, 10 states have created conservation equipment funds from which districts could acquire items, usually at reduced costs. These are shown in Table 4. Most of these funds were established at the end of World War II or soon after, when equipment formerly available to districts from the Soil Conservation Service was in short

TABLE 4. STATE CONSERVATION EQUIPMENT FUNDS, BY YEAR ESTABLISHED AND
      AMOUNTS[a]

State	Year	Amount
California	1948	$1,000,000
Connecticut	[b]	25,000
Florida	1947	50,000
New Mexico	1949	20,000
Rhode Island	1944	25,000
South Dakota	[b]	125,000
Texas	1949	5,000,000
	1961	55,000[c]
	1962	30,000[c]
Vermont	1947	30,000
Virginia	1944, 1952	303,312
West Virginia	1947	250,000

[a] Based on responses to a special 1961–62 questionnaire for this study sent to state conservation committees, which is explained in Appendix B.
[b] Not reported.
[c] For new districts.

supply. In addition, many districts acquired equipment from SCS on loan or by grant, and others with funds provided by county governments. Regardless of the original source, conservation equipment is estimated to be provided in about 30 per cent of the districts, but in widely scattered areas.

State funds, where available, and funds from local sources are used for a variety of other purposes, of course. Responses from districts receiving the special questionnaire for this study indicated that half used local funds to purchase office supplies and equipment; provide prizes, scholarships, and awards for their conservation education programs; and for communications, printing annual reports, and paying various reimbursable expenses. About 20 per cent provided insurance and surety bonds for their employees.

It is difficult to determine all the factors which produce the wide variation in the level of income and the objects of expenditure by districts. Leadership is commonly cited as important, but for this study no attempt was made to test this factor. Undoubtedly, the entire system of state and local taxation and its relationship to other ongoing programs has some bearing. Another factor about which information is limited is the amount and distribution of influence which the local supporters of SCS and districts have on the county and state level, especially in relation to the state extension services and their supporters. A few examples, however, will illustrate the significance of this factor.

Conservation districts in both Ohio and Iowa receive large annual appropriations, but, in Ohio, these are divided between state and county appropriations, and, in Iowa, they come wholly from the state. Iowa districts, moreover, do not compete with private contractors or equipment dealers. The differences between the two states may be caused by long-standing variations in the relations of their respective extension services to the state Farm Bureaus. When soil conservation districts were being formed in the thirties and early forties, Ohio's Director, H. C. Ramsower, did not oppose them. In contrast, the Iowa Extension Service's R. K. Bliss was a leading opponent, working through the Committee on Extension Organization and Policy of the Association of Land-Grant Colleges and Universities. In Ohio, the Farm Bureau was not the local, legal sponsor of extension work; in Iowa, the State Farm Bureau was. The Ohio College of Agriculture did not particularly fight the organization of districts through county agents, although informed observers agree that individual county agents sometimes did. Iowa's Extension Service offered spotty resistance to conservation districts, but it was not in a strong position after the defeat of the American Farm Bureau Federation's proposals that Congress "decentralize" SCS in favor of the extension services. The Iowa Soil Conservation Committee included one of the most powerful figures in Iowa politics, Commissioner of Agriculture Clyde Spry, who had been an organizer and a vice president of the Iowa Farm Bureau. Following congressional defeat of the American Farm Bureau's proposal in 1948—and, thus, a defeat of Iowa's Allan B. Kline, who was then its President—the fruits of an organizing campaign by the Iowa Association of Soil Conservation Districts paid off in 1949 with the largest state appropriations for districts in the nation.[18]

Explanations of the same sort can be offered in New York and Kansas, where only county appropriations are supplied, and in various other states, such as California or in Arizona, where opposition to districts from the stockmen—later augmented by certain irrigation interests—have forced districts to rely on earnings. In South Dakota, a quarter of a century of battle between the extension services and AAA, PMA, ASCS, on the one hand, and SCS and conservation districts, on the other, has kept out both state and county appropriations. But, in Oklahoma, a militant and well-organized association of districts has overcome extension opposition—as in Louisiana—to provide the second highest level of state appropriations in the country.

[18] William J. Block, *The Separation of the Farm Bureau and the Extension Service* (Urbana: University of Illinois Press, 1960), pp. 103 and 120; and interviews in Ohio, July 1961, and in Iowa, September 1961.

The intensity of farmers' demands for "permanent" soil and water conservation practices subsidized by the Agricultural Conservation Program is apparently a major factor in connection with the amount of appropriations. Where demand for ACP practices under technical supervision of the Soil Conservation Service is highest, state and county appropriations for districts tend also to be highest. It is in these areas primarily that districts provide clerks and some conservation aides.

STATE ADMINISTRATIVE SUPERVISION OF DISTRICTS

It is still too soon to tell whether the Soil Conservation Service and the National Association of Soil and Water Conservation Districts have been successful in their efforts to insure that district affairs are administered by the district supervisors. State soil conservation committees have been urged to increase their assistance to achieve this goal. In practice, "administration" in probably two-thirds or more districts is rather elementary, although equipment programs can involve matters requiring careful attention. It is necessary to keep records of receipts and disbursements, equipment, agreements and contracts, work orders, minutes of meetings, and reports.

According to replies to the special questionnaire sent to districts, virtually all now keep their own financial records—one activity which the SCS Administrator has insisted must not be performed by agency personnel. Almost the same proportion of district boards record and file the official minutes of all district meetings; only 18 (6 per cent) reported this function performed by SCS personnel. In approximately the same proportion, district boards keep record systems required by state law or their state soil conservation committees; 5 per cent reported that SCS personnel perform this function. Slightly more than half the district boards or their representatives sign contracts or agreements for conservation work to be done; 10 per cent reported that SCS personnel do this. Nearly the same distribution of function was reported for assigning the use of district equipment and maintaining it, although, again, official SCS policy forbids agency personnel to perform these two functions. About 25 per cent of the district boards (67) reported that they make out, sign, and record work orders, and 10 per cent (28) reported that SCS personnel do this. Since 60 per cent of the boards (171) reported that they have some kind of equipment, it is probably used with a minimum of formal administration.

There are two functions which SCS personnel are far more likely than district personnel to perform. Eighty per cent of the boards (226) reported that SCS maintains all records connected with farmer-district agreements; since SCS has a standard records system to account for all contacts with district cooperators, there is no sound reason for district supervisors to duplicate it. In about half the districts reporting, it is also the practice to have SCS personnel distribute planting materials, fish and wildlife stock, and similar items directly to farmers, although such distributions are made in the name of the district.

The Department of Agriculture has always required districts to make an annual report and transmit a copy to the Department. In answer to the special questionnaire, 92 per cent of the boards (258) said that they prepare annual reports. Among these, 80 per cent (204) reported that either the board or its secretary works jointly with the SCS work unit conservationist to prepare the report. Slightly over 10 per cent (28) reported that the district secretary performs this job, and 7 per cent (18) said that the SCS work unit conservationist alone prepares the report. The small remainder may be prepared by supervisors or other personnel, but all board members participate in only 52 districts. It seems safe to say that only rarely is a district's annual report prepared without very close collaboration between the district board's personnel or representatives and SCS. There is no evidence of this close association with any other federal or state agency, except the state soil conservation committee in some states. Even where the district and state committee cooperate on the annual report, information concerning the application of practices is supplied chiefly by SCS, although care is usually taken also to report ACP accomplishments based on the records of the ASCS and other agencies.

Assistance to districts and supervision by state conservation committees require staffs adequate to insure a minimum level of performance. As might be expected, there is wide variation among the states in these matters. Appropriations for the state committees vary from none in Montana to $613,912 in California for fiscal year 1962. Some states pay the state committee staff with appropriations to other agencies, especially where an extension service's state conservationist is assigned to the committee. In Georgia, for example, the salary of the conservation committee's executive secretary is paid by the Georgia Extension Service. The committee's executive secretary in Montana, where no state funds are appropriated for either the state conservation committee or districts, is a rancher who is also supervisor of a soil conservation district. The California Division of Soil Conservation of the Department of Natural Resources has the largest permanent staff of any in the country (44); $234,584 was allocated in 1961 to salaries and expenses for watershed planning parties under Public Law 566; $100,000 was for a special program of assistance for "community-type" programs in districts; and an additional $33,230 constituted the state's share of the cost of operating the SCS Pleasanton Materials Center. Thus, the net appropriation for salaries and the expenses of the State Conservation Committee and its staff to supervise districts is $146,098, an impressive total measured by the standards of other states—although no part of these funds is for direct assistance to districts. In California, the item for retirement funds alone ($27,275) was higher than the total appropriation to the New York Conservation Committee ($26,074) and all except 18 other state committees for all purposes. The California Committee's item for health and welfare alone exceeds the total state appropriation for the Massachusetts and Mississippi committees. In all, only 20 state conservation committees receive total appropriations of $24,000 per year for all objects of expenditure.

Some idea of the scope of administrative supervision of conservation districts by state committees can be suggested by the size of their permanent staffs. Seven state conservation committees have no staff of their own. Of these, the Alaska and Hawaii committees perform the minimum essential functions through other state departments, while the Vermont, Rhode Island, New Hampshire, Mississippi, and Nevada committees do without. Twenty states provide only one or two employees (usually a full-time executive secretary, who is the sole administrative officer, and a full-time clerk-stenographer). Sixteen other states provide three to seven full-time and part-time employees, exclusive of any carried on the conservation committee's payroll for watershed planning parties and related personnel for Public Law 566. Oklahoma, for example, has a total of 53 full-time employees, but 46 are employed for projects under the Watershed Act. Included in California's total of 44 employees is a chief of a division of soil conservation and 12 soil conservationists, an equipment engineer, and 8 employees in a clerical section. The remaining employees are in watershed planning parties. The soil conservationists provide direct field contact with the district boards, but do not provide technical assistance for conservation practices. Measured by the number of contacts a year with district supervisors, the California staff appears to have set a standard not contemplated, much less approximated, in other states. Even so, the California staff is concerned with district administration, and does not provide technical assistance.

In three-fourths of the states, the chief administrative officer of the state committee is a former SCS employee or, as in a few cases, a former vocational agriculture teacher or administrator. In approximately one-fourth of the states, the secretary was formerly, or is at present, an employee of the state agricultural extension service. The others have diverse backgrounds, and only one has received formal education in public administration and state and local government.

There is, understandably, wide variation in the systems of administrative supervision of districts among the states. Most of the techniques are such familiar ones as (1) advice and consultation within guidelines set forth in written policies, (2) improvement of personnel, (3) making reports and inspections, and (4) substitute administration. The control exercised over district administration varies from almost none in Hawaii, Mississippi, and Nevada to a high degree in New Jersey. The New Jersey State Conservation Committee appoints all supervisors and may remove them. It requires districts to submit annual progress reports, plans of operation, and budget requests. Any district which fails to perform these required acts does not receive state funds. If this condition persists, the State Committee may remove the delinquent district supervisors from office. The New Jersey State Committee, reorganized in 1959 to revive moribund districts, takes the stern view that the districts are legal subdivisions of the state, governed by officers who are accountable to the public in general and not just to farmers who are residents of the district. The Committee views itself as a "watch dog" of district activities, and apparently does not intend to witness the neglect of proper administration by district boards.[19] This attitude is rarely expressed, however, by members or staffs of the remaining state committees.

The prevailing view of state committee members seems to be that conservation districts are independent units of local government which should function with aid, but not "dictation," from their state committees. Usually, persuasion is preferred to outright control. The attitude

[19] In New Jersey, the State Conservation Committee has taken very active steps to align activities of the districts with a considerable variety of state conservation agencies and to recognize the growing importance of urban, suburban, and rural nonfarm interests in problems affecting soil conservation. However, no steps have been taken to alter farmer control of the agencies. The New Jersey Committee has attempted to gain some independence from the county Farm Bureaus by requiring them to nominate two persons for each supervisor vacancy. This policy was established in 1960, after the county agricultural boards had sent in only one name each, a move that was interpreted as a challenge to the State Committee, which had been completely reconstituted in 1959 by amendments to the existing law. (Data from the New Jersey response to the special questionnaire and interview with Executive Secretary of the State Conservation Committee, March 28, 1961.)

in California, despite its large state appropriations, is typical. The large field staff seeks by consultation and advice to aid in the organization and reorganization of districts, to develop cooperative relations in support of districts, and to encourage them to improve their internal administration. It does not require districts to adhere to specific procedures, but merely "encourages" districts to submit copies of their annual audits, reports of annual activities, and annual work plans. Informal, though apparently detailed, guidance is given to the management of equipment purchased by districts from state funds. A state legislative investigation of the mismanagement of district equipment demonstrated a need for this kind of supervision.

Guidance by advice and information is the least coercive form of supervision and is used by virtually all of the state committee staffs. At a minimum, this normally consists of issuing a handbook to guide supervisors in the proper performance of their duties and a periodic newsletter giving information and policy guidance. Frequently the newsletter is issued by, or in conjunction with, the state association. In three-fourths of the states, the supervisor's guide is in loose-leaf form so that current policy guides can be made readily available. A few committees, such as those in Pennsylvania and Arizona, frequently issue printed or mimeographed guides, but do not have handbooks. Missouri and Washington provide none, or very little, of this kind of supervision, and Mississippi provides almost as little. Advice and information is given during staff visits to district boards and at area or state meetings to improve supervisor performance, although care is often taken not to call these "training" sessions. All but ten states provide training meetings. Nowhere are such meetings held in conjunction with the state extension service, however, except in Idaho, Iowa, Missouri, Montana, New Jersey, Maryland, Minnesota, Nebraska, Ohio, and Wisconsin. Only Hawaii does not provide staff visits to districts. Staff visits to districts to inspect and aid in keeping records systems are made in three-fourths of the states; in addition, all but five provide standard forms for equipment records, receipts, disbursements, contracts, and other documents used in district administration; sixteen provide districts with stationery.

Districts also receive advice and assistance in planning promotional activities, although there is very little in Nevada, Iowa, Maine, and Oklahoma. Generally speaking, this assistance consists in planning tours, demonstrations, and conservation contests to stimulate interest and activity. Twenty-one state committees have programs for awarding prizes, either to outstanding district boards or to individual farm operators. Only in Missouri does the State Committee not encourage districts to enter an annual prize competition sponsored by the Goodyear Tire and Rubber Company.

Supervision by requiring specified reports and papers is common. Annual audits of district finances are required in all states except Alabama, California, Georgia, Hawaii, Iowa, Kansas, Nevada, North Carolina, and Wisconsin. California "encourages" the submission of audits; accounting for funds is probably adequate because of other forms of supervision in Iowa and Wisconsin.

A relatively important form of supervision to insure compliance with a wide range of policy directives is the requirement that districts submit a copy of the official minutes of each board meeting to insure regularity of action. The only states which do not require this sound practice are Hawaii, Mississippi, Nevada, Indiana, Kansas, South Carolina, Minnesota, Pennsylvania, and Wisconsin. According to the memorandum of understanding with the Department of Agriculture, districts are required to submit annual reports of their activities. All state committees, except California, Florida, Hawaii, Illinois, Massachusetts, Mississippi, Oklahoma, South Dakota, Texas, Vermont, Virginia, and Wisconsin, require that one copy of the annual report be sent to them also. In Mississippi, this function is said to be performed still by personnel of the SCS.

Among the 32 states making appropriations for district functions, all except five (Alaska, Michigan, Ohio, Texas, and Vermont) require that the districts submit vouchers for the payment of their expenditures. This is probably the most common form of substitute administration practiced by the state committees. Twelve of the states making appropriations require also that districts submit purchase orders for office supplies and equipment. Several states appropriate funds to the state committee for the payment of certain expenses of the districts—such as travel, supplies, postage, and similar items. In these, district boards are required to submit vouchers to the state committee for payment. The committee staff prepares warrants and submits them for payment, thereby relieving district supervisors of this administrative responsibility. Technically, this kind of financial assistance is carried as an expense of the state committee. Since its staff performs the necessary administration, it, too, is one of the commoner forms of substitute administration. It is practiced, for example, in Alabama, Pennsylvania, and Virginia. Another type of substitute administration is mimeographing district annual reports. This is done in Alabama, Alaska, California, Connecticut, Florida, Michigan, New Jersey, Oklahoma, Pennsylvania, Virginia, and Washington.

STATE SUPERVISION OF DISTRICT FINANCES

Supervision of soil conservation districts by the state committees is with few, if any, exceptions, more extensive where state funds are appropriated to districts than in those where reliance is placed wholly on local sources of funds.

The legal objects of expenditure of state funds by district boards are spelled out in varying detail among the states. In some, the appropriation acts and other general legislation specify these objects; in others, the state committees have imposed specific limits within broad legal limits. For example, Rhode Island and Texas both require only that funds be spent for the purposes of the district program, but in Texas there are several restrictions in various acts of the legislature. Districts may not purchase machinery, equipment, seeds, fertilizer, or other supplies without a written demand to a district board from ten landowners and occupiers in addition to a finding entered in the district boards minutes that there is sufficient demand to justify purchases and that the revenue expected to be returned from sale or use of the item can reasonably be expected to exceed the cost of replacement. Texas also requires that purchases be made through the state Board of Control. The districts, however, cannot legally use state or local funds to promote public interest by giving awards, prizes, or entertainment. These activities must be supported with private donations for these specific purposes. It is in this connection that the Dealer-District Conservation Program in Texas is important. In addition to limiting the objects of expenditure, state legislation in Texas imposes a variety of accounting procedures which appear to be more numerous than those commonly found in other states. In contrast, Iowa, for example, specifically permits state funds to be used for such promotional activities as are forbidden in Texas. Montana forbids the use of public funds for the payment of association dues, but Texas permits it. Ohio has a mixture of legislative restrictions and of policy decisions by the State Committee which affects the use of funds. Indiana, Michigan, and Vermont require that districts submit an annual audit as a condition for receiving state funds; Minnesota requires districts to file annual reports. In short, some measure of control over district expenditures is exerted in the 32 states appropriating funds for district activities.

The state committees which have discretion in distributing state appropriations to districts may exert some leverage on district programs, but how much is not clear. Louisiana allocates an amount to each district, giving equal weight to the number of farms and the acreage in farms. Oklahoma distributes the funds in equal amount to each district. Iowa distinguishes between administrative expenses and those in support of technical assistance. Each Iowa district receives a flat grant of $500 annually for its administrative expenses, including the expenses of district commissioners. The remaining state appropriation is distributed on the basis of "need" to the state's 100 districts. Salaries for clerks and conservation aides are paid from this part of the appropriation, and certain other expenses, such as rental and equipment, are also permitted. Needs are determined on the basis of estimates submitted by the district boards, the recent history of use of such funds, and the priority which the Iowa State Soil Conservation Committee decided to give districts having watershed projects under Public Law 566 which were (1) authorized for construction, or (2) authorized for planning. The results for the fiscal year 1961–62 (the most recent information available for this study) were decreases in the allocations of some districts and increases in those of others. Certain drawbacks are inherent in this type of situation. Local agressiveness in seeking small watershed projects varies, along with the specific nature of the problems to be solved by the project. As a result, the assignment of priorities to districts having small watershed projects may result in an undesirable diminution of funds for districts having critical conservation problems unrelated to the Small Watershed Program (which places heavy emphasis on small flood control projects, often to the benefit of urbanized areas).

Ohio, one of the other states making large appropriations to districts through its State Conservation Committee, has an extremely flexible procedure for distribution of funds. The level of local income (which includes county appropriations) is considered in relation to "need" and state funds are then distributed "proportionately." That is, each district receives a share of state funds in proportion to (1) the total available to all districts and (2) the amount needed to raise each district to a level shared more or less in common by all districts, although no amount is specified. Illinois makes a distribution, taking into account each district's history in using funds, its current local resources, and its work plan. Minnesota, which is also among the few states making large appropriations for districts, provides for two budgets, similar to those in Iowa. The first budget, for supervisors' compensation and district expenses, gives equal weight to the number of farms and to the acreage in farms in each district for expenses, and a fixed equal amount to each district for supervisors' compensation. The second budget provides funds for clerks, aides, equipment, and similar expenses, and it is based on the history of each district's past use of funds, its annual report, and its current work load.

Of these states making the largest and most significant appropriations to be distributed to districts by their state committees, only Oklahoma makes a distribution undifferentiated by some criteria of "need," although the Louisiana formula is not a very satisfactory type, since it is unrelated to the critical conservation needs in any district. The Iowa system meets openly the usual demand that all localities receive an equal share of funds by making such an allocation only for a minimum of selected purposes.

Judgments of need by state committees apparently are based upon the recommendations of the Soil Conservation Service, since the committees lack their own technical staffs. It appears that as state appropriations for small watershed projects increase, as they are likely to do, the geographical distribution of these projects probably will materially determine the areas to be treated. Virginia, for example, allocates funds from the State Conservation Committee only to districts having watershed projects to hire clerks and aides and to pay certain other expenses. Because of their involvement in the distribution of state funds, the state committees can be expected to become more actively involved in umpiring the demands of localities for these benefits as the efforts of state district associations to secure state appropriations succeed and increase. It still remains doubtful, however, that the state committees will independently determine the pace and emphasis of either the soil conservation program initiated by Public Law 46 of 1935 or the Small Watershed Program. Their dependence on federal programs, especially the Agricultural Conservation Program, appears to be much too great for any such development.

Some state committees are active in seeking appropriations for districts, but others consider lobbying to be the function of state district associations. Differences over this matter, the composition of some state committees, and weaknesses in the leadership of many state district associations have probably influenced the level of appropriations for the functions of both state committees and conservation districts.

DISTRICT PROGRAM SUPERVISION

The state conservation committees conform with the view of their functions originally conceived by the Department of Agriculture except for their administrative supervision of districts. The Department did not want the committees to develop staffs for provision of technical assistance or supervision to districts; the committees have not done so, despite a limited effort in this direction between 1937 and 1940. Consequently, although many state committees encourage and assist districts in planning programs and making agreements with technical agencies, they are careful not to displace the technical agencies, especially the Soil Conservation Service.

For example, about half of the state committees that responded to the special questionnaire reported that they either require or encourage conservation districts to review their long-range programs periodically, as SCS urges. A half-dozen committees actively help districts make annual work plans. The Michigan, Ohio, and Virginia committees reported they occasionally have recommended amendments and offered suggestions for changes in both programs and annual plans. These three committees also have suggested changes in the annual work plans of districts, but the remaining 15 state committees which either require or encourage the submission of annual work plans have never rejected one as inadequate. With few exceptions, districts are free to reconsider their programs periodically and to make such annual work plans as they see fit.

It should be emphasized also that state soil conservation committees do not maintain technical staffs which review the conservation operations undertaken in districts. This phase of operations normally is left to districts and SCS, although the situation is complicated by the ACP program. The South Carolina State Committee is an exception in that it has "technical advisory committees which meet periodically to observe the effectiveness of conservation practices and discuss evaluation. Specialists in SCS, Experiment Stations, Extension Service, and other agencies are invited to participate in the evaluations."[20] And the New York State Committee has "assisted boards where technical standards are not being met" and has checked with the SCS State Conservationist to "correct" the situation in "several instances."[21] These two, however,

[20] Response to special questionnaire sent to state conservation committees: from South Carolina.
[21] *Ibid.*, response from New York.

were the only cases reported in which the state soil conservation committees have in any way been concerned with the technical operations of SCS or any other specialized agency functioning within districts. No other state agency, it might be added, performs this function of technical review.

State committees also assist and influence districts by initiating, or helping to negotiate, memoranda of understanding between districts and agencies other than SCS, as was anticipated in the standard soil conservation districts act prepared by the Department of Agriculture in 1936. Also in the SCS-NASCD joint statement, "Working Together," it is suggested that district boards seek the assistance of state and federal agencies other than SCS.[22] In actual practice, however, only about one-third of the state committees have been active in promoting such formal agreements. In one-third of the states, districts have entered into agreements with two or more agencies, other than the U.S. Department of Agriculture; in another one-third of the states, there are no such agreements; and in the remainder, there is at least one agreement. Districts in only nine states have an agreement with the state extension service; in some of these cases, the agreement is made on the state level between the state committee and extension service, as in Indiana and Ohio. With the exception of the few formal understandings reached with state highway commissions, these agreements have been made chiefly as the result of federal interagency competition for program responsibility. Ordinarily the state committees do not seem to be in a position to do much more than encourage district boards to sign the agreements which have been worked out in each state only after the state and federal agencies have compromised their differences and agreed on the limits of SCS program activities through the districts. This is true even of agreements covering such matters as special research programs, river basin studies, and some activities under the small watershed program. In performing their limited function of encouraging these cooperative agreements, the state committees vary greatly. (District initiative in these matters is rare to the vanishing point, except in a few instances in which particular districts have asked for, and supported with their own funds, cooperative research projects with their state experiment stations.)

All of the state conservation committees except Alaska's indirectly assist districts by performing some services for the state associations of

[22] National Association of Soil Conservation Districts and U.S. Soil Conservation Service, "Working Together, S.C.D.-S.C.S.," Oct. 1959.

soil conservation districts. Two-thirds of them arrange the program of the state association's annual meetings, get speakers, and secure accommodations. About one-third publish a periodic newsletter for the association, and somewhat more assist in the publication of the association's own newsletter by providing copy for it. Nearly all offer various types of advice and assistance to the officers of the state association, and about two-thirds regularly, or occasionally, inform the public of association activities through releases to the news media. Two-thirds of the state committees also hold at least one annual meeting in conjunction with the state association. At these meetings, the issues which trouble supervisors are aired, agency representatives present their programs and help shape supervisor attitudes toward them, and supervisors are exhorted to a better performance of their duties. The meeting expenses are paid from a variety of sources, with most state committees apparently providing some financial assistance from state funds, as is normally the case with other public agencies.

SUMMARY AND COMMENT

Since 1937, when the Department of Agriculture proposed soil conservation districts, they have been financially dependent units of government. The districts have obtained funds from assessments and contributions, earnings through equipment operations, and from state and county appropriations. The struggle between the land-grant colleges and Soil Conservation Service to control administration of the Department's technical assistance for soil and water conservation has embroiled districts, and probably accounts partially for their highly variable pattern of fiscal support from public sources. Another factor appears to be the level of demand for permanent conservation practices.

Since most conservation districts are thrown upon local sources of income, some have resorted to practices which have embarrassed both the Soil Conservation Service and the National Association of Soil and Water Conservation Districts. Acting together, they have prompted state soil conservation committees and state associations of districts to improve assistance to, and supervision of, financial and other district activities.

A review of the sources and levels of district income, the objects of expenditure, and the extent and kind of state administrative supervision reveals wide variations among states, and even among districts within the same state. The unevenness of their performance is probably the reason why successive Secretaries of Agriculture have been unwilling to place significant responsibilities on districts—especially for the Agricultural Conservation Program, much of which the NACD opposes and for which it does not wish to accept responsibility. These factors provoke serious questions in consideration of the usefulness of districts in administering any future programs of resources conservation.

CHAPTER 10

District Program and Policy Functions

STATES WERE ENCOURAGED to make it possible for farmers to organize soil conservation districts, originally, as a means of performing five basic functions in a national soil conservation program. In cooperation with the Department of Agriculture, districts were to arouse farmers to the hazards of erosion and teach them methods of controlling it; to plan programs meeting the needs of their communities for improved land use and to plan operations for each farm consistent with its capabilities and needs; to provide farmers with assistance from district resources and those of state and federal agencies; to insure use of this assistance to accomplish permanent results in the public interest; and to make and enforce land use regulations where they were needed and community sentiment was prepared to support them.[1] Broadly speaking, the conservation districts were created to harmonize national program planning with the needs and wishes of the people affected by the programs.[2]

In fact, a review of the present activities of conservation districts shows that they generally perform the educational function almost to the exclusion of the other four. They still depend upon the Soil Conservation Service (SCS) to plan and review their long-range programs, annual plans of work, and individual farm plans. Nominally, districts provide farmers with technical assistance, but it is supplied almost exclusively by SCS. Between one-third and one-half of the districts provide some equipment or other special services to farmers. District boards do not, however, actively review accomplishments on the land, and not more than a dozen have enacted land use regulations.

---

[1] A concise description of the Department's official expectations from the districts, aside from what has been pointed out in earlier chapters of this study, may be found in *Soil Conservation*, Vol. III (November 1937): Dillon S. Myer, "The Next Step: Emphasis Shifts to the District Plan," pp. 126–28; and J. Phil Campbell, "Associations Lay Groundwork for Legally-Constituted Districts," pp. 132–34. The authors were the Chief and Assistant Chief, respectively, of the Division of Cooperative Relations and Planning, Soil Conservation Service.

[2] W. Robert Parks, *Soil Conservation Districts in Action* (Ames: The Iowa State College Press, 1952), p. 223.

DISTRICT AGREEMENTS WITH THE DEPARTMENT OF AGRICULTURE

The responsibilities which the conservation districts and the Department of Agriculture officially assume for the soil conservation program are specified in two memoranda of understanding. The basic one is between the Department and the district, and there is a supplemental memorandum between the district and the Soil Conservation Service. At present, no agencies of the Department except SCS have agreements with the districts, but some state and local agencies do. The memorandum between the Department and districts has been revised periodically; that used at present became effective on February 1, 1962, and supersedes one previously formulated in 1952. The present agreement continues most of the stipulations of earlier versions, but emphasizes the Department's growing interest in water conservation and the needs of a new urban and "rurban" clientele. The present agreement also contains provisions giving effect to some SCS policies developed since 1953 to stress administrative self-reliance in the districts.[3]

The Department promises to assist districts in executing their long-range soil and water conservation and "resource-use" programs within the limitations imposed by its own statutory authority and available means of assistance. Specific programs of assistance are to be supplied through those agencies of the Department having supplemental agreements with districts. Understanding and agreement must be reached on ten specified conditions:

(1) Aid is to be furnished in accordance with the applicable regulations of agencies of the Department.

(2) Educational and farm forestry work is to be carried out only in accordance with existing or future agreements between the Department, or its agencies, and state agencies (that is, agricultural extension services and state forestry departments).

(3) The jurisdiction of federal agencies over federally owned lands will not be affected.

(4) Personnel and facilities of the Department and its agencies are under its "administrative" jurisdiction.

(5) All matters requiring "administrative action or approval" by any agency of the Department will be handled through "that agency" and the Department.

[3] U.S. Department of Agriculture, Secretary's Memorandum No. 1488, Feb. 1, 1962.

(6) Districts shall use equipment and materials provided by the Department in accord with agreements made governing their use.

(7) Personnel, facilities, and funds available to districts from state, local, and private sources are to be under the "administrative" jurisdiction of the district or other cooperating state or local agency.

(8) Neither the Department nor a district is bound to expend funds in excess of amounts available to it or for a period in excess of that authorized by law.

(9) The standard memorandum and any supplements to it may be modified or terminated by either party by giving sixty days' notice in writing.

(10) Any supplemental understandings or arrangements "now in effect" are to remain in force, subject to the provisions of this new memorandum of understanding.

It is stipulated that districts will act subject to eight specific conditions. They will, either on their own initiative or "upon request," consult with departmental agencies and recommend conservation activities within their districts and they "will" prepare annual plans of work to carry out their programs. Districts are required to enter into agreements with owners and operators, "fixing the responsibilities of the parties in carrying out those plans," where the Department furnishes aid, except for "consultive type services" (which are of growing frequency in urban areas). Each district is responsible for determining "the kind, amount and priority of work to be performed by it on farms, ranches, and other land, and for seeing that the provisions of agreements it enters into with owners and operators of land are carried out." Districts are to provide such forms of assistance as they are "able to obtain," but they are forbidden to "charge for assistance made available by the Department," and are enjoined to conduct their activities in such a way that the public understands this policy. Districts are to submit annual reports to the Department through the state conservationist of the Soil Conservation Service; they are to keep their records in such condition that agencies of the Department can secure information "by examining these records." Finally, any district making "substantial changes in its longtime program" will inform the Department to avoid "possible misunderstandings."

The supplemental memorandum of understanding entered into by SCS normally adds little to these terms. It usually stipulates that SCS will furnish a work unit conservationist for each district, together with clerical assistance, transportation, and other facilities for its own personnel. Further forms of assistance may be made available occasionally. Included may be field equipment of a kind not ordinarily owned by farmers within a district, or planting materials which are not readily available. The SCS agrees to make assistance available, if it is in a position to do so, subject to its own decisions, but upon consultation with district boards. In Missouri, SCS agrees to provide only an "area conservationist" and the services of other personnel "as needed and available," but the staff in a district may be augmented upon the establishment of a small watershed program sponsored by an approved local organization within a district to "integrate" its watershed and regular soil conservation programs.

In these supplemental agreements it is "further understood" that SCS will maintain control of its personnel and determine the locations of headquarters. The SCS employees are authorized to assist districts in making surveys, preparing conservation plans for farms and "other units," explaining agreements and securing signatures to them, aiding landowners and operators to "perform operations which require technical skill beyond the experience of the individuals involved," serving as consultants to the districts, aiding them in keeping records and preparing reports and engaging in other related activities.[4]

The present detailed "model" of district functions and relations with SCS also is elaborated in a pamphlet, "Working Together," jointly prepared by the National Association of Soil Conservation Districts (NASCD) and SCS in 1959. Sound relations, in the words of this joint statement, rest upon "four fundamental requirements." These are: recognize the responsibilities of others; know the conservation job; appraise the resources needed; and report on objectives, problems, and progress.[5]

About the first of these requirements, the pamphlet says each "must recognize the clear line of separation between the respective responsibilities of the two organizations." This distinction will be reinforced best, it is agreed, if there is an annual review of the memorandum of understanding by each district board and the SCS work unit conservationist, joint initiative in the cooperative planning and action of all groups and agencies functioning in the district, initative by the district board in inviting representatives of agencies other than SCS to attend board meetings, and initiative by the board members in attending the

[4] Memorandum of Understanding between The Soil District of _____ County, The Missouri State Soil Districts Commission and The Soil Conservation Service, United States Department of Agriculture.

[5] National Association of Soil Conservation Districts and U.S. Soil Conservation Service, "Working Together, S.C.D.-S.C.S.," Oct. 1959.

meetings of other agencies. Both SCS and the board members must "recognize that the fiscal and management operations of the District are the sole responsibility of the District Governing Body, and that neither the Work Unit Conservationist nor any other Federal employee should engage in these activities."

The second of the four requirements focuses upon the district's annual work plan. In order to allocate effectively the available resources of manpower, equipment, and materials, it is recommended that an up-to-date inventory of conservation needs be kept in each district; that the work unit conservationist aid the district board in developing a "brief, realistic and readable" annual plan which sets forth goals and the actions necessary to reach them; and that progress in meeting these goals be checked monthly by both parties. The prospective annual workload for the district should include consideration of six specific items: (1) a plan for the application of practices on the lands of existing cooperators on the basis of information gathered by SCS; (2) an estimate of the number of new cooperators expected; (3) a statement of priorities of assistance; (4) an estimate of group enterprises to be serviced; (5) plans for operations and maintenance of watershed projects; and (6) consideration of shifts in emphasis among the various practices to "keep the program in balance." Boards are urged to estimate, also, the needs for soil surveys, the number of referrals from the Agricultural Conservation Program (ACP) to be serviced by SCS but administered by the ASC county committee, and work to be done in other programs under the authority of SCS and the Agricultural Stabilization and Conservation Service (ASCS).

Initial planning to include these items is to be followed by the conversion of these goals by SCS into appropriate units (such as man-years) for agency control of its annual work program. This action is supposed to be followed by a further meeting of the board and the work unit conservationist to develop "actual goals" and to decide upon ways of securing the assistance of "equipment dealers, county agents, vocational agriculture teachers, farm organizations and others as well as the press, radio and television in furthering the district's objectives." Follow-up meetings during the year are urged to make adjustments found to be necessary because of weather or other changing conditions. SCS personnel are to present periodic reports of SCS activities to the district board.[6]

A sound appraisal of the resources needed by each district involves

[6] These procedures are an almost exact duplication of those given in U.S. Soil Conservation Service, Management Memorandum SCS-5, Oct. 29, 1959.

recognition that maximum accomplishments can be achieved only with careful use of the available supplies of materials, equipment, and manpower. Therefore, the joint statement by SCS and NASCD recommends that each district board prepare an annual estimate of funds to be made available from local, county, and state funds (that is, prepare a budget), and it is urged that boards make full use of services from county and state legal and fiscal officers.

The job of reporting results is to be a cooperative one. Both board members and SCS personnel are urged to assist representatives of all news media to cover "newsworthy events," to stimulate public attendance at board meetings, to develop public consciousness of conservation needs through a variety of promotional activities, and specifically to call the attention of "urban people" to the benefits of a district program—"wildlife, recreation and water supply." The SCS is to supply district boards with information that may be used for public reporting and the board is to "make a special effort to give particular recognition and credit to all agencies and groups contributing to the advancement of the district program."

RELATIONS WITH AGENCIES AND THE PUBLIC

There are no typical soil conservation districts; they differ within states and among states. Similarly, there is wide variation in the quality and extent of district administration, financing, equipment programs, special services, and in relations with other agencies and the public. It is possible to describe, nevertheless, some activities commonly undertaken and to estimate the proportion of districts which engage in them.

The NASCD and SCS consistently have encouraged districts to hold monthly public meetings and invite to them all private groups and public agencies which can assist in a program of conservation. Meetings provide a forum for informing the public, testing its reaction to district programs, rallying group support for districts, and providing a wider range of contacts between farmers and conservation agencies. There is no clear evidence, however, that district meetings have actually resulted in integrated planning in a very high proportion of cases.

Of all the responses from districts to the special questionnaire for this study (see Appendix B), 85 per cent indicated that the district held a regular meeting once a month. The remainder met fewer times—about 5 per cent met at intervals of two months or more. Three-quarters of the districts which responded said agenda are scheduled in advance of meetings. About 20 per cent said they give regular notice of their meetings to the general public, and less than 20 per cent reported using local news media for this purpose. But only one-seventh of the districts said that they specifically invite even district cooperators to attend. One-third of the districts reported some attendance by cooperators and other persons from the public.

All but two of the districts reported that an SCS work unit conservationist regularly attends their meetings, and nearly as many reported occasional visits by other SCS personnel. Interestingly, nearly half reported that the county agent attends at least occasionally. About 5 per cent reported regular attendance by a representative of the state soil conservation committee, and 60 per cent reported that such a person occasionally attends. Representatives of other agencies and private groups attend now and then. In about 60 per cent of the districts, a representative of the ASC county committee sometimes comes. Contractors providing equipment for conservation practices attended in about 45 per cent of the districts; and in about one-third, equipment dealers, vocational agriculture teachers, representatives of the Farmers Home Administration, state forestry agencies, and farm organizations attend occasionally. In 5 per cent, or fewer, of the districts, there is occasional attendance by a representative of the county governing body, a watershed organization, a federal agency other than SCS, or groups interested in recreation.

It was not feasible to determine by responses to the questionnaire how the boards of districts having all or parts of two or more counties within their boundaries handle contacts with SCS work unit conservationists, ASC committees, and other groups. However, four such districts were visited for field interviews. In these, the boards normally do not meet on a regular basis with the work unit conservationists assigned to all the counties situated wholly or partially within the district. Contact between district boards and SCS work unit personnel located in counties outside the headquarters county of the district was either nonexistent, according to supervisors interviewed, or it was very infrequent. Obviously, the lack of procedures for regular consultation raises serious questions about the ability of a district board to have any knowledge of conditions, much less a program, in the entire district. A similar situation occurs in those areas, particularly in the Western states, where there are two or more districts within a single county.[7]

[7] The typical situation found in the four multiple-county districts and in a two-district county which were visited for field interviews is indicated by a letter from

In summary, the responses indicated that a large majority of districts conduct their business regularly, having maximum contact with the Soil Conservation Service; much more limited contact with representatives of other agricultural agencies, farm organizations, conservation contractors, equipment dealers; and only very limited contact with the general public. This type of operation almost certainly means that district supervisors' conceptions of "conservation" will be received chiefly, if not exclusively, from the Soil Conservation Service. In addition, it is almost as certain that the discussion of conservation problems attracts relatively little public concern, so that matters are left largely to the professional agricultural workers and other persons who have a direct interest in the administration of conservation programs.

DISTRICT PROGRAM AND WORK PLANS

Since 1937, the Department of Agriculture has required each district to formulate, with technical assistance from the Department, a long-range program of goals and actions calculated to achieve its conservation needs. Formally speaking, the memorandum of understanding between the Department and the district is effective only when the district supervisors meet this requirement. In his study of conservation districts, Parks noted the steady relaxation of departmental requirements of detail in district programs and a decline in their length.[8] Formulation of the district program has been followed by the preparation of a district work plan to serve both as a condition for receiving SCS assistance and as a basis for effectuating the district program.

---

the chairman of a multiple-county district in North Carolina: "We received your recent questionnaire. We discussed this briefly at our last meeting and decided that since our district is made up of five counties and five work units, we could best answer the questions on a county basis. Please send us sufficient copies for the five counties." When the board of this district was informed of my desire to have a single report for the district, it prepared and returned one.

[8] Parks, op. cit., p. 30. The application of the "farmer-district agreement" was materially reduced as early as 1939. In December 1937, Secretary Henry A. Wallace approved a form of the memorandum of understanding between SCS and districts and another for the "farmer-district" agreement. The latter were to be effective for five years and had to be signed as a condition for receiving any assistance from SCS, even a farm plan. NARG 114 (SCS General Correspondence, 1937, File 107A). In May 1939, Hugh H. Bennett recommended to Wallace, who approved, that when a farmer wanted only a plan, he should not be required to sign one of these 5-year agreements. The farmer would enter an agreement only when the district provided him with materials, equipment or labor. See National Archives Record Group 114 (SCS General Correspondence, 1936–40; Soil, 1.3), H. H. Bennett to Secretary Wallace, May 9, 1939. (National Archives records cited hereinafter usually will be identified by the abbreviation NARG.)

There is no evidence, however, that either the Department or SCS refuses to give technical assistance in districts which fail to prepare annual programs. The heaviest pressure upon districts to prepare annual plans at present appears to come from those state soil conservation committees which have made the performance of this function a condition for receiving state funds.

A district board which systematically aims the application of practices toward well-defined land conservation goals must develop procedures which relate its annual work plans not only to its long-range program but also to the accomplishment of objectives in individual farm plans. Both district programs and farm plans ought to reflect recommendations on current practice and local needs. It is not likely that all types of practices on all classifications of land used for agricultural purposes should be treated as equally deserving of public assistance. Therefore, determinations ought to be made of areas critically in need of treatment and priorities set for their treatment, if the boards are to perform a significant policy-making function. District boards ought, also, to take the actions, within the limits of their authority, necessary to insure maximum coordination in the utilization of materials and services provided by all public agencies. If they do not, they can hardly make good any claim to recognition as primary local units of government for the accomplishment of public soil conservation goals.

It is certainly a matter for technical judgment whether, and when, district programs ought to be reviewed to bring them up to date with changes in both the system of agricultural production and recommendations of technical practices. Although supervisors of a district having a program written in 1945 or 1950, for example, might raise the question whether review and revision were in order, the initiative is likely to lie with technicians of the assisting agencies. Since somewhat more than two-thirds of existing districts have been organized for more than ten years, they might appropriately reconsider their programs. As of 1962, approximately half (143) of the districts responding to the special questionnaire for this study had done so. Among these, 97 had revised their programs since 1955. Participation by board members in the National Inventory of Soil and Water Conservation Needs—which the Department of Agriculture organized in 1957—undoubtedly stimulated some interest and, possibly, reconsideration of goals and achievements, since about 80 per cent of the districts reported that they had participated in making the Inventory. In any event, it is the present policy of the Soil Conservation Service to encourage boards to reconsider their long-range programs and make appropriate changes. There is rather widespread agreement among persons familiar with the districts, however, that supervisors tend to have little knowledge of the contents of their long-range programs and give little or no consideration to them as the foundations for annual plans of work. Supervisors interviewed for this study were inclined to think the long-range programs had no practical importance, or to profess complete ignorance of them. Generally, they limited their statement of district goals to some version of the slogan: "Use every acre according to its capability; treat every acre according to its needs." Quite common was the view that the supervisor's job is to "help farmers farm the conservation way."

The various efforts to stimulate district boards to make some kind of annual plan have been successful, if responses from the sample questioned for this study are representative of typical behavior. Ninety-five per cent (262) of the respondents said that they make such plans; and 80 per cent do so during the months from October through January, when plans must be made if they are to be related in any way to the county programs of the ASC committees.

The term "annual work plan" seems to consist in most districts chiefly of a list of annual promotional activities. Answering a request to describe the "major actions which you have planned for the current year in your plan," the Atchison County District in Kansas replied: "conservation on land; Publicity; Sponsor Watersheds."[9] The brevity of this description of the annual plan appears to be characteristic of the overwhelming majority of districts—except those in which the supervisors are so inactive that they have scarcely considered even such a minimum plan. However, some of the district boards, especially those which are required or urged by their state soil conservation committees to follow a specified form, spell out their plans in greater and more specific detail than the Atchison County District.

An example is offered by the Chester County District in Pennsylvania which mimeographs a program of monthly activities suggested by the State Soil Conservation Commission. In 1961, the Chester County District Board adopted the following annual plan: (1) Secure 50 new district cooperators. (2) In January, send annual, fiscal, and audit reports to the State Commission; elect officers. (3) Attend the annual convention of the NASCD in February; order soil stewardship material, and have the chairman see about "radio program." (4) For March, it set a deadline for the Future Farmers of America conserva-

---

[9] This quotation and the others which follow, including information about budgets, are taken from the responses to the special questionnaire by these districts or from copies of their annual reports and audit statements.

tion speech contest. (5) Action in April was to be a "check on radio program on Soil Stewardship." (6) The second week in May was set as "Soil Stewardship Week," and $100 was to be provided for the "Workshop General fund." (7) Require Future Farmers of America scrapbooks for annual statewide contest to be turned in. (8) Hold a conservation field day in the Honey Brook area in August (no activity in July). (9) In September, plan activities for Farm-City Week to be held late in November and prepare suggestions for the county conservation program meeting in October with the county Agricultural Stabilization and Conservation Committee. (10) In October, attend the annual meeting of the state association of districts. (11) In November, encourage cooperators to order trees from the state Department of Forests and Waters. (12) In December, "order Game Commission trees and shrubs; ask cooperating agencies to submit annual reports; hold annual planning session." At the annual planning session for the 1961 program, the District Board also adopted a brief budget of slightly over $1,000. All the items of expenditure, except $100 for clerical services, were to conduct the promotional activities which constituted the 1961 annual program for this District. Its income was to be derived from a donation by the county governing body and the sale of affiliate memberships to local banks. The actions planned by this Board for each month were in almost exact accord with the ones urged by the Pennsylvania Soil Conservation Commission, which issued a printed form for the districts called "Our District Program and Budget for 1961." A few other state soil conservation committees—in Arizona, for example—offer similar detailed guidance.

The response of the Fayette County District in Alabama was somewhat more general:

> To promote a conservation education program that will make it possible to enlist the support of all segments of our population in carrying out our soil and water conservation program when and where needed. To assist landowners in planning the use of their land for which it is best suited and the treatment that it needs for protection and improvement. To constantly search for farming enterprises that fit into a conservation program that will contribute to an increase in income for farmers of the District. To take leadership in special events such as Soil Stewardship Week and Soil Conservation District's Week as a means of making the general public more conscious of the need for soil and water conservation.

The Chairman of the Madison County District in Kentucky reported the 1961 annual plan to consist of "(a) 25 New Cooperators; (b) 25 New Basic Conservation Plans; (c) 15 Revisions of Old Plans; (d)

Increased application of erosion control structures such as sod water-ways, diversion channels, terraces, contouring . . . ; (e) Better public relations—thru newspapers, radio, etc." And the Marshall County District in Iowa planned several promotional activities for 1961, but also included two activities rarely reported by most districts. It conducted an "inspection tour of practices by district commissioners" and held a meeting with the conservation contractors in the county.

Supervisors were asked to check on a list of nine activities the specific promotional activities which they hoped to carry out in 1961. Responses from 80 per cent indicated plans to distribute conservation literature to schools, libraries, and the public; 75 per cent reported tours and demonstrations for farmers; 50 per cent included attendance by leaders from urban areas. From 50 per cent to 60 per cent planned to prepare and distribute publicity materials for local radio or television stations, participate in the Goodyear Awards program, and conduct conservation contests and awards programs for such rural youth groups as the 4-H clubs and chapters of the Future Farmers of America. About 40 per cent of the districts were sponsoring conservation workshops or scholarships for local schoolteachers to attend during a few weeks in summer. Surprisingly, 30 per cent reported that they included urban youth groups in their awards program; the most likely explanation of this high percentage is that district boards frequently include Boy Scouts in promotional activities. This practice has carried over from 1935, when Bennett received Henry Wallace's permission to launch a conservation promotion program through youth groups.[10] From responses to the questionnaire, it appears that county agents participate in district educational programs in between one-half and two-thirds of the districts.

When the district promotional activities become routine, the annual planning meeting is not likely to stimulate an analytical or critical state of mind among supervisors—as responses to the questionnaire showed.

---

[10] NARG 16 (Erosion File), memorandum, H. H. Bennett to Secretary Wallace, April 18, 1935. Bennett's initial proposal was to work through the U.S. Commissioner of Education and the state education departments to reach adult farmers through the vocational agriculture teachers. He further suggested that a program of "summer institutes and training courses" be established for rural schoolteachers through the state departments of education. This tie with the vocational agriculture teachers and their youth group, the Future Farmers of America, has been most useful to SCS in competition with those state extension services which have been "uncooperative." Every effort is made to work through the schools and such groups as Girl Scouts and Boy Scouts. Among adults, the state bankers associations and the rural life divisions of the major churches have been cultivated persistently.

Only 30 per cent reported that they periodically review their conservation promotion programs with private groups and governmental agencies which can contribute to it. This same low percentage of districts reported any analysis of their promotional activities to determine which kinds were most effective in securing the application of conservation practices to the land. Only 10 per cent reported that they had ever conducted studies of the factors which deter farmers from becoming active cooperators. And only 17 districts said they had ever heard of the existence of such studies made by anyone else.

Apparently most supervisors consider the "annual plan" to consist primarily of promotional activities sponsored by districts, but guided by professional agricultural workers. Official SCS policy apparently anticipates that annual district plans will be of this type, since SCS has informed its field personnel that districts "should include items in their work plans that are solely the responsibility of districts."[11]

SCS PROGRAM PLANNING

It is difficult to determine the extent and kind of influence the district boards have on the annual operating plans and long-term goals of such technical agencies as SCS. Most of the available evidence points to the conclusion that resistance by supervisors to specific practices or techniques advocated by the technical agencies occasionally forces agencies to modify their operations. At the same time, there is not much evidence that supervisors engage in conscious and formalized planning or review of annual or long-term plans proposed by the technical agencies. In the atmosphere of easy informality which permeates most district board activities, a skillful professional worker can influence a good many decisions in such a way that few of the members will realize it has been done. Certainly a most important element in the SCS-district relationship is the initiative assumed by the technical agencies in planning annual and long-term operations.

Officially, SCS insists that it merely helps districts to execute their own programs and annual plans. It takes the position that SCS does not have annual, national, state, or district program goals which are to be achieved with the consent and assistance of supervisors. As one SCS state conservationist remarked during an interview: "When people ask me what our annual program is, I have a perfect answer. We don't have

---

[11] U.S. Soil Conservation Service, Districts Memorandum SCS-5, Dec. 12, 1956.

one; we merely help the districts carry out theirs." This claim would reflect seriously on the responsibility of SCS as an arm of the Department of Agriculture charged with the achievement of national objectives, if it were wholly true. There would be no foundation in fact for the SCS claim that its program ought not to be "decentralized" to the state extension services, whose chief claim for such a change rests on the assertion that extension workers are best able to shape conservation programs to suit varying local conditions.

As a matter of fact, SCS has an elaborate set of procedures for planning annual programs of work, supervising the application of practices to the land, and reviewing accomplishments by its work units assigned to the districts. The SCS Administrator and members of his Washington staff have face-to-face meetings with state conservationists and their program staffs each August during the annual convention of the Soil Conservation Society of America. One feature of these meetings is the discussion of annual plans of operations, although this is not necessarily the primary activity. During the early fall months, each state conservationist and his program staff hold a meeting at which they work out goals and plans of operations with the assistance of specialists from the SCS field staff in Washington. Normally one of the "area representatives" of the SCS Administrator also attends this meeting. Policies, problems, and objectives are discussed so that areas of special emphasis can be given special consideration. Following this meeting, the state program staffs meet with the area conservationists for a similar discussion which is intended to enable area conservationists to help each work unit establish its goals and annual plan of operations, preferably by late November.

Preparation of the "annual plan of operations" by each work unit, area, and state is based on an official SCS directive.[12] The plan of operations is defined as one fixing the "goals" of work units and the "objectives" of area and state conservationists, together with the "general timing (month or season) of the actions to be taken." It is an SCS policy that the annual plan of operations for the work unit will specify goals defined as "tangible accomplishments in surveys, planning, application and maintenance of conservation work and facilitating activities." Work unit goals are required to be set in "measurable units" such as the number of new cooperators to be sought, new plans written, old ones revised, miles of terraces constructed, acres of range reseeded, woodland fenced, or other quantitative measures. In cases where a

[12] U.S. Soil Conservation Service, Management Memorandum SCS-3, March 19, 1959.

statement of objectives is appropriate, such as "improve the working relationship with the _____ soil conservation district," one should be included. Objectives are improvements of various behavior patterns and relationships which are not susceptible of quantitative measurement. Only those practices and activities which require special effort for accomplishment on the part of the work unit staff are to be included in the annual plan. Goals "which do not require aggressive action" are to be part of a "long-time schedule."

These procedures for planning annual goals can easily become a system of quotas to be met by the work units, and they are usually so viewed by field personnel who are given the chance to talk with a guarantee of anonymity. The SCS, however, insists that only "realistic goals" be set by taking into account the size and composition of each work unit staff, resources locally available, physical, social, and economic conditions, changing agricultural technology, trends toward urbanization, and other relevant factors. Area conservationists, who are responsible for supervising work units, are directed to see that the establishment of realistic goals does not impair sound working relationships with district boards. It is suggested that to the "maximum extent possible," area conservationists "will influence the development of realistic goals by working closely with Work Unit Conservationists prior to the joint district-Service planning meetings." This procedure is to enable the work unit conservationists to know in advance of the meeting with the district board the goals which his area conservationists considers to be "realistic." To put the matter as the typical work unit conservationist sees it, when he discusses his annual goals with the board and attempts to secure their approval, he will know exactly how much he will be held accountable for by the area and state conservationists. The SCS memorandum dealing with this matter continues by saying: "It should seldom be necessary for an Area Conservationist to directly contact a district governing body for the purpose of securing their approval of the revision of jointly developed goals." To insure the accomplishment of realistic goals in the work units, area conservationists are enjoined to "keep in mind national, state, and area objectives in assisting the WUC and the district to develop realistic goals." The achievement of these goals and objectives "will, of course, be taken into account in evaluating work performance."[13]

Beginning in the calendar year 1960, area conservationists were made responsible for evaluating each work unit under their authority

---

[13] *Ibid.*

within the following guidelines: (1) appraise the level of accomplishment in relation to each unit's realistic goals; (2) use careful judgment
in cases where some goals were exceeded and others not met; (3)
consider factors beyond the control of the work unit staff; (4) use other
grounds considered proper by the state conservationists. State staffs
were directed to make a careful "onsite study of goals and accomplishment before citing the work unit for high production or taking
necessary administrative action if the production is substantially below
the goals."[14]

Some indication of the influences of district boards on annual
operations plans can be gleaned from the fact that, of the 277 districts
responding to the special questionnaire, 65 per cent (183) met in 1961
to review the deficiences of their operations in the previous year so as to
consider measures of improvement for the program year. Only 80
districts, however, reported that they planned to take specific actions as
a result of such discussions, and only 62 met during the year to review
and revise their plans to meet unexpected circumstances. Of these
boards, 60 per cent (167) reported that they had adopted priorities of
technical assistance for the year. About 25 per cent (82) of the boards
gave priority to specific practices, and the remainder generalized their
aims in terms such as "promote watersheds" or "get more practices on
the land." It is quite probable that, because many boards have informal
procedures, members do not consciously adopt priorities and review
the SCS plan of operations. For example, the Fayette County District
(Alabama) reported that it "has not set any priorities to cooperators."
In its annual report for 1960, however, it said that emphasis in two areas
in the county was being placed on shifting from row crops on hilly land
to draining more productive bottom lands. This report was a newspaper
story in which the work unit conservationist was quoted. Pennsylvania's
Chester County Board reported that "practical farmers get priorities
over small acreages which request only ponds; permanent practices
get top priority." The Atchison County Board in Kansas put terracing as
its first priority, and planning second. In Iowa, Marshall County Board
put terracing first, and strip cropping second. It is fairly certain that
boards which have consciously set such priorities have given careful
attention to the annual plan of operations set by their respective work
unit conservationists. The statement of "priorities" which was completely in accord with the practices of supervisors interviewed
in the districts around the country, however, is this questionnaire
response from the Fulton County District in Illinois: "Take things as

---

[14] U.S. Soil Conservation Service, Advisory Notice W-1156, Dec. 30, 1960.

they come." In one district where interviews were conducted, the board members were asked to give their approval of SCS assistance on one project and to deny it on another. In the first instance, approval was given to aid the Boy Scouts by giving technical guidance for work on a camp pond. The board supported SCS in the second case by refusing to stake out a pond on farm land which had been purchased by a local professional man who was believed to be planning to turn the land into a residential subdivision. The board's action was particularly welcome to SCS, since a local banker had called to support the request for this work. The members of this board had more or less consciously adopted a crude set of priorities by denying SCS technical assistance for projects not connected with farming. With growing urbanization, however, SCS has worked strenuously to alter the attitudes of farm supervisors toward such work, so as to continue to function in areas in which land use is shifting from agricultural to urban or suburban purposes.

Among the districts in which field interviews were conducted, there was no evidence that any of the boards review SCS goals or annual quotas in any formal way. In one, there was some agreement by the members that certain practices were being emphasized at the cost of others which they considered desirable. Uniformly, however, they felt that existing procedures and farmer attitudes left them little occasion to influence actual priorities in the application of practices. They generally are not willing to take the initiative personally in inducing farmers to apply any particular practices. Supervisors were asked to identify the areas in their districts in critical need of conservation. Usually they could do so, except in the multiple-county districts. When they were asked what actions they had taken or planned to take to correct such situations, however, their answer was "nothing." With one exception, they did not consider it to be their proper function as supervisors to go out and directly exhort neighbors to change their ways. Since, in some of these districts, there has normally been a backlog of requests for assistance not filled because of the lack of means, the supervisors felt that the whole question was academic. Nevertheless, when district boards were asked whether they review and approve the technical plans of cooperating state and federal agencies before practices are applied to the land, only 52 per cent said "yes." Of the remainder, 122 said "no," and 10 did not answer. In response to the question whether they had ever required any of the cooperating technical agencies to modify their technical standards to meet local conditions, only 20 (7 per cent) answered "yes"; two-thirds of them answered "no"; the rest (46) did not answer.

A rare example of critical analysis was displayed in the 1961 annual report by the McLean County District in Illinois. The Board had decided in 1958 that it could no longer justify the use of SCS personnel to stake out tile drainage on lands needing this practice; it would, instead, leave the drainage work to private contractors. Experience with this policy convinced the supervisors that drainage, irrigation, and fertilization practices should be promoted by the district, but that SCS technicians should concentrate on land needing erosion control practices. Further, SCS should leave forestry management generally to the professional foresters, and rely on the state extension services to recommend agronomic practices, especially pasture improvement and the proper use of "level crop land." The McLean County District Board was of the opinion, at the same time, that: "Better cooperation from the state university must be secured. They don't seem to quite believe in soil conservation or see the need of having [vocational agriculture] teachers trained specifically in soil and water conservation." This Board's conclusions are of particular interest because its previous annual report was sent to other districts in Illinois and their views solicited in return. Replies received by the Board showed that opinions of farm plans varied from "considering them absolutely useless to being more important than getting conservation on the land . . . The only really sour note we ran into was that some directors suspected that some SCS officials consider farm plans a 'sacred cow' to be used always and no questions asked." These strictures on farm planning seem to be typical of the attitudes of supervisors. The Soil Conservation Service takes the position that the application of conservation practices is the primary objective of all its activities. At the same time, it emphasizes the "special importance" of farm planning which it views as "essentially a prerequisite to the primary job . . . Districts and SCS need to avoid unnecessarily getting into other activities that sidetrack them from these objectives."[15] Still, a number of supervisors remains convinced that the technician's time properly should be used for the application of practices and not for planning or other "facilitating" activities.

Some thoughtful supervisors are also doubtful of the efficacy of their own promotional activities. The McLean Board found in its survey of other Illinois districts that "there wasn't any coordination between an extensive promotional program and a good program of accomplishments on the land." As one director put it, "We sell conservation to everyone—the preacher, the banker, the school teacher, everyone except the farmer on the land."

[15] SCS, Districts Memorandum SCS-5, *op. cit.*

THE INDEPENDENCE OF DISTRICTS

In order to check responses from the district boards, the executive officers of the state soil conservation committees were asked in the special questionnaire to describe how district governing boards review and, if occasion should arise, require amendments to the annual operating plans of agencies giving them technical assistance. Out of 34 responses, all except one state official reported that the SCS annual work plans are discussed in an atmosphere of informality. Differences of view, if they occur, are aired on the spot. One state committee in the Northeast reported that there is "not much, if any" review. Only two states reported that there had been recent occasions when a district board had taken up its differences with SCS in writing "through channels." One committee spokesman in the Northeast said that districts review their memoranda of understanding annually and, if "shifts in emphasis or increases in the types of technical assistance given by the cooperating agencies are required, the District makes a formal request, in writing, to the appropriate agency. If this procedure does not result in the desired change, the State Committee is asked to negotiate the problem in the district's behalf." There are scarcely more than a dozen districts in this state, however. Two of the committee secretaries pointed out that if the process of review fails to satisfy supervisors of a district, they take the matter up in the area meeting of supervisors and, possibly, carry it to the state meeting where it may become a matter for discussion. Should they find sufficient support for their position, but not receive satisfaction from the assisting agency, the supervisors take the issue to the National Association of Soil and Water Conservation Districts.

The state committees were also asked to report the specific ways in which district boards influenced the technical agencies to modify their recommendations or specifications on practices in order to suit local conditions as understood by the supervisors. Twenty-eight state committees responded. One in the Northeast said that boards do "not to any extent" influence these matters. From the Mountain states, another reported, "Many cases have been attempted but not very successfully. Supervisors will discuss and encourage certain modifications which usually end up with alternate ways of doing the job." An executive secretary in a Northern Plains state said: "The . . . districts are constantly on guard to see that technical specifications are sound. A recent example of this was [when] an Area Engineer arbitrarily decided that a tube be asphalt dipped and because of the action of a district board this policy was rescinded. . . . Also the activity of one district board has been instrumental in seeing that terrace specifications . . . were modified to more nearly meet the needs of the local farmers. These are not exceptions but situations going on almost constantly." Again in the Northeast, a state committee reported that there have been "several instances where the district has asked the SCS to modify farm plan recommendations." On Public Law 566 small watershed projects it "is not uncommon . . . to have specifications modified to meet individual needs." The executive secretary in one of the major farm states of the Midwest reported five examples of modifications brought about "through practical experience of district cooperators as expressed through the boards." He further observed that "most changes can be traced to basic research, however."

From one other Northeastern state came this thoughtful response: "This is without question a weak and significant point. We have been so involved in overcoming opposition to soil conservation districts . . . and organizing new districts, and have placed such great emphasis on promotional and educational activities, that technical specifications and practices have received far less attention than they deserve. We have taken steps in recent planning efforts with districts at annual planning sessions to secure a more adequate review of such specifications and practices so that the district directors will become familiar with their duties and responsibilities in this field and become more likely to exercise initiative."

The state and national associations of district supervisors are viewed by hostile critics as mere rubber stamps for SCS and lobbies to pressure for state and federal appropriations. In some instances, this characterization is correct. But other state associations and, certainly, the National Association, provide two-way communication between supervisors and the technical agencies. Actually, the available evidence indicates that the associations are very important channels through which supervisors can force both technical and administrative changes. Recently, for example, one state association forced the Soil Conservation Service to transfer its state conservationist because the association leaders felt that they could not work smoothly with him. Personnel changes on the local level at the request of district boards are rather common, according to information gathered in interviews. The effectiveness of state associations has varied widely, however. In 1956, Waters Davis, former NASCD President, said that state associations varied from excellent to "very poor" and that NASCD should act to improve them. Where they are weak, he charged, "the original fault was in the State Conservationists of the Soil Conservation Service itself." The state conservationist, Davis said, "didn't see the concept of the people themselves doing the job that people themselves can do. So he just turned all his Area Conservationists and Work Unit Conservationists loose. I was noticing those Plowing Matches for instance. There was an SCS guy explaining each one of those practices that was going on. If a Supervisor had explained those and at the same time explained who he was, you'd have sold Soil Conservation Districts and the people's concept. But with this SCS guy doing something a Supervisor, if he's any good, can do for himself, the SCS—well I won't say they are overstepping the mark, but at the same time they are being something more than the junior partners of Soil Conservation Districts.[16]

The annual meeting of the National Association provides an important occasion for state and area supervisor associations to make demands for program and administrative changes, although many of these are discussed in relative privacy and are not formally recorded. Others are offered as resolutions for discussion as part of the agenda. At the national meeting in 1961, for example, the Kansas Association expressed its opposition to any amendment to Public Law 566 "which would permit the assumption of the contracting obligations of local sponsors by an agency of the Federal government." In contrast, the Arkansas Association asked the NASCD to support an amendment permitting the "Federal government to do the contracting, if requested by the local organizations." The Arkansas supervisors explained that "most local organizations have experienced difficulty in providing the necessary finances and personnel to properly award and administer contracts." They were thereby often "forced" into "untenable and dangerous positions" by the existing requirement that local sponsors be the contracting organization. When this issue was brought to the floor for action by the Council of the NASCD, there was a spirited debate which revealed sharp differences among the members, but the Kansas position carried. In this instance, disagreement between SCS and some of the district leaders was rather thoroughly aired in a meeting that was semipublic.[17]

Differences of this sort have led some people who are intimately

[16] National Association of Soil Conservation Districts, "Minutes of the Meetings of March 19–22, 1956," p. 18.

[17] NASCD, "Resolutions Index," mimeographed, Memphis, Tenn., Feb. 5–9, 1961. The author was an observer at this annual convention of the NASCD.

familiar with the relationships between SCS and districts to assert that the districts are not rubber stamps, and that they are independent of SCS. There is certainly a tendency which can be detected at meetings and in interviews for district supervisors and the employees of many of the state committees to be very sensitive about this matter. The remarks by Waters Davis in 1956 show that many supervisors resent the idea that districts are the junior side of the "partnership" with SCS. Many expressed differences of opinion as signs of district independence of SCS domination. This kind of discussion of district influence in SCS misses a more fundamental point, however. District boards have influence, but almost no real power, over the Agricultural Conservation Program administered by the Agricultural Stabilization and Conservation Service. And this program is crucial in determining how much agricultural soil and water conservation is accomplished in the United States.

SUMMARY AND COMMENT

An analysis of the role of districts in the Agricultural Conservation Program can more easily be undertaken if the districts' countributions to soil conservation are kept in mind. To this end, four factors seem significant:

(1) The districts have actively promoted programs to make farmers aware of erosion hazards, improved land use, and the availability of assistance from the Soil Conservation Service. To a great extent, however, districts identify soil conservation methods and objectives with the SCS and only infrequently with other federal or state agencies.

(2) The Department of Agriculture insisted in 1937 that states organize districts, rather than voluntary soil conservation associations, under state law, since the districts could receive appropriations and provide assistance for public purposes and voluntary associations could not. Districts were expected to provide labor, materials, and equipment for all farmers and to carry out land use regulations when they were deemed necessary.

So far not more than a dozen districts have enacted land use regulations. Districts receive appropriations in variable amounts from 60 per cent of the states, some counties, and from contributions and earnings. Perhaps as many as 500 to 600 districts provide conservation aides and 1,000 hire clerks who relieve technicians for direct assistance to farmers. Probably 900 to 1,000 districts provide conservation equipment appropriate to the needs of farmers at favorable rates, when it is not available through other sources. These services are useful and undoubtedly have contributed to improved land use, although the performance of most districts falls considerably short of the Department of Agriculture's original expectations.

(3) District program planning consists chiefly of annual plans to carry out promotional activities. As many as three-fourths of all districts may conduct annual conservation education programs. Most funds derived from local sources are apparently spent for supervisors to attend association meetings, to pay dues to state associations, and to conduct prize contests or provide conservation scholarships for teachers to attend conservation short courses.

(4) District supervisors occasionally influence the technical and administrative operations of the Soil Conservation Service, but it is doubtful that their decisions have much effect on other agencies or on the kinds and amounts of practices which farmers install on their land.

CHAPTER 11

The Districts' Part in the
Agricultural Conservation Program

THE HIGHLIGHTS OF the rivalry between the Production and Marketing Administration (PMA) and the Soil Conservation Service (SCS) for dominance over the soil conservation program were narrated in Chapter 6. The competitive urges in the agencies were restrained to some degree, first in 1949, by the provision which Representative Jamie L. Whitten of Mississippi inserted in the agricultural appropriation bill authorizing PMA county committees to transfer up to 5 per cent of their program funds to SCS for technical assistance for PMA-subsidized practices. Next, Secretary of Agriculture Charles F. Brannan took the second important step to reduce friction when he issued Memorandum 1278 in February 1951. This directive gave specific responsibilities to the Soil Conservation Service and the Forest Service for participating in the Agricultural Conservation Program (ACP). They, together with the state extension services, were joined with other interested state agencies and the state PMA committees to form an organization known as the ACP development group. After the Agricultural Stabilization and Conservation Service (ASCS) replaced PMA, the Secretary directed this agency to invite the governing boards of soil conservation districts to advise the county committees which are responsible for recommending practices and the amount of federal subsidy for each in county conservation programs.

These changes had four basic consequences which have profoundly affected the role of soil conservation districts in the administration of the Department of Agriculture's soil conservation programs. The first of these is that district governing boards serve only in an advisory capacity to county ASC committees which are completely responsible for administering the Agricultural Conservation Program on the county level. Second, since ACP subsidizes the whole range of farm conservation practices, and districts have no funds for such work, districts are left with little power to control farm conservation. Third, since ACP subsidizes all of the "permanent" practices installed under the technical direction of the Soil Conservation Service, the district boards have no authority over this phase of the Department's program. This situation is ironic insofar as spokesmen for the districts have created the impression that their program stresses "permanent" practices, and "genuine soil conservation" in contrast with ACP, which is accused of frittering away federal funds on practices that ought to be part of good farming operations. And, fourth, there is little, if any, basis for the idea that districts, the Soil Conservation Service, or the Agricultural Stabilization and Conservation Service have distinct and separate programs or unique conservation missions to accomplish.

RESPONSIBILITIES FOR PROGRAM PLANNING

Memorandum 1278 provided for annual program planning on the county and state levels, without reference to the regional offices of SCS which still existed in 1951. At least in theory, this directive made the county the primary unit in the planning process. It made the state the primary level for the technical phases of program planning by providing for both the ACP development group and the necessary technical advisory committees representing all state agencies having an interest in, and contributing to, the Agricultural Conservation Program. The Secretary of Agriculture has declared that representatives of ASCS, SCS, and the U.S. Forest Service are "equally and jointly responsible for developing the state program." In recommending practices and setting specifications for them, the agencies are enjoined to "reach unanimous agreement. . . . No one of the responsible agencies has veto power to the end that no further action can be taken."[1] The Administrator of ASCS will decide cases in which disagreement persists. Present procedures reserve for the Department of Agriculture authority to approve state programs within the framework of a national program which is authorized annually in its "Handbook." The national program changes little over short periods of time and, consequently, the practices authorized for federal subsidy have remained relatively stable during the past decade, except for wildlife measures added in 1962.

The county ASC committees are completely responsible to the Secretary of Agriculture through their state committees for administer-

---

[1] U.S. Department of Agriculture, "Agricultural Conservation Program Service Handbook," 1-ACPS, mimeographed, May 1, 1960, as amended, par. 84; see also pars. 84–93 for general information.

ing the Agricultural Conservation Program. Soil conservation district governing boards may serve only in an advisory capacity to perform certain program planning functions.

Current regulations require that the annual ACP plans of both the county and the state programs be prepared in two steps. The first consists of recommendations of practices to be included; the second, of decisions that approved practices are to be included in the programs and subsidies (cost-sharing) paid at rates specified for these practices. Each ASC county committee is directed to invite the governing board of any soil conservation district organized within its boundaries to share in the preparation of county program recommendations. Others to be invited include representatives of SCS, the county agricultural agent representing extension, a U.S. Forest Service representative having jurisdiction over "farm forestry," the county supervisor of the Farmers Home Administration, and representatives of other agencies having conservation interests. Evidence gathered in field interviews with ASCS state officers and the county committees indicates that local dealers in seeds, equipment, and other materials used for conservation practices usually are invited to attend in an unofficial, but sometimes influential, capacity. Final responsibility for making the county recommendations, however, rests with the county development group. It consists only of the ASC county committee, the local SCS technician, and the Forest Service representative; the governing bodies of soil conservation districts are excluded from this group.[2]

Among districts responding to the special questionnaire for this study (see Appendix B), 80 per cent reported that one or more of their supervisors, together with the SCS work unit conservationist, attended the last annual meeting of the ASC county committee at which preliminary recommendations were prepared. Only 12, less than 5 per cent of the 277 responding, reported that all supervisors attended this meeting. Forty-three (15 per cent) reported that one or more district supervisors were also members of the ASC committee in their respective counties, and 25 districts (10 per cent of responses) reported that only the SCS technician represented the district. Seven reported that no one represented the conservation district.

In one Southeastern state, where the proportion of districts responding to the special questionnaire for this study was very low, the executive secretary of the state conservation committee reported as

---

[2] *Ibid.,* par. 83.

follows: Out of 99 counties, representatives of conservation districts in 48 participated in formulating the ASC committee request for ACP funds; 70 participated in selecting practices for cost-sharing; 48 helped determine rates of cost-sharing; and 59 coordinated these plans with their district programs. Information on this subject was somewhat incomplete, in that only 76 of the 99 counties responded to this inquiry from their state committee. According to the report from another Southeastern state committee, in which response to the districts questionnaire also was extremely low, only three districts in the state were known to participate in annual interagency planning meetings. In one Western state, where responses to the district questionnaire were low, the state committee reported that no meetings of this sort are held. In another, also in the West, it was reported that about half of the districts regularly participate. Still another in the West reported that about one-fourth of the districts are active in this respect. Apparently, at least one member in 65-70 per cent of the district boards attends one or more meetings of the ASC committee to discuss the county program.

Evidence gathered from field interviews is that most supervisors view attendance at this ASC committee meeting as a chore. Sometimes a supervisor attends because he has some influence with members of the ASC committee and can help the SCS technician carry a decision in favor of, or against, the inclusion of particular practices. Or the supervisor can otherwise defend the SCS technician against some hostility from the ASCS side—especially any growing out of the 5 per cent transfer of funds. Insofar as experience in the districts where field interviews were conducted is typical, it can be said that most district supervisors know little or nothing about ASCS regulations and have almost no idea of the role they could play in formulating and influencing the county programs.

Where districts are organized on a multiple-county basis, or a district contains parts of more than one county, no effort seems to be made by district boards to deal as a whole with the various ASC county committees concerned. In the four multiple-county districts visited in Virginia, Georgia, Louisiana, and Texas, all negotiations over county programs, cost-share rates, and the 5 per cent transfer were normally conducted by the SCS work unit conservationists responsible for each county and the ASC committee of that county. Occasionally, they did have the assistance of a district supervisor, if one lived in the county. Where more than one district was located in a county, the practice was for the senior work unit conservationist of SCS to negotiate with the ASC county committee. In such circumstances, it is evident that the conservation district work plan bears little, if any, substantial relation to the county conservation programs of ASCS. Even in districts organized on the basis of single counties, it appears to be doubtful that as many as 15 to 20 per cent of the boards give regular and comprehensive attention to formulating the ASC county program and relating it to the annual plan of work of the district. Of course, since ASC county and state programs remain rather stable in respect to authorized practices— although not necessarily as to rates of cost-sharing—district supervisors may justifiably feel that their participation in the formulation of county programs is largely a formality.

These conclusions are supported by responses to inquiries on the questionnaire. District boards were asked which practices authorized in their state ASC programs, if any, their respective ASC county committees excluded from their programs. Of those districts which answered, 62 (22 per cent) reported that there were no exclusions; 135 (50 per cent) specified practices which were omitted, and 32 of these (11 per cent) gave specific reasons for the exclusions. Districts were also asked which practices, if any, were included by the ASC county committee for the 1961 program year that the district board considers unwise, unsound, or unnecessary for cost-sharing. In answer, 71 per cent (191) said "none"; only 9 per cent (25) specified ACP practices which they disapproved; and the remaining districts did not answer. When the boards were asked what practices, if any, their ASC county committees refused to include for cost-sharing, 75 per cent (209) replied "none." Only 5 per cent (13) reported any such refusals, and the remainder did not answer the question. District boards were asked whether they had any suggestions to make for improving the ASCS program and working relations among the agencies involved. "None" was the answer of 50 per cent (144). However, 20 per cent (57) offered suggestions, most of which involved details of operating procedure; 6 of these recommended more funds and 7 more "education." The remaining districts (30 per cent) were silent. When asked what proportion of the conservation practices installed in their district received some ACP cost-shares, 31 per cent (90) reported that it was between 80 per cent and 100 per cent; 16 per cent (45) reported that it was between 70 per cent and 80 per cent; and 40 per cent (108) reported that it was below 70 per cent. The remaining 13 per cent did not answer. Interestingly enough, 65 per cent of these respondents reported that they answered this last question on the basis of an estimate made by SCS. Only 33 (12 per cent) obtained the data from their ASC county committees and 18 (6 per cent) from their own district records.

Each county program includes the practices which its ASC committee has selected from the authorized state program. Strictly speaking, county committees cannot be forced to include any practice which they do not want, and occasionally they are adamant in their refusal to include some authorized practices. According to information given in field interviews, the attitudes on this problem vary among ASC state committees. Some have taken the firm position that the technicians of the land-grant colleges, the SCS, the Forest Service, and other related agencies may think that they know what practices farmers ought to want, but that county committeemen will exercise the final authority to include or exclude particular practices in their programs. In other states, pressures are sometimes exerted by exhortation and other actions by the ASC county office managers and supervisors of several counties. Again, it should be noted that the state soil conservation committees have no functions to perform in administering the state and county programs of ASCS.

Although it is true that the county committees have the power to exclude nationally approved practices from their programs (and also to request special practices subject to approval annually), the ASC state development groups are not without influence over the county programs.[3] First, of course, is the formal power to approve each program. The power to fix cost-sharing rates and to determine most of the specifications for practices adds very much to the practical influence of the state development groups. Most of them use this authority to stimulate some practices and to retard others. For example, in 1961, field interviews with personnel of both ASCS and SCS revealed that both agencies were working to secure wider farmer acceptance of some new "wildlife conservation" practices in ACP for 1962, especially in the East.[4] At the same time, SCS was trying in some states, at least, to cut down on the construction of "storage-type reservoirs," or farm ponds, as they are generally known. In California, a change in the political composition of the ASC state committee was expected to result in a sharp increase in the rate at which costs of certain practices valuable on irrigated land would be supported by ACP. In Iowa, the

[3] *Ibid.*, pars. 126–60.

[4] At the Senate hearings on the budget of the Agricultural Stabilization and Conservation Service for 1964, Administrator Horace D. Godfrey remarked: "Just recently, at the insistence of many people, we have included in ACP some practices to benefit fish and wildlife. We have made quite an effort during the last 2 years to get these practices included in county and State programs and to get them used on farms." *Department of Agriculture Appropriation Bill for 1964*, Senate, 88 Cong. 1 sess. (1963), p. 552.

new ASC state committee members said they hoped to increase the proportion of ACP funds used to stimulate "permanent" practices, especially terraces.

Congress has, however, by a rider to the agricultural appropriation, placed some limits on involuntary changes in a county's program, if they have the effect of making participation by farmers more difficult in one program year than in the previous one. The prohibition against "restrictive changes" forbids a state committee to omit practices or reduce the percentage of cost-sharing for practices in any county, unless such changes have been recommended by a county committee and approved by the state committee. This prohibition was intended by the House Agricultural Appropriations Subcommittee where it originated to protect at least one of its members from suffering adverse effects from changes in the county programs within his congressional district during an election year. As long ago as 1948, an election year, Representative H. Carl Andersen, Republican of Minnesota and a member of the Appropriations Committee, was quite disturbed over cuts in the allocation of ACP funds to the states. When the Under Secretary of Agriculture, Norris E. Dodd, reported the sum allocated to Minnesota, Andersen said: "Now I am getting into pay dirt. Minnesota received only $3,038,000?"[5]

RESPONSIBILITIES FOR PROGRAM ADMINISTRATION

The ASC county committees are responsible in general for administering ACP; soil conservation districts have no direct administrative responsibilities, although they may be consulted on one or two matters.

If a farmer wants to install a practice such as improvement of permanent cover in his pasture, he goes to his county ASC office and makes a formal request. Since this practice is one for which the ASC county committee alone is responsible, the committee decides whether the practice is needed and, assuming that funds are available, authorizes payment for it. Most farmers will install this kind of practice without technical guidance from any agency, although they must follow the specifications established by ASCS in the ACP's "Service Handbook."

[5] *Department of Agriculture Appropriation Bill for 1949*, House, 80 Cong. 2 sess. (1948), Part 2, pp. 170–82 (especially p. 171). See also "Agricultural Conservation Program Service Handbook," *op. cit.*, pars. 161–74 (especially par. 165 for allocation of funds).

If a permanent practice, such as standard terraces, is desired, the farmer applies to the ASC county committee. The ASC committee then requests the local SCS technician to make an on-site inspection of the farmer's premises to determine whether the practice is needed and practicable and where the terrace should be located. If the technician determines that terracing is not needed, he reports to the ASC and the matter is ordinarily ended. If the technician decides it is needed and practicable, he reports to the ASC committee, which then authorizes the practice at the rate of subsidy which has been approved for this practice in the county program. The SCS technician is responsible for laying out the job, once it is approved by the county committee. He must also supervise installation, inspect completed work, and certify to the ASC county committee that the practice has been properly installed. This done, the farmer receives his payment. At no point in these procedures does the governing board of a soil conservation district have any power of decision.

If a farmer wants one of the two forestry practices subsidized by ACP for which the U.S. Forest Service has technical responsibility, the procedure followed is essentially the same as in cases for which SCS has technical responsibility. Needless to say, soil conservation district boards have no authority over personnel of either the U.S. Forest Service or any state forestry agencies which have been designated by the Forest Service to provide technical supervision of the subsidized forestry practices.[6]

The regulations direct ASC county committees to enter into "mutual agreements" with the technical agencies and soil conservation district governing boards to schedule technical assistance in such a way that farmers are assured they will receive timely aid in carrying out practices for which cost-sharing is provided. "The determinations on *priority of ACP work* to be done under the transfer agreement or the designation of farms on which such work is done, is the responsibility of the county committee." It is also provided, however, that in those counties in which ACP funds are transferred to SCS (up to a maximum of 5 per cent of the total funds available for cost-shares), the recommendations of the SCS technician and the district governing board shall be given "full consideration" so that SCS will discharge its dual responsibilities to ASCS and the districts to "accomplish the

---

[6] "Agricultural Conservation Program Service Handbook," *op. cit.*, pars. 196–99. In cases where either of these agencies is unable to assume its usual responsibility, the ASC county committee may employ suitable persons to perform this function. *Ibid.*, par. 200; see also pars. 353–54.

objectives of both." Particularly pertinent is this admonition to SCS and district boards: "In servicing ACP practices the technician shall make no distinction between farmers who are cooperating with a local soil conservation district and farmers who are not. Practice referrals are to be serviced and in line with the time the farmer intends to carry out the practice, allowing only for adjustments in the order of servicing necessary to make the most efficient use of time and travel."[7]

Regulations for the ASC county committees also state firmly that no farmer shall be required to pay for the technical services rendered for ACP by agencies of the Department of Agriculture, nor is he required to make any deposit of funds to guarantee that he will complete practices. In addition, he is not "required to sign a cooperative agreement with any local soil conservation" or other special district "as a prerequisite to obtaining the necessary technical services furnished by the Department." Regulations do state, however, that there is "no reason" why an SCS technician may not suggest to a farmer receiving ACP assistance the "advantages of cooperating with a local soil conservation district." Districts are forbidden to make "assessments against" or require "deposits from a district cooperator for technical services furnished by the Department."[8]

The ASC county committees have been directed to "invite" district governing boards to discuss annually whether or not to transfer to SCS up to 5 per cent of the county's ACP funds to enable SCS to service the practices for which it has technical responsibility. Such transfers are negotiated on the county level, but they are made on the state level. These funds are used by SCS chiefly to provide part-time aides who work under the direction of SCS technical personnel. It is solely the responsibility of the ASC county committee to decide whether to transfer the full 5 per cent, or any portion of it. If the decision is in favor of the transfer, the county committee and the SCS technician are responsible, jointly, for recommending the transfer to the ASC state committee. Even though a county committee decides against a transfer, the SCS technician is "obligated to discharge his technical responsibilities . . . to the extent his resources will permit without disrupting his discharge of regular duties under other assignments." Any ASC county committee which refuses to transfer funds is directed to "adjust its program in light of the ability of the technician to service approved ACP practices without the additional funds."[9]

[7] *Ibid.*, par. 203 (emphasis added).
[8] *Ibid.*, par. 205.
[9] *Ibid.*, pars. 214–15.

Soil conservation districts can neither forbid SCS technicians assigned to them to service ACP requests approved by the ASC county committees nor can they force county committees to make a transfer to expedite this work. They may participate in the negotiations only upon invitation from the county committee. No specific sanctions are provided for cases in which a county committee does not extend the invitation or make the transfer. The county committees, however, are subject to obvious forms of administrative persuasion by the ACP state committees and higher authorities in ASCS. Transfers are not made in all counties, but the number has gradually and generally increased since the 1950 program year in which this procedure was initiated at the direction of the Subcommittee on Agricultural Appropriations of the House of Representatives.

These transfers of funds were made originally under considerable duress by many ASC county committees; for example, only 96 made them for the 1950 program year, and 98 did so in 1951. The significant factor in forcing the change was Secretary Brannan's Memorandum 1278, which in 1951 vested responsibility for the technical supervision of certain "permanent" practices in SCS. In the 1952 program year, transfers were made in 1,153 county committees. There was a similar dramatic increase to 2,380 in 1953; by 1957 the total had reached 2,881; and in 1960 was 2,878.[10] Even so, there are still between 100 and 150 counties which fail or refuse each year to make this transfer. Missouri accounts for the bulk of these, although Georgia has about 20, and Florida and Minnesota both have about 12 counties which do not transfer. Each county committee's recommendations relevant to such a transfer are reviewed on the state level by a joint ASC-SCS (or other) committee which is responsible for working out, if possible, reciprocally acceptable agreements.[11]

The annual negotiations of these agreements is frequently the subject of some friction between ASCS and SCS and their local partisans. For example, during the hearings on the SCS budget estimates in 1962, Donald A. Williams, Administrator of SCS, testified under questioning by Representative Andersen of Minnesota that SCS had recommended to the Department of Agriculture for "conservation operations" for 1963 a total of $119,144,000 "which included a suggestion that a direct appropriation of $10 million be made to the Soil Conservation Service

[10] U.S. Department of Agriculture, Agricultural Stabilization and Conservation Service, *Agricultural Conservation Program, Statistical Summary, 1960,* mimeographed, Table 24, pp. 100–01.
[11] "Agricultural Conservation Program Service Handbook," *op. cit.,* par. 223.

in lieu of ACP transfers." Andersen asked Williams if he meant that SCS wanted this amount to be appropriated to it directly instead of working out the agreements with ASC on the county level, and Williams answered: "That is correct."[12]

Again, during the hearings on the SCS budget estimates in 1963, Chairman Whitten lectured Williams, reminding him of the reasons for the transfer.

> I think this committee can claim some credit for efforts to get each to see that the other service was one of the chief supporters of its service. Each without the other would have a hard job functioning. . . .
>
> May I say that every two or three years the Association of Soil Conservation Districts has come to me as chairman insisting that this 5 per cent be given to SCS. I have pointed out that my experience in Congress makes me know that it would not be two years before SCS would not be getting any 5 per cent. They would get just so much a year, and that is it. I have plenty of evidence to support that contention.
>
> On the other hand, I have had many folks concerned with the ACP program come and say, "Why should we give that crowd 5 per cent? Why don't you just let us do our own work?"
>
> The point I make is that the reason for the 5 per cent is to join the two together. It keeps the ACP from setting up a new group of technicians in competition with SCS which remains the technical arm of ACP. But it also means that you get paid if you do this work. If they do not want you to do the work, they do not pay you. And if they pay you, you have to do the work.[13]

At the 1962 hearing on the SCS budget, Administrator Williams testified that SCS was reimbursed during the 1962 fiscal year from ASCS for "1,320.6 man years of the value of $8,297,263." He claimed that this amount was only 58 per cent of the total cost to SCS for servicing these requests and that, therefore, SCS "absorbs" 42 per cent of this cost. Williams did not mention the actual amount of the reimbursement from ASC, but this figure was reported in the testimony of Horace D. Godfrey, Administrator of ASCS, to be $7,983,000. The difference between the "value" which SCS placed on this service and the actual reimbursement is only $314,263.[14] The extent to which SCS

[12] *Department of Agriculture Appropriations Bill for 1963*, House, 87 Cong. 2 sess. (1962), Part 2, pp. 944–45.

[13] *Department of Agriculture Appropriation Bill for 1964*, House, 88 Cong. 1 sess. (1963), Part 2, pp. 1033–34.

[14] *Department of Agriculture Appropriation Bill for 1963*, House, *op. cit.*, Part 2, pp. 1035–75. Part of the confusion possibly arises from the fact that the SCS reports

provides technical service for the Agricultural Conservation Program is emphasized by the fact that during the 1962 ACP year SCS serviced approximately 35 per cent of the total national expenditure for all ACP practices.

The extent of the technical responsibility which SCS assumes for ACP varies somewhat among the states. First of all, this responsibility may be for any one, some, or all of four steps required for installing a practice: (1) determination of need and practicability, (2) site selection, (3) supervision of installation, and (4) certification of performance. For example, during the 1964 program year, SCS was responsible for all four of these steps for ACP practice G-4 (a wildlife habitat management practice) in ten states, but in North Dakota it was responsible for only the first step of G-4. Another reason for variation is that the "ACP National Bulletin" specifies the practices for which SCS must assume technical responsibility in all states. The Department's regulations of the Agricultural Conservation Program also stipulate, however, that ASC state committees may negotiate additional assignments of responsibility with SCS. For example, during the 1961 ACP program year, SCS had technical responsibility for contour strip cropping (ACP practice A-5) in Virginia, but in Pennsylvania it did not. The state handbook provided, instead, that either SCS or the Pennsylvania Extension Service was responsible for field strip cropping (ACP practice A-6). This practice is similar to A-5, but technicians differ over the wisdom of using one rather than the other practice in a situation in which either might be applied. On the other hand, ACP regulations require that the Forest Service be responsible for designating the agency which will be technically responsible for initially establishing trees (ACP practice A-7) and improving a standing forest (ACP practice B-10). During the 1961 program year in Virginia, the State Division of Forestry provided technical supervision of these two practices; in Louisiana, the Soil Conservation Service was designated as the responsible agency. According to reports, there was sharp skirmishing over the issue in the latter case, and SCS was accused of wresting this assignment from the Louisiana Forestry Commission to the dismay of the U.S. Forest Service.[15]

---

its operations on the basis of the fiscal year, while the ACP program administered by ASCS corresponds normally to a calendar year, although an ACP program year can start in September rather than January of the year for which it is numbered.

[15] U.S. Department of Agriculture, Agricultural Conservation Program Service, "Agricultural Conservation Program, Handbook for 1961, Virginia," p. 4; *Ibid.*, "Louisiana," p. 4; based further on field interviews in October, 1961. Other data supplied by ASCS in letters to the author, May 8, 1964.

Since the ASC committee is responsible for checking farmers' performance of practices, except in cases where SCS or a forestry agency has been given this responsibility, district boards have no share in this important phase of program administration.[16] Inasmuch as they are even forbidden to limit the amount of time SCS technicians devote to servicing ACP, it is apparent that, as farmer demand for ACP permanent practices increases, the opportunity for district boards to direct the work of SCS decreases.

Some farmers—perhaps as many as 10 per cent—install practices without ACP subsidies. When they do so, they are authorized to receive technical assistance from SCS, if they are district cooperators. In such cases the Department holds district boards responsible to some degree for achieving permanent results, but apparently most of them accept this responsibility only in a very limited and ambiguous way. Of the districts responding to the special questionnaire, 90 per cent (250) reported that they require the farmer-district agreement to be signed by the cooperator and at least one member of the board. Only about 11 per cent of them, however, reported that the agreement is discussed by the farmer-cooperator and the entire membership of the board before it is signed and accepted by both parties. When the district boards were asked how they assure themselves that practices are properly installed and maintained under these agreements, 47 per cent reported that they leave the matter to SCS; 11 per cent reported that they take some kind of further action on their own; 16 (5 per cent) said they do nothing; the remainder did not respond. In response to a question as to whether the board, or a member delegated by it to act, made on-the-spot reviews of the work of the technical assisting agencies, 41 per cent answered "yes"; 42 per cent said "no"; and the remainder did not respond. There was very limited response to the important question: "What steps, if any, does the board take to correct unsatisfactory work by any of the cooperating agencies?" Two districts reported that the situation was reported to SCS; 2 said they asked for a change of personnel; 16 said that there was no need for action; and 25 said they did nothing. The overwhelming majority did not answer.

Either the situation is satisfactory to most boards, or they did not care to commit themselves in writing on an issue about which all concerned tend to be sensitive. When this matter was discussed with supervisors in field interviews, they most frequently characterized their relationship with SCS as being analogous to a school board's relation to

[16] "Agricultural Conservation Program Service Handbook," *op. cit.*, pars. 400–427.

the superintendent. At the time I visited one of the districts, relations obviously were strained. Several supervisors were very angry because at least two jobs installed under SCS supervision with district equipment had been highly unsatisfactory to the farmers concerned. The district supervisors felt that the fault actually lay with SCS because of technical incompetence. Interviews with a widely scattered sample of supervisors and professional workers familiar with SCS and district relations revealed that incidents of this sort are not uncommon. In one case where a district had become wholly inactive, the explanation offered by well-informed individuals was that the SCS technician had been incompetent and the district board could no longer withstand the disapproval of farmers. The result was that this district was abolished by legal procedures when it was attacked by opponents.

It can be argued with considerable force that district supervisors have neither the time nor the competence to inspect actively the work done in their districts. Only technically trained personnel who report their activities to the boards are competent to perform this function. However, the boards can claim to be nothing more than local sponsors and promoters of SCS and its programs if they make no effort to see that farmers who receive technical and financial assistance from public agencies use this aid to achieve permanent results. There is no question that the Department of Agriculture originally viewed this as a primary function which led to establishment of the boards. The progressive deterioration of the use of this function has been traced throughout this study. Today the Soil Conservation Service—and the districts—have abandoned all but the pretense that farmer-district agreements establish a responsibility of governing boards to insure that the public investment in soil conservation shall not be frittered away. The official policy of SCS is that it "regards district cooperative agreements as an important device for recording the agreement between the district and the cooperator as to what each intends to do. Such agreement can do much to prevent misunderstandings. Furthermore . . . a well designed . . . agreement will have a desirable effect in emphasizing to a cooperator his responsibility in developing and carrying out a conservation plan for his unit. *We do not consider a district cooperative agreement in the light of legal contract on which a district could base a law suit against a cooperator.*"[17]

The idea that district boards should act as public bodies, using legal sanctions when necessary to secure their goals, has been discouraged

[17] U.S. Soil Conservation Service, Farm and Ranch Planning Memorandum SCS-3, Aug. 11, 1955, emphasis added.

by the Soil Conservation Service. Instead, the key actor in this performance is the work unit conservationist, who is literally on the ground
with the farmers of nearly every agricultural county in the nation. In a
training directive issued in 1960 to its personnel, SCS summarized the
essence of its position by saying that the "main objective of your work
in soil and water conservation is to *influence* the people who control
and operate the land to become conservationists . . . Permanent accomplishments can be expected only to the degree that soil and water
conservation becomes part of the pattern of conduct of groups of
people, especially the neighbor groups. When neighbors 'think conservation,' then . . . conservation becomes the traditional way of life. It
becomes a 'major function' of the group."[18]

The fact that conservation district boards are discouraged from using
any methods stronger than gentle persuasion to induce farmers to
conserve their soil does not mean that no one is accountable for the
proper use of federal conservation assistance. The Secretary of Agriculture has placed this responsibility in the Agricultural Stabilization and
Conservation Service. The ASC county committees have been expressly
directed to determine that all practices installed by farmers with ACP
subsidies and the technical assistance supplied by the U.S. Department
of Agriculture have been established according to departmental regulations and are maintained for their normal lifespans. The ASC committees are responsible for all practices, whether they are "permanent" or
"enduring," such as terraces, or less "enduring," such as a winter cover
crop. Where federal funds are used to install a practice, the ASC
county committees are completely responsible, and there is no reason to
devolve additional responsibility upon districts which are not subject
to the effective direction of the Secretary of Agriculture.

ACP: EXTENT AND DISTRIBUTION
OF CONSERVATION PRACTICES

Since it was created in 1935, the Soil Conservation Service has
insisted that it has a unique mission to plan and execute a national

[18] U.S. Soil Conservation Service, "Working Effectively with Groups of People,"
mimeographed (1960), especially pp. 17 and 22–23. The National Association of
Soil and Water Conservation Districts, in a more recent statement, reiterates this
point of view. It asserts that "under the American system of private property
ownership, the nation's efforts toward resource conservation and development
depend, in a singular way, on the full and voluntary participation of private
landowners and operators." NACD *Tuesday Newsletter,* June 18, 1963.

program of soil and water conservation. Because the Department of Agriculture required the states to establish soil conservation districts as a condition for receiving technical assistance administered by SCS, there is a prevailing impression that the districts, too, have a distinctive soil and water conservation program and a unique mission. The Agricultural Stabilization and Conservation Service engages in a similar kind of image-building. Officials of ASCS and its supporters insist that ACP is the only conservation program which reaches all farmers. Spokesmen for the National Association of Soil and Water Conservation Districts (NACD) counter with the claim that ACP is a subsidy for farm operations in the guise of soil conservation. Both camps try to obscure the fact that Congress and the President hold the Secretary of Agriculture responsible for these allegedly different programs. Consequently, some questions are in order: Are there two distinctive programs which differ from each other in some way? If there are two, do they have different objectives requiring different means for their accomplishment? And are the distinctions, if any, desirable, especially for the future?

At the budget hearings for 1964, ASCS Administrator Godfrey showed Senators a chart "to illustrate the fact that is not generally recognized and that is that ACP is the only ongoing conservation program which applies to all of the farmland."[19] Donald A. Williams, the SCS Administrator, told the same group of Senators that the "soundness of providing cost-share funds for payment to land-owners and operators based upon a technically sound plan has been proven over and over again."[20] He added: "Soil conservation districts have been the parent organization, providing the stimulus and leadership for growth of the national conservation program during the last 2½ decades."[21]

During the hearings in 1963, Administrator Williams asserted that SCS had assisted farmers to convert 21.5 million acres of cropland to other uses. When Senator Spessard L. Holland, Democrat of Florida, challenged him to clarify his claim, Williams inserted in the record a statement disavowing that SCS was exclusively responsible for these land use changes. He said SCS had studied farmer actions in sample soil conservation districts and found that "it is impractical to measure separately the impact of technical assistance and the various methods of financing soil and water conservation work on private lands. The

[19] *Department of Agriculture Appropriation Bill for 1964,* Senate, *op. cit.,* p. 544.
[20] *Ibid.,* p. 182.
[21] *Ibid.,* p. 184.

records of applied conservation work in districts have been kept on the basis of completed land treatments, rather than on the factors or sources of funds contributing to decisions of landowners to go ahead with their conservation work."[22] The 21.5 million acres converted to "less intensive" uses, SCS reported, were in districts and 90 per cent was on the land of district cooperators, but this figure did not include other conversions made without SCS technical assistance.

When SCS analyzed 2,000 examples of cropland conversion by district cooperators, it found that 316 used only the technical services of SCS; 1,406 used ACP cost-shares; 159 made contracts to receive subsidies under the Great Plains Conservation Program; 299 used long-range contracts then provided by the Soil Bank Program; and 125 used loans from the Farmers Home Administration.[23] This analysis indicates that 70 per cent of the district cooperators making cropland conversions used ACP and 29 per cent more used one of the other forms of federal financial aid. If the proportion of practices involved in converting cropland to trees or grass is 70 per cent among district cooperators, there is no reason to believe that it is lower for the permanent practices, such as terracing, which require SCS technical assistance as a condition of receiving ACP payments. Consequently, the volume and distribution of ACP permanent practices for which SCS has technical responsibility greatly influence the work load of SCS, and are far more influential than any decisions made by most soil conservation district supervisors.

Since SCS does not officially report the extent of its technical assistance to farmers for ACP, it is necessary to rely upon ASCS statistical reports for this kind of information. These reports also provide a clear index of the volume, location, and character of the demand for permanent practices, although it gives no clue to areas of critical need. There is likely to be variation between any two consecutive years in the total national volume of demand for permanent ACP practices, as indicated by Table 5.

From 1960 to 1961, the table shows a decline in the total number of referrals to SCS, but a sharp increase in the number of completions. Since there are approximately 4.5 million farms in the United States, SCS technicians determined the need for ACP practices for one farmer out of every 11. The SCS technicians serviced ACP practices to completion for one farmer out of every 18 in 1960, and one out of every 12 in 1961. Almost 70 per cent of the total increase in the number of

[22] *Ibid.*, p. 183.
[23] *Ibid.*

TABLE 5. CHANGES IN TOTAL DEMAND FOR PERMANENT AGRICULTURAL CONSER-
VATION PROGRAM PRACTICES SERVICED BY THE SOIL CONSERVATION SERV-
ICE, 1960 AND 1961[a]

Item	1960	1961	Change 1960 to 1961
Total number farms using ACP cost-shares	1,029,000	1,217,000	+ 188,000
Number ACP requests referred to SCS	407,130	401,773	− 5,357
Number requests certified by SCS as completed	244,596	373,795	+ 129,199
Total ACP funds transferred to SCS for service	$7,260,000	$7,983,000	+ $723,000

[a] Based on *Department of Agriculture Appropriation Bill for 1963*, House, 87 Cong. 2 sess. (1962), Pt. 2, p. 843; Pt. 4, p. 1821; and *Bill for 1964*, House, 88 Cong. 1 sess. (1963), Pt. 2, p. 1035; Pt. 3, p. 1717.

farmers participating in ACP during the 1961 program year was accounted for by the installation of "permanent" practices serviced by SCS.

Figure 4 shows that there is a wide variation among states in the total volume of demand for permanent ACP practices. Ten states in 1962 accounted for one-half of all requests by farmers for ACP practices supervised by SCS.

There is also wide variation among localities in the extent to which particular practices are useful and are actually applied. For example, Table 6 shows that in 1962 about 36 per cent of the counties in the United States included standard terraces (ACP practice C-4) in their programs and 25,185 farmers undertook such practices. Of these farmers, however, 85 per cent were in six states and only 18 per cent of the counties in the nation.

During the 1962 program year, in contrast with Iowa's 2,061 farms which had installed terraces, the other Central states had very small programs. Except for Iowa, the largest number of farms installing standard terraces that year was in Illinois with 327; next, Minnesota had 229; Wisconsin, 94; Indiana and Ohio, 35 each; and Michigan, only 12. And in each of six states only one farm installed standard terraces with ACP assistance. These were: Arizona, California, Massachusetts, New Hampshire, New York, and North Dakota. The contrast between neighboring states is further indicated by Alabama with 1,013 farms on which terraces were installed, and Tennessee with only 31.

Another rough index of the relative emphasis given to certain types

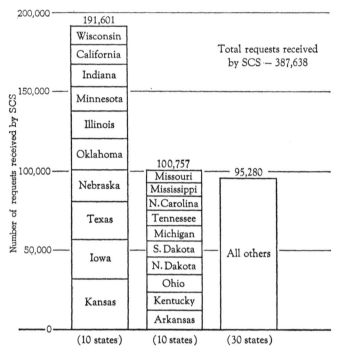

FIGURE 4. Distribution of demand for permanent Agricultural Conservation Program practices serviced by the Soil Conservation Service, 1962

TABLE 6. PERCENTAGE OF TOTAL STATE AGRICULTURAL CONSERVATION PROGRAM FUNDS USED FOR INSTALLING STANDARD TERRACES ON FARMS IN SELECTED STATES, 1962[a]

Location	Number counties authorizing terraces	Number farms installing terraces	Per cent of total state ACP funds used for terraces
Iowa	87	2,061	4.48
Kansas	99	8,446	30.24
Missouri	85	1,810	2.70
Nebraska	82	3,715	15.54
Oklahoma	64	3,423	9.74
Texas	136	2,025	2.29
Total in 6 states	553	21,480	
Total for nation	1,103	25,185	2.57

[a] Data on ACP practice C-4 (standard terraces) are from Agricultural Stabilization and Conservation Service, *Agricultural Conservation Program, Summary by States, 1962* (Washington), pp. 6, 8, 10, 15, 17, 19, 38, and 55.

of practices is the percentage of ACP funds spent in each state for each of five categories of practices. The group of practices designated by the prefix C is primarily for the conservation or disposal of water. During the 1962 program year, Utah spent 82.95 per cent of all ACP funds for these water-connected C practices; Arizona followed with 81.77 per cent; California, 79.5 per cent; Nevada, 79.45 per cent; and Idaho, 73.44

TABLE 7. NUMBER OF FARMS IN UNITED STATES COMPLETING SELECTED PERMANENT AGRICULTURAL CONSERVATION PROGRAM PRACTICES SERVICED BY THE SOIL CONSERVATION SERVICE, 1962[a]

ACP designation	Type of practice	Number U.S. farms completing practices
A-5	Contour strip cropping	5,222
A-8	Tree planting for erosion control	11,062
B-5	Livestock wells for grazing management	7,987
B-7	Dams, pits, ponds for grassland management	38,070
C-1	Sod waterways to dispose of runoff	28,060
C-4	Terraces to control erosion and conserve water	25,185
C-5	Diversion terraces	10,312
C-6	Storage dams to control erosion	2,922
C-7	Structures to protect inlets, outlets, and channels	11,750
C-8	Protect stream banks and channels	2,037
C-9	Open drainage ditches	26,524
C-10	Underground drainage systems	40,666
C-11	Shaping land for drainage	2,853
C-12	Reorganizing irrigation systems	11,648
C-13	Leveling land for irrigation	10,926
C-14	Ponds to store irrigation water	4,112
C-15	Concrete lining for irrigation ditches	8,240

[a] Data are from Agricultural Stabilization and Conservation Service, *Agricultural Conservation Program, Statistical Summary, 1962* (Washington), pp. 12, 18, 25, 27, 35, 38, 40–43, 45, and 47–52.

per cent. In contrast, Alabama spent 7.46 per cent; Florida, 5.39 per cent; Georgia, 5.24 per cent; and Tennessee only 2.92 per cent.[24]

If the number of farms on which SCS serviced an ACP practice is used as a measure, the importance of water conservation practices in SCS operations is evident. The tabulation in Table 7 for the 1962 ACP program year includes some practices for the conservation of water in ACP's A, B, and C groups of practices. The breakdowns of totals for selected practices shown in Table 7 reveal wide variations among

[24] Agricultural Stabilization and Conservation Service, *Agricultural Conservation Program, Summary by States, 1962* (Washington), pp. 6, 8, 10, 15, 17, 19, 38, and 55.

states. For example, Pennsylvania and Wisconsin together account for nearly 50 per cent of A-5, contour strip cropping, which is a very simple and effective practice; and North and South Dakota account for more than 60 per cent of A-8. In North Carolina, where lingering images of the eroded Piedmont would suggest that terracing would be a major practice, terraces actually are installed on only 179 farms, but either open-ditch or underground drainage practices were completed on 5,605 farms in 1962.[25]

Another way of illustrating the distribution of ACP practices is to look at selected major practices in the five states having the greatest number completed with SCS technical assistance. The identity of these states—four Plains states and one Central state—and the extent of ACP practices are shown in Table 8.

TABLE 8. SELECTED AGRICULTURAL CONSERVATION PROGRAM PRACTICES SERVICED BY THE SOIL CONSERVATION SERVICE IN FIVE STATES[a]

ACP designation[b]	Iowa	Nebraska	Kansas	Oklahoma	Texas
B-5	934	937	222	410	942
B-7	735	940	1,297	3,404	5,874
C-1	3,305	3,142	4,780	2,920	1,277
C-4	2,061	3,715	8,446	3,423	2,025
C-5	755	572	1,184	1,042	1,032
C-6	607	685	325	133	17
C-7	4,047	98	290	2	120
C-9	810	252	516	176	303
C-10	9,305	176			51
C-12		307		48	2,809
C-13		1,264	267	76	1,075

[a] Based on Agricultural Stabilization and Conservation Service, *Agricultural Conservation Program, Statistical Summary, 1962*, pp. 25–50, *passim*.
[b] For type of practices to which ACP designations refer, see body of Table 7.

In addition to these variations in the application of permanent ACP practices, there are similar differences in the amounts of practices for initial establishment of ground cover, improvement of permanent cover, limited protection of cropland, and improving wildlife habitat. There is also an extremely wide variation among states (and, in all likelihood, individual farmers) in the average amount of annual finan-

[25] Agricultural Stabilization and Conservation Service, *Agricultural Conservation Program, Statistical Summary, 1962* (Washington), pp. 12, 18, 25, 27, 35, 40–43, 45, and 47–52.

cial assistance given in the Agricultural Conservation Program. These differences, shown in Figure 5, are caused by several obvious variables, such as differential costs among practices (for example, concrete irrigation ditch lining and liming cropland) and others. Nevertheless, it is significant that the range of average assistance for the four program years 1959–62 varied from a high of $1,140 per farm in Arizona to a low of $80 in North Carolina. The contrast between neighboring states with similar conditions of terrain and types of farming is sometimes very great, also, as a comparison of Arizona with Utah reveals. Arizona's average of $1,140 is approximately five times greater than Utah's average of $223. Such disparities raise questions about the criteria used for distributing ACP funds, especially since the high average payments in 10 of the 13 Mountain and Pacific states reflect a heavy national investment in production increments on irrigated cropland.

The total number of farms with permanent ACP practices installed with the technical assistance of SCS are shown for the nation in Figure 6 and by regions in Figure 7. The Plains states and Mountain and Pacific states, as Figure 7 shows, accounted for 46.9 per cent of all practices applied in 1962 with SCS technical assistance and ACP subsidies. The Soil Conservation Service allocated 43.9 per cent of its personnel "assisting districts" to these nineteen states. These are areas of high demand for water conservation practices, and they have produced three of four recent presidents of the National Association of Soil and Water Conservation Districts: Waters Davis, from Texas; Nolen Fuqua, Oklahoma; and William Richards, Nebraska.

Table 9 reveals that in the 15 states with the largest number of ACP practices completed with SCS assistance, 8 had annual state appropriations to soil conservation districts of $80,000 or higher in 1962. Four states (all involving a past struggle by the Extension Service and SCS) appropriated none. Two of the other states making small appropriations were also well-known battlegrounds for the two agencies. Six of the 10 states highest in the number of ACP practices permit their appropriations to be used for district conservation aides. Nine of the top 10 states provide state appropriations for soil conservation districts to hire clerks. Only six other states which are not in the top 15 applying permanent ACP practices do so. Only two outside of the top 15 provide conservation aides. It was pointed out in Chapter 10 that there is apparently some reciprocal influence of ACP and state appropriations for district conservation aides and clerks. Louisiana, which is not in the top 15 applying permanent ACP practices, appears to be the major

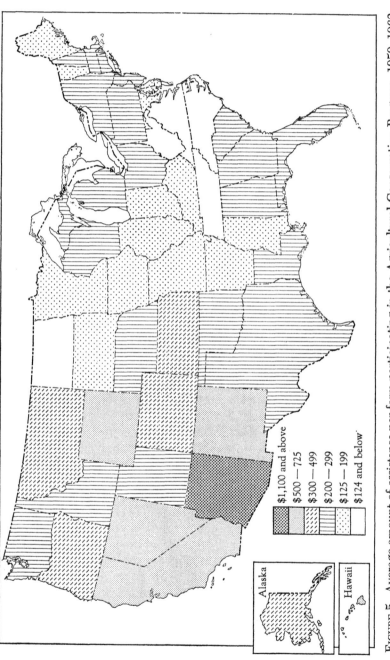

$1,100 and above
$500 — 725
$300 — 499
$200 — 299
$125 — 199
$124 and below

Alaska

Hawaii

Figure 5. Average amount of assistance per farm participating in the Agricultural Conservation Program, 1959–1962

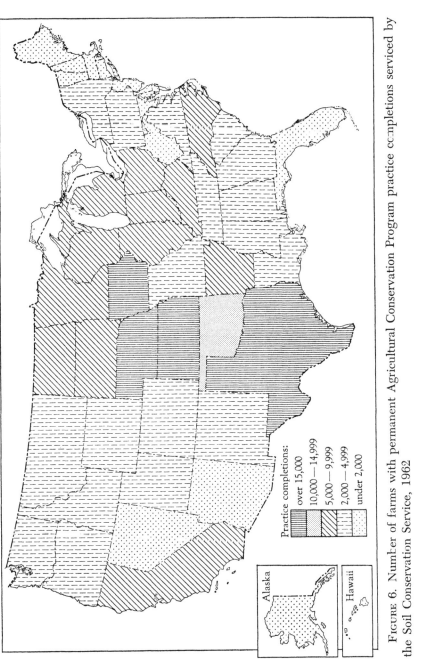

Practice completions:

- over 15,000
- 10,000 — 14,999
- 5,000 — 9,999
- 2,000 — 4,999
- under 2,000

Alaska

Hawaii

Figure 6. Number of farms with permanent Agricultural Conservation Program practice completions serviced by the Soil Conservation Service, 1962

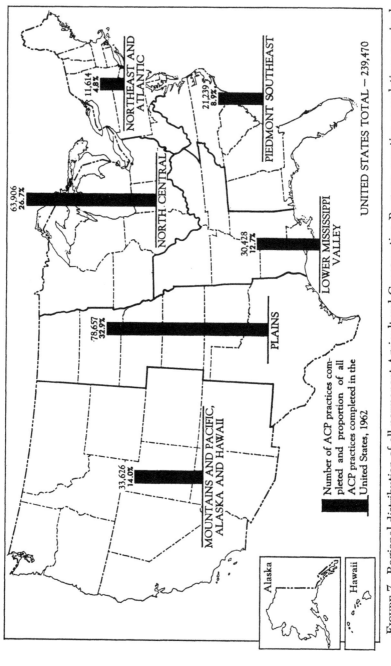

FIGURE 7. Regional distribution of all permanent Agricultural Conservation Program practice completions serviced by the Soil Conservation Service, 1962

TABLE 9. HIGHEST AND LOWEST RANKING OF STATES BY THE NUMBER OF AGRI-
CULTURAL CONSERVATION PROGRAM PRACTICES IN RELATION TO SOIL
CONSERVATION SERVICE STAFF AND STATE AND COUNTY APPROPRIATIONS
TO DISTRICTS

States by ACP rank	Number ACP practices completed with SCS assistance 1962	Man-years of SCS staff allocated to districts 1962	State appropriations to districts[a]	County appropriations to districts[a]
1. Kansas	19,530	352.2	$ 0	$134,561
2. Iowa	17,562	339.3	500,000	0
3. Texas	16,757	908.7	95,000	0
4. Nebraska	15,294	312.6	80,000	110,000
5. Oklahoma	12,678	353.2	337,328	0
6. Illinois	9,731	295.8	136,250	25,000[b]
7. Indiana	9,139	233.3	25,000	10,000[b]
8. Minnesota	8,811	225.1	195,840	42,361
9. California	7,833	331.4	0	23,514
10. Ohio	7,728	230.0	147,400	195,425
11. North Dakota	7,325	182.4	15,000	c
12. South Dakota	7,073	185.1	0	0
13. Arkansas	6,937	282.3	173,600[d]	c
14. North Carolina	6,779	243.7	0	52,248[e]
15. Kentucky	6,622	246.7	74,000[d]	c
41. Maine	650	42.8	f	c
42. Nevada	533	53.0	0	0
43. Florida	419	133.9	0	14,686
44. Massachusetts	297	40.6	1,275	1,450
45. Connecticut	206	27.1	f	0
46. New Hampshire	194	25.2	0	0
47. Delaware	173	23.3	c	c
48. Alaska	109	5.0	f	c
49. Hawaii	94	24.8	0	0
50. Rhode Island	44	9.1	f	0

[a] Based on responses to a special 1961–62 questionnaire for this study sent to state conservation committees, which is explained in Appendix B.
[b] Estimated by state soil conservation committee.
[c] Not reported.
[d] Includes funds for operation of state soil conservation committee.
[e] Incomplete report from 76 of 99 counties.
[f] Appropriations are made to state committees for some district purposes but not reported separately.

exception: it appropriates approximately $375,000 a year to provide for clerks, aides, and other expenses for districts. None of the 10 states with the lowest number of permanent practices is known to have made a significant appropriation to districts.

ACP OR "DISTRICTS PROGRAM"?

In view of the available data, what basis is there for concluding that the Department of Agriculture supports two distinctive and separate soil and water conservation programs? If it were found that it does so, would this policy necessarily be desirable in the future in the face of a sharp decline in the total number of farmers?

The effect of Memorandum 1278 of February 1951 has been to bring ASCS and SCS together to a great degree. Representative Whitten has called SCS the "technical arm" of ASCS, and to a limited extent this is correct. However, ASCS is still free to obtain recommendations and specifications for practices, except for permanent and forestry practices, from the state colleges of agriculture—and the ASC committees do so in most states. Also, ASCS is, of course, equally free to accept or reject SCS recommendations and specifications for practices that are not permanent. If ASCS uses specifications set by one of the colleges for a practice subsidized by ACP, a farmer can follow a different specification set by SCS for that practice only if he does not use ACP funds to install the practice. Here is a source of interagency friction and a basis for the claim that the district program is "different."

Officers of the NACD and SCS spokesmen have always claimed the district program is distinctive because it inspires permanent soil and water conservation practices. They contrast this program with ACP which, they allege, subsidizes ordinary farming operations. The ACP work is highly concentrated, however. In 1962, approximately 80 per cent of all ACP practices installed with SCS assistance were in twenty states; only one of these, North Carolina, was in the East. In these twenty states, there is little likelihood that there is any real distinction in either the character of the work or the accomplishments of the ACP or districts.

In areas where the volume of ACP practices serviced by SCS is low in proportion to the total number of farms, the technicians are available more frequently to render assistance on the farms of district cooperators. Similarly, technicians would have more time to serve coopera-cooperators where the proportion of farmers who are district cooperators is low in proportion to the total number of farms, if they did not also have to service ACP requests for farmers who use ACP, but refuse to become district cooperators. In areas like Iowa and Kansas, where profitable practices are installed at a heavy annual rate, the competition among farmers for SCS assistance is probably intense and accounts, no doubt, for the NACD contention that there is a continuing and "serious shortage of Soil Conservation Service technicians."[26] This situation, in which district cooperators must share with farmers who are not cooperators the available time of SCS technicians, serves as another foundation for the claim that the district program is "different." Figure 8 shows the relationship between the total number of farms, those where ACP practices are applied with SCS assistance, and the number of farms with basic conservation plans prepared by SCS.

If the Department of Agriculture were to terminate the requirement that SCS service ACP practices and confine technical assistance to cooperators with soil conservation districts, as some spokesmen of the NACD advocate, what proportion of all farmers would be reached? In the face of the economic and institutional obstacles to acceptance of soil conservation by farmers shown by Held and Clawson, what kind of farmers would install the relatively expensive practices? Would they concentrate on input yielding a relatively quick return—such as drainage and irrigation improvements now subsidized so that all farmers can install them?[27]

In 1964, the majority of farmers and ranchers were not district cooperators. The Soil Conservation Service listed 4,417,899 "operating units," in districts, but, of these, only 1,944,000 were district cooperators whose holdings totaled 618 million acres. Figure 9 shows that 31.9 per cent of the operating units were covered with conservation plans and 32.1 per cent of the land area was included.

There are marked regional variations in the proportion of operating units covered with basic plans shown in Figure 8. Variations among states are much more striking, however. For example, New Jersey and Maryland both adopted districts enabling legislation in 1937. But on July 1, 1962, in New Jersey, only 19.9 per cent, of the operating units and 14.8 per cent of the agricultural land were planned. On the same date, there were plans in Maryland for 37.7 per cent of the farms and 35.2 per cent of the land. In the Southeast, where districts were organized most easily in the thirties and early forties, 52.2 per cent of Georgia's units had plans, but Virginia had plans for only 27.3 per cent,

[26] *Department of Agriculture Appropriation Bill for 1964*, House, *op. cit.*, Pt. 2, p. 1239. See also pp. 255–79, 774–80, and 1237–49.

[27] R. Burnell Held and Marion Clawson, *Soil Conservation in Perspective* (Baltimore: The Johns Hopkins Press for Resources for the Future, Inc., 1965), Chapter 10.

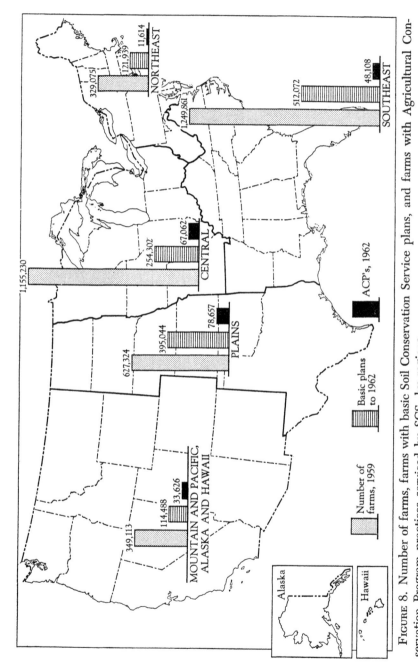

FIGURE 8. Number of farms, farms with basic Soil Conservation Service plans, and farms with Agricultural Conservation Program practices serviced by SCS, by regions

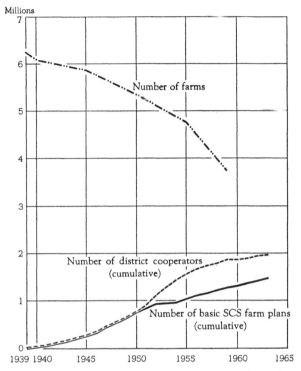

FIGURE 9. Number of farms, 1939–1959, and number
of district cooperators and farms with Soil Conservation
Service farm plans, 1939–1964

North Carolina for 24.1 per cent, and Tennessee for 19 per cent.
Oklahoma led the nation with 62 per cent on its units planned, and, as
might be expected from previous data, the Plains states led all others
regionally. The Central states had the lowest percentage of plans for
units—23.3 per cent—despite the fact that operating units which were
numerous outside of districts in Missouri and Indiana were not counted
in this report. The figures for units with plans in Missouri were 13.3 per
cent; Ohio, 30 per cent; Iowa, 29.8 per cent; and Arizona, 40.5 per cent,
but the Arizona units included only 18.8 per cent of agricultural land.[28]

The distinctive characteristic usually claimed for districts is that
cooperating farmers agree to use their soil and water resources according to a complete farm conservation plan made with "district

[28] *Department of Agriculture Appropriation Bill for 1964*, House, *op. cit.*, Pt. 2,
pp. 938–47.

assistance"—meaning SCS assistance. This practice is a legacy from the New Deal, when M. L. Wilson convinced Secretary Henry A. Wallace that such planning was essential to keep policy-making and administration sufficiently democratic to balance the increasingly centralized power directed by a national planning process. The Committee on Extension Organization and Policy of the Association of Land-Grant Colleges and Universities resisted Wilson's ideas so successfully that district program planning became an educational activity used to stimulate interest in organizing districts. It never became part of a process of continuous adjustment of local to national plans. Farm planning was not exclusive with SCS and the soil conservation districts, however, as the Missouri Balanced Farming Rings showed. The extension services provided farmers with individualized "farm management plans." These, as Louisiana's Director of Extension, H. C. Saunders, pointed out to Congress in 1948, were in competition with SCS farm plans because soil conservation could not properly be separated from the whole range of farm management. Not only have both agencies engaged in farm planning, but also both have been described as serving virtually the same clientele. Once again, this is an indication of what is meant by calling the district program "distinctive."

Based on the percentage of operating units with conservation plans as late as 1964, nearly two-thirds of all farmers still would not have been eligible for the Department's technical assistance, if farmers were required to be district cooperators as a condition for receiving the aid. And this has been a problem in finding a solution to the interagency controversy from the start. For example, in 1948, when Representative Clifford R. Hope, Republican of Kansas, offered his bill to reorganize the soil conservation agencies, he proposed to administer payments for permanent practices in one class, and all others in a second class through the soil conservation districts. But this solution was bitterly protested by Representative Stephen Pace of Georgia who said 70 per cent of the farmers in his area would get nothing out of Hope's conservation program. Congressmen who are most concerned with farm policy know the Agricultural Conservation Program has provided benefits to a large number of widely distributed farmers annually; in 1961, 1,217,000 participated.[29] It is probable that this large group of beneficiaries includes also most of the individuals who are served by SCS "through" soil conservation districts.

[29] *Ibid.*, Pt. 3, p. 1718.

SUMMARY AND COMMENT

As the present working relations between ASCS and SCS and the respective functions of the ASC county committees and the soil conservation districts are reviewed, it becomes clear that Secretary Brannan's Memorandum 1278 was an astute solution to a troublesome political problem. He satisfied some members of Congress who want the Department of Agriculture to provide maximum conservation benefits for the several million farmers who vote in congressional elections. He also satisfied some people who prefer that SCS provide technically sound assistance through districts to farmers who without federal subsidies would install some practices—even if the practices required considerable capital investment.

Borrowing some imagery from the farm, we can say that the Department has hitched an unmatched pair in the traces and made them pull together. One agency is viewed as the partisan instrument of the political party which controls the Administration. During consideration of the 1964 appropriation, a spokesman for ASCS said his agency tried "to see that it [ACP] would be spread around the way it is needed the most" during "a political year, and some people attributed to us intentions that we certainly did not have."[30] The National Association of Soil and Water Conservation Districts, mindful of the fact that a bipartisan group of congressmen saved SCS from "decentralization" in 1948 and 1953, insisted in 1962 that key administrative personnel of SCS "be maintained under the civil service system . . . We will vigorously oppose any actions which would subject appointments in the Soil Conservation Service—local, state, or National—to partisan political pressures."[31] As Representative Hope so perceptively remarked in connection with the proposed reorganization of SCS in 1953, districts are supported by the "better and more influential" farmers.

Presumably the Secretary of Agriculture has vested real responsibility for administering the Agricultural Conservation Program in the ASCS because a line agency is accountable to him in ways in which the conservation districts are not. In practice, however, the ASC county committees seem frequently to develop a base of power largely independent of the Secretary and rooted in their own clientele. Nearly

[30] *Ibid.,* p. 1719.
[31] *Department of Agriculture Appropriation Bill for 1963,* House, *op. cit.,* Pt. 2, p. 944.

twenty years ago, Charles Hardin observed that farmer committees exercise highly decentralized control over personnel and policies. It tends to "become vested in chairmen (and others) of state PMA committees." Hardin predicted that the outcome would probably be an "agricultural field organization built upon a network of state and local farmer-elected committees which effectively interlock with influential persons in state and local farm organizations . . . responsible neither to the Secretary . . . nor to any effective general electorate."[32] Later, Reed L. Frischkneckt—an adviser at the time to Secretary Ezra Taft Benson and his associates—concluded, on the contrary, that the "national administrative agency has maneuvered . . . use of state, county and community farmer committees."[33] In 1962, Morton Grodzins dissented from the majority of a committee appointed by Secretary of Agriculture Orville L. Freeman to study the farmer committee system. Like Hardin, Grodzins criticized the committees on several grounds, not the least of which was that in "a rural community . . . powerful people have a great opportunity to punish their local opponents with a wide range of economic, social and political weapons. The linkage in many counties between political (or farm) organizations and ASC committees is also prejudicial to justice. Where this relationship exists it at least implies that the dominant organization in the county can prevent certain people from holding membership in the committee; at most, it means that the organization consistently receives for its adherents special consideration in committee adjudications."[34]

Grodzins' argument was not merely speculative. In March 1965, the U.S. Civil Rights Commission gave authoritative support to critics of the Department who have contended that, since 1942 or even earlier, most Negro farmers, as well as white farmers with small operations, in the South have been deprived of many of the direct benefits administered by the farmer committees and the soil conservation districts.

[32] Charles Hardin, "Reflections on Agricultural Policy Formation in the United States," *American Political Science Review,* Vol. 42 (October 1948), pp. 881–905.

[33] Reed L. Frischkneckt, "The Democratization of Administration: The Farmer Committee System," *American Political Science Review,* Vol. 47 (September 1953), pp. 704–27. At the time this article was published, Frischkneckt was serving as an adviser to Secretary of Agriculture Ezra Taft Benson, Under Secretary True D. Morse, and Assistant Secretary Earl J. Coke on matters pertaining to the reorganization of the farmer committee system and the Soil Conservation Service. Several memoranda which he prepared are in National Archives Record Group 16 (Soil Conservation Service, Organization 1, 1953–55).

[34] Morton Grodzins, "Separate Statement by Morton Grodzins From Review of the Farmer Committee System," U.S. Department of Agriculture, mimeographed, November 1962.

These farmers and the land they use, in the Commission's words, are relegated to "a separate, inferior and out-dated agricultural economy" in an "arc of poverty which sweeps from Maryland to Texas—the largest geographic and social concentration of the poor."[35] The Commission found this condition not only in the state extension services—where it was already a matter of public knowledge and little corrective action—but also in the Agricultural Stabilization and Conservation Service and ASC committees, the Farmers Home Administration, the Soil Conservation Service, and the soil conservation districts.

Conservation policy is necessarily a political—and party—issue and not merely a technical or administrative one in the narrow sense of these terms.

[35] *New York Times,* March 1, 1965, p. 1.

CHAPTER 12

Soil Conservation or
Multiple-Purpose Districts?

IN 1935, HUGH H. BENNETT discussed the federal government's new erosion control program before the House Committee on Public Lands. He told the congressmen the program, in his judgment, ought to be administered cooperatively with such conservancy districts as the Muskingum district, organized on the watershed of a tributary of the Ohio River. In 1937, Under Secretary of Agriculture M. L. Wilson told the Association of Land-Grant Colleges and Universities that the Department's new conservation objective of regulating land use was to include the "recalcitrant minority" who would not voluntarily conserve their soil. This objective, he said, could not be achieved through counties because they lacked the "necessary scientific traditions" and their boundaries had "political significance, but often very little if any . . . in land-use management." Instead, the most suitable instrumentality, he contended, was a soil conservation district organized under state law over a "naturally bounded area like a watershed." In 1938, Secretary Henry A. Wallace said that his Department's objective was to channel all of its resources—research, education, and "action"—to farmers by focusing on "watersheds."[1]

What, in the course of the last quarter of a century, has happened to this idea as expressed by Bennett, Wilson, and Wallace—the early supporters of the conservation program? Were Gaus and Wolcott correct or wrong in calling this creative venture unwise, and suggesting that it be tried only on an experimental basis until some results were in and evaluated? For, as they said in 1940, at issue was this fundamental question: Can the existing units of government, especially counties, be made into effective instrumentalities for solving basic conservation problems, or must they be displaced, by-passed, or reconstructed to serve special purposes?[2]

[1] See Chapters 1 and 2 for the details.

[2] John M. Gaus and Leon O. Wolcott, *Public Administration and the United States Department of Agriculture* (Chicago: Public Administration Service, 1940), pp. 383–84.

A generation later this question is still with agricultural leaders. The National Association of Soil and Water Conservation Districts (NACD) is now urging that soil conservation districts be converted to multiple-purpose districts. If in the future this should be done, would the multiple-purpose districts displace or absorb the functions of all existing agricultural agencies on the county level? Would they displace or absorb other existing special districts concerned with resources administration? Would they perform some necessary functions better than counties, if the counties instead were granted appropriate powers from the state legislatures which must create any new or enlarged conservation districts?

Before considering some of the implications of these questions, it may be useful to see how soil conservation districts have been organized, note present trends and ask how they serve, if at all, to "coordinate" units of national and state government.

TRENDS IN THE NUMBER AND SIZE OF DISTRICTS

On July 1, 1964, there were 2,928 soil conservation districts in 48 states, 11 subdistricts in Alaska, 8 work areas in Connecticut and 18 in the territories. The total land area within districts was 1,739 million acres; of this amount, almost 1,060 million acres were in farms. Since the Census of Agriculture in 1959 reported 1,124 million acres in farms, about 98 per cent of this land is situated in districts. Approximately 2.5 million acres are organized in six other special districts with which the Soil Conservation Service (SCS) has cooperative relations.[3] There has been a steady growth in the number of districts since 1937, but not quite all agricultural lands are located within them. The annual rate of increase held at a nearly constant rate between 1937 and 1948, and since then declined gradually. The total number of districts should stabilize in the near future, although there may be slight changes even after all agricultural lands are in districts, since reorganizations could change the total numbers. In the past, as Figure 10 shows, the total area of land in districts and of land in farms has increased at rates more or less similar to the increase in districts. Since 1947, however, newly organized districts have tended to include within their total areas

[3] U.S. Department of Agriculture, Soil Conservation Service, "Soil Conservation Districts: Status of Organization by States, Approximate Acreage, and Farms in Districts," mimeographed, July 1, 1964.

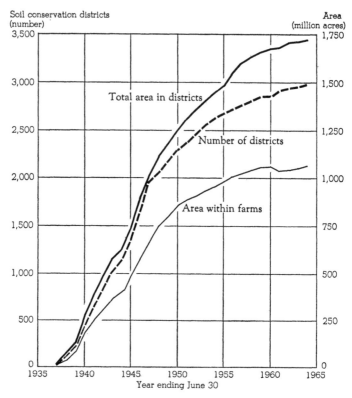

FIGURE 10. Number of soil conservation districts, total area, area within farms in districts, 1937–1964

proportionally smaller acreages of land in farms as the total number of districts increased.

Gross national data of these sorts, however, can obscure such significant variations as uneven regional growth rates, which have limited the usefulness of districts in administering the Department of Agriculture's soil conservation programs. The maps shown in Figures 11 through 14 illustrate such differences over the past quarter century. The conservation districts were first organized most widely in the Southeast and Plains states; they spread northward and westward, principally after World War II, when increased technical assistance for drainage and irrigation practices was added to the original erosion control program. Major increases in the area within districts have occurred in the Central states and Mountain and Pacific states in recent years. In 1962, for example, the net national increase was 16.2 million acres.

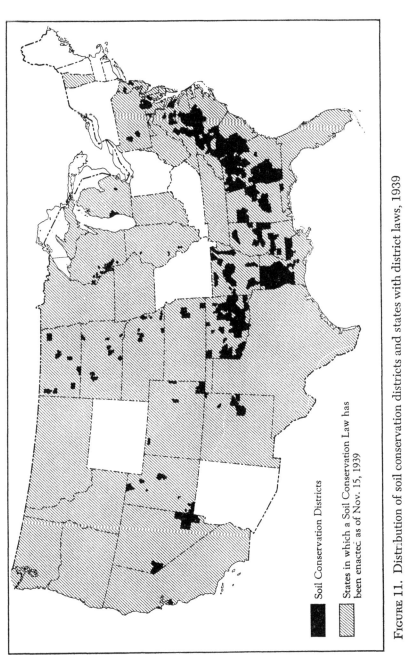

Soil Conservation Districts

States in which a Soil Conservation Law has
been enacted as of Nov. 15, 1939

FIGURE 11. Distribution of soil conservation districts and states with district laws, 1939

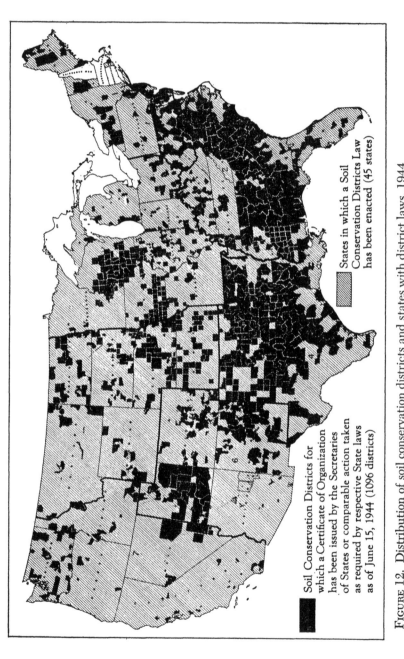

Soil Conservation Districts for which a Certificate of Organization has been issued by the Secretaries of States or comparable action taken as required by respective State laws as of June 15, 1944 (1096 districts)

States in which a Soil Conservation Districts Law has been enacted (45 states)

FIGURE 12. Distribution of soil conservation districts and states with district laws, 1944

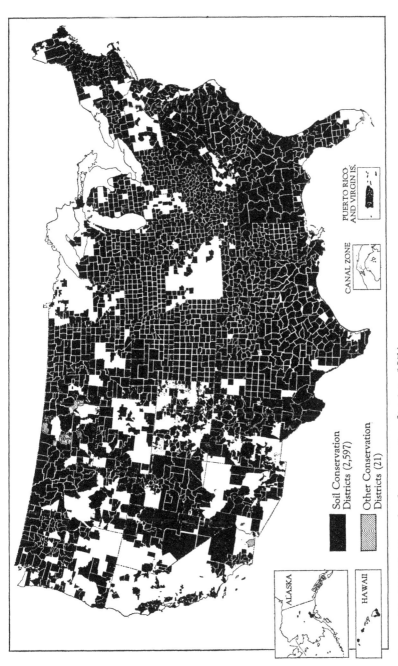

FIGURE 13. Distribution of soil conservation districts, 1954

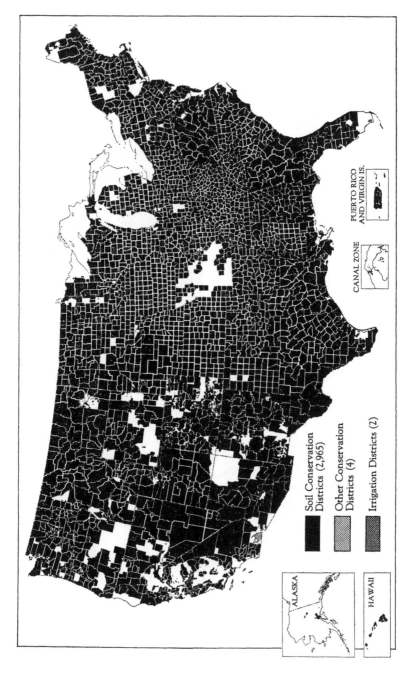

Soil Conservation
Districts (2,965)

Other Conservation
Districts (4)

Irrigation Districts (2)

ALASKA

HAWAII

PUERTO RICO
AND VIRGIN IS.

CANAL ZONE

Figure 14. Distribution of soil conservation districts, 1964

Of this total, 6.9 million acres were in the 11 Mountain and Pacific states. In the Central states, 3.3 million acres were added, with Missouri alone accounting for an increase of 2.3 million acres.

Since reorganizations of existing districts influence both the gross and net number of districts, they need to be considered when increases are reported. In 1963, for example, the Soil Conservation Service reported to Congress that 55 "new" districts had been organized in 1962; the net increase, however, was 19, distributed among 9 states (Missouri and Pennsylvania had 5 each; Indiana and Michigan, 2 each; and 1 each for Idaho, Minnesota, New York, Oregon, and South Dakota). Twenty-four "new" districts were created in 3 states by subdividing existing multiple-county districts (Alabama, 4; North Carolina, 19; and Virginia, 1). Three states had net decreases (Arizona, 6; Illinois, 1; and Washington, 1). Transfers, consolidations, and additions of areas affecting existing districts occurred in 8 other states (California, Colorado, Hawaii, Montana, New Mexico, North Dakota, Texas, and Wyoming) without changes in the net number of districts. The statistical picture of these changes is confused further by the fact that there can be a net increase of the area in districts, but a decrease in the net number of districts in a state in a single year. Arizona, for example, had a net loss of 6 districts, but an increase of 413,435 acres in the area within remaining districts. In another example, there was no change in the total number of California's districts (160), but the land area within them increased 1.3 million acres after 4 new districts were organized and 30 transfers, additions, and consolidations were effected.[4]

Despite the confusion which these details create, four trends in the changing number and size of districts are clearly emerging. In the Southeast, multiple-county districts are being divided into single-county districts, with virtually no net change in the total area included within them. In the Central and Western states, three other trends are evident: Areas which have never been within districts are being organized into new districts; territory not previously within districts is being added to existing districts; and transfers among existing districts are being made along with consolidations merging small existing districts into larger ones. As a result of these reorganizations, most districts in the future are likely to be organized along county boundaries except, perhaps, in scattered areas of the Mountain and Pacific states. This trend certainly is in contrast with the Department of

[4] *Department of Agriculture Appropriation Bill for 1964*, House, 88 Cong. 1 sess. (1963), Pt. 2, pp. 1060–63.

Agriculture's original wish to have districts organized on the basis of watersheds. It is also contrary to the practice implied by the small watershed legislation, Public Law 566. If most districts are to become conterminous with counties, there is some question why additional, but separate, units are needed as multiple-purpose conservation districts. In 1963, the Administrator of the Soil Conservation Service reported to Congress that his agency encourages both subdividing and consolidating districts "from the standpoint of good operations."[5]

The current organization of districts for the first time in some areas may be taken as evidence of their popularity, but in most of these places it is more likely to be a sign that long-standing opposition is only now giving way. Many factors have been involved, but in most states where the organization of districts has been slow during the past twenty-five years, legal provisions relating to organizing them have worked to the advantage of the opposition.

LEGAL IMPEDIMENTS TO ORGANIZING DISTRICTS

The standard soil conservation districts act prepared by the Department of Agriculture contains simple procedures to organize districts with a minimum of obstacles. Twenty-five land "occupiers" (as distinguished from owners) may file an organizing petition with the state soil conservation committee. The state committee is to hold a public hearing on the proposal within thirty days, and then to conduct a referendum in which all land occupiers within the prospective district are eligible to vote. The negative vote of a simple majority stops further proceedings; otherwise, if the proposal is judged to be feasible and practical, the state committee completes the organization of a district.[6]

Differences with the Department of Agriculture over fundamental issues and opposition from some of the state extension services led many legislatures to depart from the suggested standard act. For example, only ten states permitted twenty-five mere land "occupiers" to petition for a district. Most states required that petitioners be "landowners." Virginia required not only a majority of the landowners, but also that those who own a majority of the acreage in the proposed district initiate the proposal. Oregon required either 25 landowners or

[5] *Department of Agriculture Appropriation Bill for 1964*, Senate, 88 Cong. 1 sess. (1963), pp. 250–51.

[6] U.S. Department of Agriculture, *A Standard Soil Conservation Districts Law* (Washington: Government Printing Office, 1936), pp. 7–15.

the owners of 70 per cent of the land to start the petition. Colorado required 25 per cent of the landowners. In Kentucky, no district could be organized in a county lacking a county agent. Until 1954, Arizona prohibited organization of any district on range land. Missouri still forbids any district larger than a county and more than one in each county to be organized. It also requires the signatures of 25 "land representatives" in each township of the area to be organized, and the "land representative" is a taxpayer who must own land, or be the agent of a landowner, in the proposed district. The effect is a blocking action since in a typical county of 16 survey townships 400 signatures are required on the petition.[7]

Provisions dealing with the referendum on a proposed district also differed materially from the Department's recommendations. Twenty-four states permitted only "landowners" to vote in the referendum. Most of these states also required more than a simple majority to carry the proposition—sometimes as high as 75 per cent. Indiana requires a favorable vote by 60 per cent of all eligible voters in the proposed district. Missouri requires approval by 67 per cent of the "land representatives." Wyoming established a complex system of weighted voting, which put land occupiers in one class and owners in another for voting purposes. Voters of each class are entitled to one vote for each acre occupied or owned, but a majority of each class must favor the proposal for the district to carry. Representatives of partnerships and corporations owning land may vote in the owner class. Legal provisions of this sort impose many burdens and hazards on those who propose to organize a district; they are clearly not intended to ease their growth. Several state laws contain a provision which does, however, lighten the labors of the organizers. Anyone who wishes not to be in a district may declare himself out at an appropriate time; some laws, such as Wyoming's, permit a landowner to withdraw from a district after being in it for a year. It is interesting to note that the few states which allowed occupiers to vote in the referendum also required only a simple majority to carry the proposition.

The width of the gulf between the Department of Agriculture and most state legislatures over the standard act in the thirties and forties

[7] The Missouri State Soil Conservation Committee also has discretion to determine whether the vote cast in the organizing referendum constitutes a "substantial expression of opinion" favoring the proposal. The Missouri Committee reported in 1962 that it has "encouraged a district to include all farm land in a county." (Response to special questionnaire sent to state conservation committees for this study: from Missouri.) Appendix A presents citations to state laws relating to soil conservation districts.

can be measured to some extent by the fact that only nine states followed the recommendation permitting a petition by twenty-five land occupiers and a referendum carried by a simple majority of them to start a district with state committee approval. The fact that most state legislatures preferred more stringent requirements is evidence of a widespread disinclination to bind landowners by the actions of their tenants.

Four states also adopted another policy which the Department wished to avoid when the standard act was framed. Wisconsin, New York, California, and Pennsylvania vested power in their county governing boards—rather than state committees—to create districts. In Wisconsin and New York, this procedure has not generally been a serious barrier to organizing districts, but proposals to organize them have become involved in partisan political conflicts. Both Pennsylvania and California, however, offer interesting insights in this connection.

In Pennsylvania, county boards were given power to create districts only after the state House of Representatives appointed a committee to investigate charges that the state soil conservation committee and state extension service had jointly opposed proposals to organize districts. A majority of the Pennsylvania legislative committee sustained the charge, but a minority claimed that in no instance had a majority of eligible voters favored any proposal rejected by the state committee. The minority also repeated claims that districts would increase taxes and permit the federal government to regulate farming operations.[8] After a coalition of farmers, sportsmen's groups, watershed interests, and a key member of the state Senate exerted pressure in the Pennsylvania Legislature, however, the law was changed in 1945. The Director of the Pennsylvania Extension Service reluctantly recognized defeat by proposing that his agency retain primary responsibility for "conservation education" and that the Soil Conservation Service concentrate upon flood control and watershed operations authorized by the Flood Control Act of 1944.[9] He later agreed, however, that both agencies would give technical advice on the "broad aspects of soil conservation" and that staff members of "each of the two agencies are to be strictly enjoined not to disparage or controvert the program of the

[8] Pennsylvania State Soil Conservation Commission, *Teamwork*, mimeographed, September 1960. See also "Committee Appointed by the Speaker of the [Pennsylvania] House of Representatives in Accordance with Resolution No. 15—February 4, 1941, Minority Report."

[9] Dean S. W. Fletcher, Pennsylvania State College, to Miles Horst, Pennsylvania Secretary of Agriculture, Aug. 22, 1945.

other."[10] Finally, in 1950, the two agencies signed a formal memorandum of understanding by which the State Extension Service agreed neither to oppose nor to support the organization of districts, but to permit the State Soil Conservation Commission to promote them.[11] During the next decade, the number of districts in Pennsylvania increased very slowly; only since the Watershed Protection and Flood Prevention Act of 1954 (Public Law 566) was enacted has the pace of organization there quickened.

California, which also vests authority in county boards to act on petitions to organize districts, has had only a slow growth in districts because of deep divisions between the Soil Conservation Service and the California Extension Service. The districts enabling law in California prescribes very complex and exacting organizing procedures which depart materially from the standard act. The sponsors of a district submit a preliminary proposal to the county governing board, which is then required to ask the state Division of Soil Conservation to investigate and report on the practicability and feasibility of the district and to recommend boundaries. The proposal must next be examined by the county boundary commission which may hold hearings, after which it shall report and make recommendations to the county governing board. At this stage, the county governing board considers the two reports it has received, and advises the sponsors of the district of any changes to be made in the proposed district boundaries. Once agreement on the boundaries is reached, the district sponsors must circulate a petition which can be qualified by either of two methods: (1) If there are 200 or more landowners involved, the signatures of 100 owners are required; if there are fewer than 200, then 51 per cent of the landowners must sign. (2) The alternate method requires the signatures of 51 per cent of the landowners who own at least 51 per cent of the land, including public lands (although these public lands may not be included in satisfying the acreage requirement). If the petition passes subsequent scrutiny for legal adequacy within a 60-day period, the county governing board must conduct a public hearing, after which it may deny the petition if it sees fit, or it may approve the petition.

If the county board favors the California petitioners for a district with its approval, and fixes the boundaries to exclude any landowner desiring exemption, the board then can hold an election of district

---

[10] Fletcher to Horst, Jan. 11, 1946.

[11] Memorandum of Understanding signed by the Pennsylvania Extension Service and the U.S. Soil Conservation Service, December 1950. (Copy in the files of the Pennsylvania Department of Agriculture.)

directors and a vote on the proposal to organize. All landowners are eligible to vote, and for the district to be organized, a majority must favor the proposition. But instead of an election, a simpler method of acting is provided in cases where the sponsors use a petition signed by 51 per cent of the landowners of 51 per cent of the land. The county board may simply appoint five district directors. Even this lengthy recital omits numerous detailed steps and requirements in California law which must be fulfilled if the proposition is not to fall from the weight of its own imperfections. Not more than two years can elapse, for example, from the start to the completion of the process. It is not difficult to understand, therefore, why California has 12 districts with fewer than 50 farms; and why the smallest district in 1964 had only three farms within it; and why, with 162 districts organized, approximately two-thirds of the Central Valley remains unorganized.

Active opponents of conservation districts in California have had little difficulty using these complex legal provisions—and the fact that districts may levy taxes—to discourage organization of districts. Informed persons within, and outside, the California Extension Service agree that Extension Director H. B. Crocheron, his successor J. Earl Coke, and many county agents vigorously opposed districts. After Coke went to Washington to become Assistant Secretary of Agriculture in 1953, he continued to receive frequent reports from California. For example, the county agent in Merced County contended that an SCS employee was stimulating a district in the county. He complained to Coke that only four people had signed the organizing petition, the purpose was to drain land which would then appreciate in value by $100 per acre, the district was an unnecessary burden on taxpayers because it could levy taxes and, furthermore, he had five extension specialists in his office prepared to handle the job without a district. He bitterly denounced the Soil Conservation Service because it proposed to "hand feed these fellows, thereby breaking down their initiative and self dependence."[12]

Occasionally, farmers and ranchers in California also opposed districts actively. A prominent rancher launched a campaign in 1953 to discontinue the San Mateo District and a large orchardist complained to Secretary of Agriculture Ezra Taft Benson that the seventeen employees of the Soil Conservation Service in Tehama County were

[12] National Archives Records Group 16 (Soil Conservation Service, Organization 1), William H. Alison, Director of Merced County Extension, to J. Earl Coke, March 24, 1953, and attached file. (National Archives records cited hereinafter usually will be identified by the abbreviation NARG.)

promoting districts which were attractive to farmers "because the cost is all paid by the federal government."[13] A member of the Madera County governing board charged that an SCS technician had refused to certify for payment some work subsidized by the Agricultural Conservation Program (ACP) on his ranch. He claimed that he was being victimized, because in the "meantime, the Soil Conservation people tried to form a soil conservation district in the mountains, which was turned down by the Board of Supervisors, of which I was a member. From that day on, my trouble started." After an investigation of these charges, the Secretary of Agriculture rejected them, but this story is the kind which is long remembered in the countryside.[14]

DISTRICT BOUNDARIES: WATERSHEDS OR COUNTIES?

The varying boundaries of different districts result from several factors: not only special provisions of state legislation, but also the practices of state soil conservation committees or other organizing bodies seeking solutions to a variety of problems. Districts in most Western states, for example, have been established with little concern for county boundaries, and not always on watersheds. The very large size of counties; irregular terrain, often fixed by high mountain ranges; variations in climate, rainfall, and land use; the national origins of occupants—all of these and similar factors encouraged sponsors to organize districts on the basis of particular "problem areas" where farmers were ready to be organized to receive technical assistance from the Soil Conservation Service. The areas of many of the districts still are not contiguous in Colorado and New Mexico. (See, for example, Figure 15.) Often such districts exist because ranchers have been suspicious of any agency which may be used to regulate land use. It is for this reason that several Western states permit landowners to exclude their lands from any district in which they would otherwise be included. New Mexico's district legislation provides for supervisors to be elected from zones in which either farming or ranching is the

---

[13] NARG 16 (SCS, Organization 1), James Mills to Secretary of Agriculture Ezra Taft Benson, April 28, 1953.

[14] NARG 16 (SCS, Organization 1), C. C. Clark to Ezra T. Benson, July 5, 1955; reply from Under Secretary True D. Morse for Benson, Aug. 19, 1955. Based also on interview with Francis Lindsay, Dec. 9, 1961. Mr. Lindsay was formerly an employee of the Soil Conservation Service and later a member of the California Legislature, where he served in the Assembly as Chairman of the Committee on Conservation, Planning and Public Works. In this position, he was able to exert leverage in favor of soil conservation districts, starting in the early fifties.

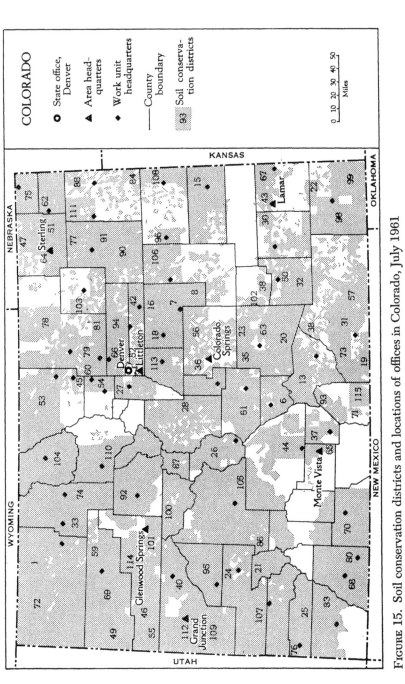

FIGURE 15. Soil conservation districts and locations of offices in Colorado, July 1961

predominant land use. Farmers apparently have tended to favor the technical services available to improve the efficiency of irrigation on cropland, but there is general agreement among informed observers that ranchers have not overcome their suspicion of districts with which they must enter into a written agreement.

The prevailing situation in the Mountain and Pacific states is districts which have been organized on areas much smaller than counties and with watershed boundaries only incidentally considered in fixing boundaries. An officer of the Hawaii Department of Agriculture summed up some of the reasons as: "(1) Initially many large landowners, sugar plantations, and ranchers were not interested in the District movement; (2) Landowners or farmers with common interests joined to form districts; (3) In a few cases boundaries were based on existing Agricultural Extension Service districts. Watersheds were not the units of organization because watershed activities started after most of the Soil Conservation Districts were formed."[15] The distribution of existing farmer or rancher support of the Soil Conservation Service's program probably has been the major factor determining boundaries in most cases.

Many of the earliest districts to be formed in other regions were organized on watersheds, but this effort was soon abandoned outside the Southeast and county boundaries were used for districts. In Minnesota, for example, it was found that using the watershed involved such difficulties as making accurate lists of landowners and informing people of voting places and district boundaries which ignored more familiar political jurisdictions. In Nebraska, the experiment with watersheds was given up early because the members of the State Soil Conservation Committee thought that organization across county boundaries unnecessarily confused relations with units of county government. In Rhode Island, districts were organized to coincide with existing boundaries of the area covered by the local Farm Bureau and county agents. Since the early state soil conservation committees were usually composed of a majority of representatives of the agricultural colleges which had long worked with the county as the basic unit, it was rather natural that they favored district organization on a county, rather than a watershed, basis. A spokesman for the Michigan State Committee, however, stated a fundamental consideration very neatly

---

[15] Response to special questionnaire for this study sent to state conservation committees: from Hawaii. These conclusions are supported also by information gathered in interviews during 1961 with the personnel of the state committees in Arizona, California, Colorado, and Wyoming.

by saying that districts in that state were "soil oriented rather than water—watersheds were not considered."[16] That is, agricultural leaders, especially in the colleges and extension services, viewed soil conservation as "soil management" carried out on individual farms as the result of voluntary decisions by operators. They did not think of planning and developing the soil and water resources of watersheds as "natural" units to include even the construction of minor flood control or other water management structures. In addition, not all SCS personnel were entirely favorable to the watershed concept. The result was a clear pattern of districts organized as single counties, with few exceptions, in an arc across the northern portion of the country from Maine into the Great Plains and southward into Oklahoma.

The pattern in most of the Southeastern states was distinctive and remains so, although changes are under way. Most of the districts organized outside of the area affected by the Tennessee Valley Authority (TVA), and prior to 1945, consisted of two or more—often four or five—complete counties. However, Florida and Mississippi have been exceptions generally, despite the early organization of very few on the watershed or multiple-county basis. Occasionally, the areas initially organized omitted part of one or more counties where resistance to the districts was known to be strong. Sometimes it is said that the multiple-county districts followed watersheds, but this was largely a talking point that seems to have gained relatively limited local acceptance. The multiple-county organization was used in this region partly because the Department of Agriculture pushed for it and the extension services in the South were not in a strong enough financial or political position to resist. In Louisiana, for example, Dean J. G. Lee, of the College of Agriculture, gave up the idea of having county agents serve as members of the districts boards as they did in some other states, and agreed to true "watershed" districts, when the regional director of SCS insisted on the soundness of the principle.[17] In a few other Southern states, SCS strategists thought that this system would minimize the number of county agents who might serve as one of the two appointed district board members. In part, too, it was an economy measure to reduce the total number of supervisors who might require per diem compensation, payments of expenses, or both. From the Extension Service's standpoint, this type of organization had the advantage of minimizing the number of supervisors per district—a desirable result if it was assumed

[16] *Ibid.*, response from Michigan.

[17] Interview with H. C. Saunders, former Director of the Louisiana Extension Service, Oct. 6, 1961.

that the soil conservation and terracing associations organized by the county agents might continue to play a leading role under the guidance of the agricultural colleges.

Starting in 1945, almost without exception, the areas in the Tennessee Valley and on the coastal plains were brought in as single-county districts. Where former multiple-county districts have been subdivided recently to become single-county districts in the Southeast, several reasons have been offered for the change. One is the hope that an increase in the total number of district supervisors will "involve" more individuals in conservation policy. They will be expected to cultivate favorable responses from county governing boards and state legislatures when funds or new laws are requested. Also, Congress is expected to react to reports of an increase in the total number of districts by providing more funds for technical assistance. In addition, it is difficult for individual supervisors to know or care much about conditions in the counties outside their own. In some districts, county agents are supervisors instead of technical advisers to district boards, and in some multiple-county districts they represent counties. Perhaps most important of all is the fact that most agricultural agencies are organized locally to function on the county level, especially the Agricultural Stabilization and Conservation Service (ASCS) which administers the ACP for which the Soil Conservation Service has important responsibilities. If there were to be genuine coordination of planning and operations by the two groups, the district supervisors must be residents of the counties in which they have to negotiate with the ASC county committees. Of course, where districts include parts of two or more counties, this problem is compounded, just as it is when two or more districts are located within one county.

There is also another reason why neither the multiple-county districts nor those organized on the basis of watersheds have been entirely satisfactory. The Soil Conservation Service does not necessarily distribute its own personnel and establish lines of administrative authority to correspond with the boundaries of districts. Normally, SCS work units are set up with one or more to a county, depending upon the work load, special projects, and the like. Each work unit is under the direct administrative supervision of an area conservationist, whose territory normally comprises several counties. Where there are multiple-county districts, as in several Southeastern states, instances occur in which not all of the work unit conservationists serving counties within such districts are supervised by the same area conservationist. Where this is the case, a board must deal with work unit conservationists who not only may be under more than one supervisor but are assisting one or more other boards. The Boards which must share the time of technical personnel face certain difficulties in trying to establish control over work priorities and schedules of assistance to farmers in their districts.

DISTRICTS IN THE SMALL WATERSHED PROGRAM

While all these factors have encouraged the organization of districts to coincide with the boundaries of single counties, the Small Watershed Act, Public Law 566, has revived interest in a unit of government organized on natural drainage areas. During the last decade, states have passed legislation permitting the organization of sub-watershed or watershed conservancy districts. Usually the sub-watershed districts are authorized to be created upon the initiative of one or more soil conservation districts, and after approval by a public referendum. These districts also normally are given the power to levy taxes or to assess benefits in order to raise the local share of funds required of sponsors under Public Law 566. Sub-watershed districts are expected to be organized on the basis of watersheds to include the outer limits of projects, just as soil conservation districts were once expected to be organized.

Officials of the Department of Agriculture have encouraged this return to the watershed as the unit for organizing a conservation district, but the Department has not advocated this kind of local sponsor exclusively for projects under Public Law 566. The Department has also encouraged states to authorize broader powers for counties, cities, other corporate municipalities, and various types of special districts including, but not confined to, soil conservation districts.[18] It is now the Department's policy to work with a wide variety of local governments as sponsors of small watershed projects. This policy is a departure from the Department's earlier practice of restricting SCS cooperative relations to soil conservation districts. It appears to reflect several of the Department's needs.

First, by working through a variety of local governments and state units, the Department can make use of their varied legal powers to undertake multiple-purpose projects. If the Department had relied exclusively on the soil conservation districts as sponsors, the number

---

[18] See, for example, Public Law 87–170, 75 *Stat.* 408, 16 U.S.C. 1002.

and type of the watershed projects would probably have been limited. Instead, by July 1, 1964, the Department had received 2,137 applications, and had approved 1,002 for planning.[19] Local sponsors need the power to construct, operate, and maintain works of improvement; and they need fiscal powers and, frequently, the power of eminent domain, or at least authority to acquire rights, interests, or easements in lands. Sponsorship of a project should exist in the area or it should be possible quickly and easily to organize it.

Second, the Department is interested in working with a broad spectrum of interests to benefit from these projects. In recent testimony, the SCS Administrator, Donald A. Williams, has insisted that "conservation is for all the people, not just for the farmers or ranchers. I think you will recall, Mr. Chairman, there is nothing in the legislation authorizing the soil conservation program in this country saying it is a farm program or saying it has to be done by farmers or ranchers." Representative Jamie L. Whitten, a member of the Appropriations Committee and a Democrat from Mississippi, responded: "I can recognize as fewer and fewer people are left actually farming and with more and more of us in other activities, it becomes more and more important to retain the support of the farmers in Congress, in the press and elsewhere."[20] In supporting the ASCS estimates for ACP wildlife practices, Whitten said that members of Congress recognize the great number of people interested in hunting and fishing. "With a declining membership in Congress who have any farm knowledge, background or direct interest, it gets to where you have got to have programs which have some appeal to that group in our nation."[21]

When amendments to Public Law 566 were under discussion before the House Committee on Agriculture in 1962, much attention was given to features intended to ease the construction of projects with recreational features. Representative W. R. Poage, Democrat from Texas, commented that a "great many people . . . will be attracted by recreational features . . . I think that you will pick up a clientele that you do not have at the present time." With this, the SCS Deputy Administrator, Gladwin Young, agreed.[22] Charles B. Shuman, President

[19] *Department of Agriculture Appropriation Bill for 1966*, House, 89 Cong. 1 sess. (1965), Pt. 3, p. 366.
[20] *Department of Agriculture Appropriation Bill for 1964*, House, *Op. cit.*, Pt. 2, pp. 1028–29.
[21] *Ibid.*, Pt. 3, p. 1727.
[22] *Food and Agriculture Act of 1962*, House Committee on Agriculture, 87 Cong. 2 sess. (1962), Serial AA, Pt. 2, p. 288.

of the American Farm Bureau Federation, opposed the provisions of Title I of this portion of the Law, on the ground that the measures "appear to be designed primarily to appeal to urban people."[23]

The policy of having a variety of local sponsors possessing sufficient legal powers, fiscal resources, and administrative skills was also based on the fact that Public Law 566 originally provided that, after July 1, 1956, no project could be initiated by the Secretary of Agriculture without such a local sponsor. This provision was intended by the enemies of Public Law 566 as an effective block to an extension of the "pilot" watershed projects initiated through the $5 million rider in 1953. Opponents undoubtedly recognized that soil conservation districts could not ordinarily qualify as sponsors, and they probably confidently expected that the Small Watershed Program would be confined to agricultural flood control.[24]

It was not only opponents of the district boards who favored sponsorship by some other unit. There was a very considerable reluctance on the part of some experienced SCS personnel to entrust most district boards with responsibility for construction contracts and for maintaining structures—even where they had legal power to do so. District boards, they feared, lacked the administrative skills to handle contracts responsibly, and there is always that unpleasant possibility that considerations of patronage, nepotism, and venality will influence the decisions of some boards. There is still a good deal of doubt that districts can be relied upon to maintain structures once SCS has turned over a completed project to the local sponsor. Many district supervisors do not want to assume this new responsibility. Members of Congress, too, have expressed concern that watershed projects featuring recreation facilities could pass from public to private hands.[25]

Districts have not been abandoned in the administration of the new multiple-interest watershed projects; SCS encourages joint or multiple sponsors, and one of them normally is a district. Districts are said to be included to insure that proper measures for land treatment are installed on areas affecting the works of improvement. Normally, one of the other sponsors has adequate legal powers, fiscal resources, and administrative skills to serve as the local contracting organization which signs the project agreement to operate and maintain the works installed by

---

[23] *Ibid.*, pp. 444–45.

[24] *Amendments to Watershed Protection and Flood Prevention Act*, Hearings before the Subcommittee on Conservation and Credit of the House Committee on Agriculture, 84 Cong. 2 sess. (1965), Serial JJ, p. 32.

[25] *Ibid.*, pp. 296–98.

SCS. And, understandably enough, some district supervisors regard these procedures with alarm, since the relative importance of districts has been diminished in the Small Watershed Program. They fear that SCS employees are being diverted from assisting districts to assisting watershed projects which are of little direct assistance to farmers. Other district supervisors, however, have welcomed joint sponsorship, since it relieves them of responsibility for performing unfamiliar and complicated administrative tasks. Even without such responsibility, some have found that they sometimes serve as lightning rods to catch the heat of their neighbors' ire over unwelcome projects.[26]

Districts can perform some essential functions to sponsor small watershed projects. They have been exhorted to promote projects and many do so. A district's leadership can be helpful in making the right contacts with both farm and urban interests affected by a proposed project. They have also been used to obtain easements, rights of way, and interests in lands required to impound water. When this work is done quietly and effectively it can reduce, or even eliminate, costly litigation and the costs of acquiring legal rights and interests in land by public proceedings. Persuasion by friends and neighbors can sometimes avoid unnecessary resentment and opposition, although not always. Some supervisors also have good connections with both county and city governing bodies.[27] These relationships are of increasing importance as counties and cities sponsor small watershed projects.

In the ten years since the Small Watershed Act was passed, the Department of Agriculture's need for sponsors of local projects has outweighed consideration of the kinds of local governing units which ought to cooperate with the national government in the administration of resources programs. The Department has encouraged a proliferation of special districts in the face of a rapidly declining farm population.

[26] In one state, two supervisors of a district said in interviews that some animosities had been stirred among neighbors by a proposed project and some people were reported to have received threatening letters through the mail. These supervisors said that "the F.B.I. is now investigating this." Another former supervisor said he had refused to run for re-election because of angry opposition from some of his neighbors to features of a small watershed project. In another state, I asked a supervisor what he and his colleagues had done to sponsor a project located partially within his district and he replied "nothing." He went on to say that for some reason a large hole developed behind a dam and the water which had been impounded ran out. When I asked him what responsibility he and his board assumed for this situation, he said: "Why, hell, none! That's SCS's hole—let 'em fill it up."

[27] One SCS state conservationist remarked in an interview that a large proposed project in his state had been delayed because a "smart lawyer" had "lined up about 200 people" who were unwilling to give easements voluntarily.

Many of these districts have the power to tax or to assess and collect benefits. By encouraging this trend when it failed to secure general acceptance of soil conservation districts organized on the basis of watersheds, the Department has complicated all questions about the proper functions and organization of these districts and of others in the federal system. In the light of past experience, it is difficult to see how soil conservation districts can be converted to multiple-purpose districts in the midst of this confusion.

INTERAGENCY RELATIONS IN THE DISTRICTS

Multiple-purpose resource development as the product of coordinated program planning and execution has been accepted as a desirable goal for half a century. The idea that the watershed is the natural unit upon which to organize such programs has been received with equal, and often uncritical, acclaim. The areal base upon which traditional units of government have been organized has frequently been rejected as unsuited to the rational use of physical resources such as land and water, and the search has continued for more appropriate units of government. The movement to organize soil conservation districts has profited greatly from this search. Governing boards and the public are encouraged to view districts as the proper agencies for planning and carrying out soil, water, and wildlife conservation.[28]

In about one-third of the states, some districts have entered into formal agreements with agencies other than SCS. This is especially a pattern in the West, where districts include not only private land but also public land under state or federal management. Most of these agreements are with either the Bureau of Land Management of the Department of the Interior or with the U.S. Forest Service, although there are some with other federal agencies.

Public lands under federal management for grazing and timber production were included in soil conservation districts because it was argued in the thirties that if districts were organized on the basis of watersheds they would provide units of government through which the various federal agencies would cooperatively attack the twin problems of erosion and flood damage. This idea gave considerable impetus to the decision to foster the creation of soil conservation districts. It is doubtful that many districts perform this function, although some are

[28] National Association of Soil and Water Conservation Districts, *Tuesday Newsletter,* April 17, 1962.

reported to have been useful. They have solicited and received funds used for some measures of range improvement. Private and public lands are normally so intermingled in a single ranching unit under private management that all parcels of land ought to be planned and used to assure optimum utilization of the range and prevent erosion. Nevertheless, SCS makes basic conservation plans only for the private, or "deeded," lands on the ranches of district cooperators and supervises the installation of ACP practices for which it is responsible. It has had no responsibility for work on the public lands since 1940.

The extent of joint conservation planning by SCS, the Forest Service, and the Bureau of Land Management is not clear. Efforts which were made several years ago to develop "pilot" projects, "sponsored" initially by a Nevada soil conservation district, have not been continued. Suspicions were aroused that these projects were intended to "accelerate" range improvements by the Forest Service and Bureau of Land Management on the public lands of ranchers who had become district cooperators. Since the federal funds available for soil and moisture conservation on public range lands have been limited, fears were privately voiced that this was a scheme to serve selected ranchers who previously had not seen the advantage of becoming district cooperators.[29] The Bureau of Land Management recently has initiated new agreements to be made with districts. The conservation district boards, however, have no legal power to affect the decisions of the Bureau of Land Management or the Forest Service.

Agreements between districts and the federal agencies managing public lands have been principally for the purpose of using such district-owned facilities as heavy construction equipment. Frequently, the agencies have had relatively small construction projects, and the district equipment can be used for these when the agencies lack the necessary equipment or private contractors do not bid on jobs. Such agreements have been advantageous to the districts insofar as they

[29] NARG 16 (Soil 1), Warren T. Murphy, Field Representative of the Secretary of Agriculture, to Herbert M. Peet, Administrative Officer, Office of the Secretary of Agriculture, Feb. 2, 1953. Similar projects were initiated in other soil conservation districts in the West. The Public Lands Committee of the National Association of Soil and Water Conservation Districts has urged the National Advisory Board Council (consisting of rancher members of the grazing district boards which function in the Bureau of Land Management) to increase cooperation with districts. The National Advisory Board Council has, in turn, urged the Bureau of Land Management to "enter into cooperative working agreements with soil conservation districts where lands administered by the BLM are within Soil Conservation District boundaries." NACD *Tuesday Newsletter,* Nov. 19, 1963; also March 19, 1963.

permit year-round use of equipment to help return the cost of their investment.[30]

District agreements with state agencies are fairly common, especially with state conservation departments. Most of these deal with either forestry or wildlife programs. Districts in six states also have agreements with their highway departments which are intended chiefly to encourage the departments to prevent unnecessary erosion along their rights of way. These highway departments do not provide districts with any direct services and the districts have no power over their actions.

Agreements with the state forestry agencies are found chiefly in the East, where small woodlots constitute about 97 per cent of the total of such holdings in the United States. Such agreements have been negotiated in some states, but not others, and they vary in important respects. The Maine Forest Service agrees to give "priority to the needs of district cooperators in the use of Forest Service personnel within the district, and encourage others to become district cooperators." The North Carolina Department of Conservation has stipulated that its agreement with districts "shall not in any way limit the Division of Forestry in discharging its service equitably to all forest landowners whether or not they have entered into cooperative agreements with the district or whether or not their property is included within the district." In both states, the districts have agreed to do the following: assist in promoting public awareness of the need for better forestry practices, especially as part of "coordinated" land use planning; make needed information available to the forestry agencies; cooperate in the distribution of planting stock, by estimating the needs of district cooperators and aiding them to procure planting stock; provide such equipment, personnel, funds, and other facilities as they may have to assist in woodland conservation; and aid in the control of fire and other elements destructive of woodlands.

The services promised by the forestry agencies in these two states are similar in general, but significantly different in detail. Both agree to assist district boards in programs of public "education" for developing public awareness of forest conservation problems and needs; to "facilitate" the procurement of planting stock; to render on-site technical

---

[30] For example, the Bureau of Land Management and the San Miguel Basin District in Colorado agreed that the District would furnish one tractor, TD18a, with dozer, ripper, carryall and operator according to Contract No. 14–11–008–24–90, July 6, 1960. This machinery was to be operated on lands administered by the Bureau in specified areas, and at rates determined in the contract. (Copy of agreement supplied from Administrative Files, Bureau of Land Management, Denver, Colorado.)

assistance to landowners; to provide marketing information; to aid in developing markets; and to cooperate with districts in fire prevention programs. The North Carolina Division of Forestry also will "consult with the district on the needs of district landowners and operators for specialized forestry assistance, and provide such assistance to the extent available." When it prepares "timber management plans" for district cooperators, it will give the district copies of them. The Maine Forest Service specifies that its on-site technical assistance to district cooperators will include training in the correct interpretation of their farm plans; marking trees for cutting; and providing limited marketing service, including the location of markets, prices, contract forms, and other procedures and information. Apparently the actual services rendered by each agency to landowners and operators do not differ greatly. The important difference seems to be that the North Carolina Division limits its contacts largely to the district board of supervisors in return for its support in promoting sound forestry, whereas the Maine Forest Service commits itself to work particularly with district cooperators in return for the district board's active support.[31]

On the surface, these differences may appear to be slight in practice, but they have more meaning in the light of the struggle between SCS and the U.S. Forest Service over their respective roles in promoting "small farm" or woodland forestry from 1936 to 1951. Each agency is committed to work through state or local agencies: SCS through districts, and the Forest Service through divisions or departments of forestry. The state forestry agencies, as much as the districts, need the support of clientele, but in several states they face rivals who have been successful in restricting the scope and character of free public technical services available to small farm and woodland operators. Since 1951, SCS has been forced to accept a secondary role in providing technical assistance for forestry practices, but it has not altogether ceased to assist district cooperators with forestry practices. Since 1953, when Donald A. Williams became Administrator, SCS has encouraged district boards to seek agreements that the state forestry agencies will provide cooperators with on-site technical assistance, preferably with first priority. The Forest Service has steadily encouraged the development of strong state forestry divisions, but has had uneven success, depending upon a variety of factors influencing events in each state.

[31] Memorandum of Understanding Between the_____Soil Conservation District and the Maine Forest Service, mimeographed; Memorandum of Understanding Between the_____Soil Conservation District and the Division of Forestry, North Carolina Department of Conservation and Development, mimeographed.

Where the state forestry agencies have small staffs of "service forest-ers," SCS has the opportunity to provide assistance through its own personnel—as in Louisiana, for example. In states such as New York, Pennsylvania, Virginia, and South Carolina, strong forestry agencies are in a position to provide sufficient services so that SCS is forced into a subordinate role.

In order to accelerate sound practices, SCS has stimulated the interest of district supervisors in agreements with the stronger state forestry agencies, since there is little reason to believe that SCS will again be assigned national responsibility in the Department of Agricul-ture for small woodland programs. This policy fosters the belief that districts are not exclusively dependent upon SCS for technical assist-ance. It also gives the forestry agencies some interest in supporting dis-tricts. One of the primary needs of these state agencies is to expand the size of their public clientele, and the districts provide additional access to the numerous private operators of small holdings. Where districts operate tree-planting equipment, they help the state forestry agencies provide direct services. Foresters who deal with district boards can also maintain surveillance over the forestry services provided by SCS.

Although it may be useful for some of the state forestry agencies to work formally or informally with district boards to reach farmers, most districts provide little more than moral support—moral support which is very welcome in preference to outright hostility. Certainly, they do not exercise control over programs. Districts are not likely to perform other functions for the forestry agencies in the near future, both because of interagency rivalries and because not all professional foresters agree with district leaders who advocate completely voluntary private decision-making in forest management.

The district agreements with state fish, game, and wildlife agencies appear to be of the same type. In Minnesota, there is a standard agree-ment whereby the districts undertake to promote in several specified ways public consciousness and support of the program of the Division of Game and Fish, Department of Conservation. The Division has promised to provide "technical and other services, to the extent avail-able . . . in the District." It is "mutually agreed," however, that "this memorandum will in no way limit the Division in discharging its serv-ices to persons in the District, who have not entered into cooperative agreements with the District." The Virginia Commission of Game and Inland Fisheries has agreed to give district cooperators "supervision and advice" in undertaking wildlife conservation measures; to "interest farmers" in soil and water conservation; and "if the farmer is not in contact with the Soil Conservation District," to "facilitate these contacts." In return, the districts have promised to help the Commission's technicians to locate farmers who are interested in wildlife conservation; to include such measures in all farm plans; to contact the Commission's personnel when demonstration farms are being planned; and to help the Commission distribute planting materials. The Virginia agreement is silent on the issue of priority of assistance to district co-operators.[32] The district boards have no legal power to block the state wildlife agencies; again, their friendship is preferred to their hostility, since the support of farmers is essential in maintaining the supply of wildlife.

So far as the "assisting" state and federal agencies are concerned, these agreements formally recognize each party's interest in providing maximum access to a public clientele under conditions of mildly regulated competition. It is an example of the rule of reciprocal accommodation by which it is possible for agencies to function. Because of the decentralized nature of interagency conflicts, two agencies may engage in both conflict and cooperation, but at different times and places depending upon the advantage which each calculates will follow the one course or the other.

This conclusion is illustrated by the experience of the leaders of the National Association of Soil Conservation Districts in 1954. They met with the Federal Director of Extension and tried unsuccessfully to find a "basis of cooperation" between districts and the state extension services. In September, Waters Davis, then President of the NASCD, wrote to Clarence M. Ferguson, Director of the Federal Extension Service. He reviewed the recent talks and included a "declaration of principles" to which he hoped Ferguson would agree. Three of these principles are of special interest. First, Davis wanted the state extension services to "recognize" that districts have "final responsibility," and are the "focal point," for assisting farmers and ranchers to conserve soil and water. Second, Davis insisted that "education" and "technical assistance" are separate and distinct activities; and said districts welcome aid from the state extension services "in the education phase."

---

[32] Memorandum of Understanding Between the_____Soil Conservation District and the Division of Game and Fish, Minnesota Department of Conservation, mimeographed; Cooperative Agreement Between the_____Soil Conservation District and the Virginia Commission of Game and Inland Fisheries, mimeographed.

The third of these proposals was a bold and direct threat to the cornerstone of the political power of the Extension Service. Davis urged that district supervisors "have an active part in guiding and assisting Extension workers in the new 'farm business planning' work of Extension." District governing boards "should actively help Extension workers select farms and ranches for the new service," according to priorities which Davis recommended. Farms operating with an SCS conservation plan would have first priority. In descending order would be farms on which a complete system of soil and water conservation is "in process of installation" and, last, those farms for which SCS conservation plans have been made.[33]

This incident, like negotiations between state conservation departments and districts, shows why the districts have not become "multiple-purpose" districts except in a formal way. Soil conservation districts have insulated SCS programs and administration to a considerable degree from disruption by competing agencies, especially the state extension services, and have given SCS access to farmers. In performing these functions, however, the districts have become so thoroughly and inseparably identified with SCS that competing agencies are reluctant to work through districts if there is a possibility that the district boards could direct agency programs in a way to favor SCS. Ironically, the districts have served as a means to isolate the competing conservation agencies from each other rather than to coordinate them.

If the conservation districts ever do become multiple-purpose resource units, as their leaders are now urging, exclusive program authority broad enough for this purpose will have to be given a single national agency, and the agency will have to administer its program through the districts. Bennett expressed this point clearly in his letters to Rexford Guy Tugwell in 1933. Bennett pushed with all his great energy and determination to make the Soil Conservation Service the sole agency responsible for national agricultural conservation and land use programs, but the divisive force of the land-grant agriculture colleges and other agencies in the Department of Agriculture was more than he or any Secretary of Agriculture could or would overcome.

In the future, the real questions will be whether the Department of Agriculture and soil conservation districts can achieve this overriding objective and whether they are the proper agencies to carry out a far broader program of watershed protection in an urban age.

---

[33] National Association of Soil Conservation Districts, "Minutes of the Meetings of Oct. 18–21, 1954," mimeographed.

SUMMARY AND COMMENT

Soil conservation districts are still being organized, although nearly all the land used for agriculture is now included within them. The total number of districts will continue to fluctuate for a decade or more, however, as the result of reorganizations. Most of the reorganizations probably will be to subdivide multiple-county districts or to consolidate small districts into larger units. Because coordination with other agricultural agencies is facilitated where districts are established to coincide with county boundaries, the trend of reorganization is in that direction, although this trend will not be so strong in many of the Western states. This development is occurring at a time when the Department of Agriculture is urging that special districts be organized on the basis of small watersheds to sponsor projects authorized by Public Law 566.

The present irregular areal pattern of district organization was caused by such factors as provisions of state law reflecting varying attitudes in the localities when districts were proposed in the late thirties and early forties. In a few states, notably California, Missouri and Pennsylvania, provisions for organizing districts were used effectively by opponents to prevent the rapid growth of districts. In other areas, especially the West, many districts were organized only on a few farms and even on areas which were not contiguous. A fear that tenants might bind landowners to undertake conservation operations led many state legislatures to limit participation in referendums on the organization of districts to landowners and not merely "occupiers" of land, as the Department of Agriculture originally recommended.

Just as the early hope for districts organized on the basis of watersheds proved to be disappointing, so the idea that soil conservation districts would be multiple-purpose units coordinating the work of numerous state and local conservation agencies has not borne much fruit. This failure, if it can be properly called that, is caused by many factors. It arises primarily from the unwillingness of agencies other than SCS to make working agreements with districts which might permit the district boards to determine which farmers were to receive the agencies' services. This problem persists, despite the fact that SCS and the districts have been special-purpose agencies, precisely because they are capable of broadening the range of their activities to become multiple-function agencies. Indeed, this has been the aim of the Soil Conservation Service, since the report of its Watershed Committee in 1937.

The fundamental problem is not technical, but political. Each government agency serves only a part of the public. One must build its support, for example, from some landowners or farm operators, and not others. In the crucial first, promotional stage of developing a new program, especially if participation is voluntary, each agency seeks out "innovators" and "early adopters" who are likely to accept its program. Since these two types—especially the early adopters—tend to be community leaders, they are looked upon as being enormously valuable if an agency succeeds in associating itself with them.

Community leaders of this sort are, in effect, the "primary" supporters of the agency. They use its services directly. Usually the agency's services are valuable, or it brings to the individual's attention some profitable new technology which puts him ahead of competitors who are, perhaps, less perceptive or less venturesome. In some cases, primary supporters are given preferential treatment; Hugh Bennett, for example, recommended this policy in 1934. The Soil Erosion Service and its successor, SCS, followed this policy on the demonstration projects, where as Bennett testified, farmers were sometimes hired on their own farms on the demonstration watersheds. They were "the boys who really got the gravy," as Georgia's Senator Richard B. Russell once remarked (see discussion of SCS popularity in Chapter 2). In field interviews conducted in the early sixties, it was usually possible to identify these innovators and early adopters as district supervisors or as persons called upon to testify in public to the good works of "districts." In return, they are honored in various ways. Recently, for example, a dam constructed for a small watershed project in a central Virginia county was named for a man whose property was used partially in one of the demonstration projects established by the Soil Erosion Service in 1934. Leaders of this sort sponsor new federal agencies by vouching for their legitimacy.

All special-function agencies, such as forestry and wildlife which can reach only a narrow range of the public, engage in such practices to develop primary supporters. All agencies, special or general, attempt to develop secondary supporters such as bankers, preachers, and teachers in the rural communities. The soil conservation districts are urged to sponsor "soil stewardship week" every May. They are encouraged to sponsor scholarships for teachers to attend conservation workshops. Children are tempted with prizes to paint posters and write essays in support of conservation. These activities serve both to dramatize the community interest in conservation and, also, to identify the agency as legitimate. Thus, two interests are served: the private and personal interest of the primary supporters and the general interest of the community. There is a tendency for each agency to have its own special supporters, although some individuals are identified prominently with more than one.

The Soil Conservation Service, the Agricultural Stabilization and Conservation Service, and the Extension Service all have limited, but differentiated functions. The ASCS, however, touches the whole range of the farm population because its primary function is to control the production of selected commodities as a form of intervention in the price structure. It still administers the Agricultural Conservation Program for a complex variety of reasons described in Chapter 11, but there are also other reasons. Production controls can be used as a stick; conservation subsidies, as a carrot. Furthermore, production controls are related to proper land use adjustments. Many thoughtful critics of the Soil Conservation Service say that it has concentrated on the physical methods of conservation and ignored the economic and institutional obstacles to sound land use. For this reason, John M. Gaus and others opposed soil conservation districts in 1939. They did not believe that they would help to solve the problems of landownership, tenancy, taxation, and related matters, which were the fundamental issues that counties should solve within a framework of national planning.

The state extension services also have general functions, although they do not regulate farmers because they do not want to. They left part of this job to the state departments of agriculture with their agreement of 1923, and they left the rest of it with the Agricultural Adjustment Agency in 1942, when the break between the Extension Service and AAA over this issue became final. County agents, nevertheless, are generalists who have brought to farmers advice and assistance from the specialists in the colleges covering the whole range of farm management and operations. They have stressed techniques for the most efficient and profitable production. This was one cogent reason why the leaders of the colleges always contended that soil conservation was necessarily a part of good farm management and ought not to be planned apart from it. The Soil Conservation Service does not dispute this position, but argues that sound farm management must start with the soil and water base upon which production must occur; the first consideration in planning ought, therefore, to be an analysis of land capability. In making farm conservation plans based on land capabilities, SCS planners inevitably are constrained to plan for all phases of farm operations, and it is at this point that the old cycle of interagency conflict starts turning. Extension leaders, like administrators of many special-function agencies, will work with districts in only a very limited fashion because they fear that the Soil Conservation Service and districts will ultimately become general-purpose agencies capable of displacing them. It was this same fear, to a degree at least, which moved the Army Corps of Engineers and the Bureau of Reclamation to oppose the small watershed legislation.

CHAPTER 13

Retrospect and Prospect

A REVIEW OF past experience is not always a guide to the future, but it can help in identifying some relatively persistent problems. Decisions of the Secretary of Agriculture in the thirties concerning the organization and objectives of the Department's soil conservation programs still limit the practical range of choices open to his successors. One basic aim of the policies adopted thirty years ago was to create a novel blend of power and responsibility; it was to be neither wholly centralized nor decentralized. The optimum use of human and natural resources depends upon sound national planning, according to the view which prevailed in the Roosevelt Administration. Some of the Department of Agriculture's leading policy-makers were equally certain that successful national planning and administration require enough decentralized authority and action to fuse the knowledge of the specialist with the experience and interests of farmers affected by agricultural programs. The novelty of these ideas in the thirties seemed to demand new organizations to develop policies and procedures which could be understood and accepted by the public.

Acting upon these convictions, the Secretary of Agriculture first created a system of committees composed of farmers to administer the new agricultural adjustment programs. He then conceived and promoted soil conservation districts organized under state law to cooperate with the Department in administering the new land use adjustment programs on a permanent basis. The Department also organized other state and county committees on which farmers served as a means of blending local and national leadership in a continuous process of planning and initiating agricultural policy. These new organizations were established usually with the aid, but sometimes over the strong opposition, of the state agricultural extension services, which had been decentralized in the traditional sense when they were authorized to receive federal funds in 1914. While the farmer committees, soil conservation districts, and other local committees also were decentralized, their relationship to the Department of Agriculture was fundamentally different from the ties which linked the extension services to the De-

partment. This new system of local government was created specifically to assist the Department in successfully administering its new
action programs. The new local organizations were elements of the
"new decentralization."

It is not surprising that there were differences between the new and
old organizations, with each having distinctive identities and institutional drives. But it is striking that intense conflicts erupted from the
outset over the control of soil conservation activities. To some extent,
these differences were the products of survival tactics developed by
leaders on both sides, but the real source of conflict lay deeper within
the constitutional and political environment in which the organizations
functioned. The new decentralized organs were created to blend
national planning with local interests, but at the same time to harmonize local action with the policies of a national administration which
was seeking to shift the balance of both centrifugal and centripetal
forces in American government to a federal center. To some degree,
these efforts have been successful, but, on the whole, there has been a
striking and persistent diffusion of power, which is the fundamental
starting point for consideration in formulating resources policy.

POSITIVE ACCOMPLISHMENTS AND THE JOB AHEAD

In 1936, the Department of Agriculture's broad objectives were to
plan and execute long-range policies of land use adjustment which
would conserve the nation's soil and simultaneously improve the rural
standard of living. The primary purpose of this study has been to report
on the means by which the Department has tried to accomplish these
objectives, rather than on the ends. Therefore, since Held and Clawson
have devoted their RFF study to an assessment of the Department's
success in accomplishing its soil and water conservation objectives, the
following brief summary draws heavily on their findings.[1]

From the point of view of accomplishments, three conclusions stand
out. First, it is particularly fitting to recognize Hugh H. Bennett as an
amazingly forceful evangelist in the cause of erosion control, and to say
that the spark which he struck aroused great public concern for
conserving agricultural lands. Many others contributed to this result,
but he must be given much of the credit for turning his personal
crusade into a major program in the Department of Agriculture. Today,

[1] R. Burnell Held and Marion Clawson, *Soil Conservation in Perspective*
(Baltimore: The Johns Hopkins Press for Resources for the Future, Inc., 1965).

conservation is a popular public policy and undoubtedly it will be supported in the future, even rather generously, because people are aware of the necessity for long-term conservation of resources. Care is required, however, to assure that these programs are suited to national needs and are in the public interest.

Second, soil conservation has become a well-established public policy which is the primary concern of the Soil Conservation Service (SCS) and soil conservation districts; a secondary, but regular, activity of the county committees of the Agricultural Stabilization and Conservation Service (ASCS); an expanded part of the regular educational programs of county agricultural agents; and a portion of the work of state conservation departments dealing with forestry and wildlife.

Third, although the information upon which to base a judgment of the physical accomplishment of soil conservation during the last thirty years is extremely limited—if not confusing and vague, as Held and Clawson have emphasized—there are evident accomplishments. These improvements have been stimulated by many agencies of government, not by a single agency with an exclusive mission to save soil. Many of these changes are also the result of private, rather than public, decision-making. Very real changes in land use have occurred in such areas as the Southeast, where the shift from cotton production to forests has added immeasurably to erosion control in an area which was severely eroded in the thirties. These changes in farm technology and the shift of cotton production resulting from management decisions by farmers have contributed in great measure to achieving some of the land use adjustment which Secretary of Agriculture Henry A. Wallace and many of his subordinates advocated. In the Great Plains, however, there has been no comparable shift of land out of crop production, and erosion is still a severe hazard on lands unsuited to crops. In that case, the Department's support of wheat prices at high levels has probably retarded appropriate adjustments of land use.

In contrast with these broadly stated accomplishments, there are some shortcomings which even the agencies dedicated to soil conservation implicitly acknowledge in pointing out the extent of their unfinished business.

First, past efforts to estimate the scope of the job to be done, and to fix a timetable for accomplishing it, have been too optimistic, even vague. National conservation planning in the middle thirties rested on a hasty reconnaissance survey of erosion, and an estimate of the time to bring erosion on most private agricultural lands under control. Both were made by the Soil Erosion Service under Bennett, who estimated that, within ten years, controls would be started on all lands suffering seriously from erosion; that, in twenty years, reasonable controls would be established; and that, in about thirty years (by 1964), controls would cover "practically all" of the better lands. With hindsight, it is possible to say that these goals were unrealistic when they were set, but the urgency of the task appeared obvious to almost anyone in 1934 who had seen the huge dust clouds rolling out of the Plains.

Although more than 98 per cent of the land in farms is now incorporated within soil conservation districts, and about half of all farm operators now have conservation agreements with soil conservation districts, only about one-third of the farmers have basic conservation plans. Professional conservationists agree that a substantial proportion of all farmers are still unwilling to accept the technicians' judgment either that erosion is serious or that proposed corrective measures are practicable. Some farmers are convinced that they lack the means of acting on the advice and exhortation directed to them. They accept cash subsidies offered through the Agricultural Conservation Program (ACP), but two-thirds of all farmers still are unwilling, or unable, to select their practices as part of the progressive achievement of a basic farm conservation plan made by the Soil Conservation Service. There is the estimate that, at the rate farm plans have been written and revised during the past thirty years, it would take at least thirty more years to plan for all farms, even if the present number of operating farmers were to remain stable. One shortcoming of judgments based on this information, however, is that the estimate was supplied by the Soil Conservation Service and does not include plans made by the state extension services. Here again, there is a reminder of the inadequacy of the data on which to base assessments of either soil conservation accomplishments or the remaining job.

The information about physical accomplishments is so imprecise that it is only safe to say that about half of the agricultural land still needs some erosion control. The severity of erosion, the need for this land under future conditions of agricultural production, the method of treating it—none of these is clearly known. Reports of limitations on production imposed by conditions of soil, slope, climate, and similar factors are vague. Little apparent effort is made to relate present conditions to future needs of treatment as the result of changing goals. Nor can one say how much the treatment said to be needed consists of recurring practices on the same land. It is almost impossible to determine, moreover, where the job needs to be done. Thus, it is difficult to tell with any precision what proportion of land needing treatment is concentrated in areas such as the Great Plains, where less than one-fourth of the cropland and little more than 10 per cent of the range had been properly treated ten years ago. Although the Plains region regularly accounts for the highest concentration of water conservation practices applied with ACP subsidies, there is little, if any, information to show whether this means also a steady accomplishment of goals for soil conservation and land use adjustment.

Positive steps can be taken to increase the amount of useful information. Congress has made the Secretary of Agriculture, and not the bureaus in his Department, responsible for administering the soil and water conservation programs, along with related forestry activities. The responsibility for providing useful and accurate reports of program accomplishments is the Secretary's, not that of the Department's subdivisions which have a long history of rivalry and conflict. Yet, each of these agencies is permitted to report soil conservation accomplishments under its own program without indicating that this information frequently overlaps, and even duplicates, the reports of other agencies. Understandably, each claims that its own work is distinctive and in no way duplicates that of others, but there is no justification for more than a single departmental report of annual and cumulative accomplishments. The Secretary has the authority to improve this situation. With only slight changes in internal reporting procedures, the Department could issue an annual statistical report modeled on the one now made by the Agricultural Stabilization and Conservation Service. Improvements should be made to assure comprehensiveness and to show the intensity of the treatment of lands. Reports should be consolidated periodically, certainly every four or five years, to report accomplishments in relation to previous goals and to provide the occasion for formulating new aims in keeping with changing needs. Most important of all, it should be emphasized that these reports reflect activities for which the Department of Agriculture, and not its bureaus, is responsible.

PROGRAM GOALS AND PLANNING

As a basis for judging the effectiveness of conservation programs, a brief chronology helps to summarize past activities and put them in proper perspective.

From 1914 onward, the Department of Agriculture channeled funds into the state experiment stations for research and into state extension services for teaching farmers how to increase production through the application of improved technology.

In 1933, the Agricultural Adjustment Administration (AAA) adopted a county committee system to raise the rural standard of living by controlling production and subsidizing farm income.

In 1933, also, the Soil Erosion Service (SES) was created in the Department of the Interior to teach farmers ways for controlling erosion so that they could continue profitable production on lands previously damaged, or threatened, by improper land use.

In 1934, the National Resources Board was in the process of establishing four broad goals for a long-range program of improved land use. These were: (1) to curb destruction of agricultural lands by uncontrolled erosion and flooding; (2) to protect public funds and physical resources from being wasted on uncoordinated governmental programs; (3) to induce changes in land use by shifting production from marginal lands to the soils best suited for production in order to raise the rural standard of living; and (4) to develop organizations and techniques for long-range comprehensive planning and execution of multiple-purpose programs to serve the national interest directly, and private advantage only incidentally.

In 1935, the Soil Conservation Service was established in the Department of Agriculture (replacing the Soil Erosion Service), following passage of the Soil Erosion Act of 1935—Public Law 46—which gave the Secretary of Agriculture broad authority to "coordinate and direct all activities" related to programs aiming at prevention and control of soil erosion.

In 1936, for tactical reasons, the AAA system of farm subsidies was changed. The Agricultural Conservation Program (ACP), established within AAA, made payments to encourage soil conservation both by removing some lands from production and by conserving others which were retained in production.

In 1936 also, the National Resources Board approved the Department of Agriculture's proposal to recommend soil conservation districts to the states so that the Department would have appropriate means of executing a permanent and comprehensive program of adjustments and controls to improve land use.

For the whole period—1914 through 1936—each of these programs to some degree was incompatible with the others. Erosion controls can be applied to lands which are not necessarily needed for production. Some sound conservation practices increase production. Improved technology increases production, even if less land is used. If land is voluntarily withdrawn from use, it is quite likely to be the operator's poorest in productive capabilities.

From the middle of the thirties to the present, the Department of Agriculture has tried without notable success to reconcile these conflicting goals. Each of the agencies concerned with some aspect of the problem is caught up in this dilemma, and certainly neither the soil conservation districts nor the Agricultural Stabilization and Conservation county committees are in a position to achieve order where the Department has failed.

Held and Clawson have estimated that since the thirties the Department has spent more than $9 billion on soil conservation.[2] Since it has been unable to treat farms and larger areas selectively, the Department's programs probably have hindered desirable land use adjustments in cases where cropland should have been withdrawn from production. Undoubtedly, the Department has been inhibited by consideration of the political effects of this kind of adjustment. For one thing, opposition is strong in the small trading centers to measures encouraging movement of the population away from farms, as experience with the Soil Bank demonstrated. To a very great degree, the Department has followed inconsistent policies because it has not been in a position to make a defensible choice between two contradictory policies. These are (1) policies which concentrate on use and conservation of the land, regardless of who occupies it; (2) policies which subsidize the operations of farmers who need increased income in order to share at all in the bounty of American life. Moreover, for a variety of reasons, participation in soil conservation has been made completely voluntary and, therefore, it is a piecemeal program.

The National Resources Board and other advocates of soil conservation from the thirties have emphasized that clear national objectives could be achieved, and waste eliminated, only if actions were planned comprehensively and developed through organizations having authority to carry out multiple-purpose programs. Plans, of course, cannot be disembodied ghosts severed from policies. It is not surprising, therefore, that all conservation planning—national, state, and local—has been weak. This is not to say that the conservation agencies do not make plans, even a great many of them, but their relation to national goals is uncertain at best. No doubt this is to be expected since the Department and its line agencies insist adamantly that planning must start with the county committees and soil conservation districts and move upward to

---

[2] *Ibid.;* see breakdown of the estimate in their Table 5.

the assisting agencies. However, the decisions which the districts and county committees make are rarely, if ever, related consciously or otherwise to plainly articulated long-range goals. The ASC county committees (assisted by SCS technicians, county agents, and the few members of soil conservation district boards who will attend meetings of the ACP county development group) annually choose from a state handbook the practices which the ASC committee alone has authority to approve. Most soil conservation district boards annually prepare simple promotional programs which have become routine and are of unknown effectiveness because they are accepted without critical thought by the overwhelming majority of supervisors. Only rarely do supervisors consider the critical conservation problems of their areas, and they have resisted SCS attempts to treat neighborhoods or small drainage areas as a whole rather than piece by piece. They do not require farmers to use technical assistance in accord with their farm plans made with "district help," and most district boards have no idea whether or not their long-range district program bears any relation to current activities.

County committees of the Agricultural Stabilization and Conservation Service are equally disinclined to relate use of cash subsidies to either the local district's program or any other known plan of accomplishment. The Agricultural Conservation Program, the controlling one among the Department's conservation activities, has been accurately compared to a mail-order catalog or a cafeteria. Held and Clawson have published ACP maps which dramatically show how the new decentralization places control of subsidized practices upon county committees. Maps reporting applied practices reveal that in a remarkable number of cases some are authorized and used in one county, but not in the county next door. Members of the county committees are nominally subject to the authority of the Secretary of Agriculture, since they are part-time employees of the Department. But their responsiveness to the Secretary's directions has been questioned by competent and disinterested observers.

Despite the fact that there is little local program planning, and most of what there is has very limited value, the line agencies of the Department, especially SCS and the Forest Service, supply some needed technical direction to the ASC development groups. Undoubtedly, these agencies have a measure of limited control over the choices made by the development groups. Also the state agricultural extension services have influence on both the county and state level, since they are included in the group which the Secretary of Agriculture in 1951 made responsible for setting the technical specifications for ACP. It should be emphasized, however, that the annual ACP programs prepared for each state still are not plans for accomplishment of long-range, national conservation goals.

The inadequacy of local planning, including the failure to require a farmer to use his basic conservation plan in connection with applying ACP practices, can be traced to many factors, including interagency competition. By 1940 the Committee on Extension Organization and Policy of the Association of Land-Grant Colleges and Universities had made clear to the Department that most colleges objected to any work by the districts in connection with program planning, work plans, and farm plans. Some extension services even encouraged district boards to refuse responsibility for assuring that farmers used services and materials to achieve public purposes. This campaign to discredit districts was waged far and wide, and included warnings to farmers that, if districts were organized with power to make and enforce land use regulations, they would at the very least lose control over their farms.

District boards now do not assume that they are responsible for SCS technical assistance to farmers. Although supervision of the accomplishments under plans is the final step which district boards were once expected to take, the Department has placed the program on a voluntary basis and made the ASC county committees responsible for the proper use of federal funds. Under these circumstances, responsibilities of the districts for planning or supervising the results of technical work is only an academic issue.

It is difficult to speak with confidence about national program planning activities since the agencies do not publicize them. The ASCS planning is concerned primarily with making small annual changes in the national "Agricultural Conservation Program Service Handbook." Attempts have been made to alter the practices for which payments will be made, the rates at which they will be subsidized, or to change land use, but Congress has reacted unfavorably to reductions in any practices which congressmen want for their home districts. While SCS engages in an annual program planning cycle, it denies publicly doing any more than helping the districts plan their own individual annual programs. There is no published long-range plan, nor is there an annual plan for the "districts program." Congress is merely requested to appropriate funds to support districts, the Small Watershed Program, the Great Plains Program, and other minor programs. One consequence of this system of arranging the estimates is that it is possible to get a reasonably clear view of the number of districts, small watershed projects, and Great Plains contracts to be administered annually, but there is no clue to the way in which each of these conservation programs is separately effective or is related to national needs or goals.

It is tempting to suggest that the Department of Agriculture take bold steps to improve planning and to insure achievements in accordance with plans—even if it is necessary to impose conditions which do not now accompany grants of cash and technical assistance. Even if |there were reasonable certainty, however, that some simple land use regulations would be effective and would not defeat the objectives of the Department's other programs, there are great obstacles to using them. The first is the attitude of most supervisors, and probably farmers, that compulsion is "un-American." Furthermore, districts enabling legislation in more than half of the states would require amendment before regulations could be enacted and enforced. County governments could be empowered by revised state laws, perhaps, to enact regulations upon the certification of a district board, but past experience with rural zoning is not encouraging. Eventual pressures on the land may stimulate selective use of this device in the distant future, but not soon.

Other approaches probably are preferable. When the proper means of achieving the Department's soil conservation goals were discussed in 1934 and 1935, some thought was given to using contracts as a means of getting farmers to follow long-range plans of land use, and the Department's standard districts legislation reflected this thinking to some extent. There were proponents of the idea that the Department might enter into contracts with each farmer. An unsatisfactory experience with this device, however, had led the Bureau of Reclamation to encourage farmers to organize irrigation or reclamation districts under state law to enforce farmer obligations. This experience influenced the Department of Agriculture in the decision to recommend that states adopt soil conservation districts which would eventually perform the enforcement function. The Great Plains and Cropland Conversion programs are returning to the idea that fairly long-term contracts to carry out complete plans made by SCS technicians will meet the conservation needs of agriculture. As the number of farms declines and their size increases—in perhaps ten to fifteen years—it is possible that this method might be extended to other areas and eventually be made a condition for any farmer's marketing his crops. In the short run, however, most farms will probably remain under the present voluntary and piecemeal system.

One or two other techniques might be tried to assure accomplishments in accordance with sound long-range plans. A program of public purchase of submarginal, or marginal, croplands might be reinstituted, but with an important change from past practice. Cash payments could be made to the county governments in which the purchases were made. These could be in amounts in excess of those usually made in lieu of taxes and their use would be limited to supporting essential public services such as schools, roads, and others which become very costly when population declines materially. Admittedly, this policy would be of limited value so long as the present surplus of agricultural labor cannot be absorbed in other employment. A somewhat different device might be used where there is need only to withdraw lands from use for short terms or to control serious erosion which the owner is unable or unwilling to prevent. Conservation easements might be obtained on such lands and payments made to keep them in proper condition to prevent erosion, or even to restore them to a degree of productivity consistent with their capability. In cases where lands were purchased or easements obtained, they would be kept under public management, but be made available for private use consistent with national needs.

Many whole farms now operated by older farmers need to be withdrawn from production. Direct income payments for life could be made on condition that the farmer produce no crops and keep his land from creating an erosion hazard. In some cases, a degree of soil conservation would be achieved simply by abandoning the land. In other cases, simple conservation measures could be applied with federal assistance. Direct income supplements or substitutes of this sort might be preferable to cash subsidies for conservation measures which many older farmers cannot afford to apply and ought not to use on land no longer needed for production. Also, these older farmers would not be driven to seek alternate employment for which they were unsuited. Some of the worst kind of rural misery and desolation might be relieved by such a scheme. It would keep these people on the land and ease the transition of some of the rural trading centers where the results of declining population are keenly felt. This proposal certainly would not cure all of the shortcomings of the Department's present efforts, but it might be one of several techniques to be used selectively to withdraw surplus land and labor from production.

The Soil Erosion Act of 1935, Public Law 46, appears to give the Secretary of Agriculture adequate authority to undertake direct operations of this kind, subject to appropriations, so the authority of soil conservation districts probably is not required. Since state appropriations to districts are very limited, and the extent and quality of administrative supervision of their internal administration vary so much, districts would not be appropriate agencies for achieving the aims of programs of the kind just described. Their advice may be of value in reflecting sentiments in their areas, but their authority should not be stretched beyond practical limits. It is difficult to see how the Department of Agriculture's programs can be more effectively coordinated than they are now, if responsibilities are dispersed in the future more than they are at present.

COORDINATION AND COOPERATIVE RELATIONS

Between 1936 and 1940, Secretary of Agriculture Henry A. Wallace expected the Soil Conservation Service to become the Department's multiple-purpose conservation agency, with the mission of planning and developing a national program of erosion and flood control on agricultural lands. Although he was fully aware of opposition by the Association of Land-Grant Colleges in 1938, Wallace announced that the Department's facilities for improving land use would be channeled in the future to farms organized on the basis of watersheds. Ultimately, area and function were to be united in soil conservation districts organized on watersheds under state law to perform cooperatively with the Department functions believed otherwise to be both constitutionally and politically beyond its powers.

Since the Secretary announced this decision, the long experience with conservation districts has raised a number of questions. If districts did not now exist, would the Department initiate them? Would creation of districts be sound policy for the Department when the future outlook is one in which farm population will continue to decline and rural political power to wane? Should the Department encourage the states to convert districts into "multiple-purpose conservation" districts normally conterminous with counties? These questions cannot be answered with assurance until the future roles of the Department's agencies administering resources conservation programs are determined.

Neither the Soil Conservation Service nor the conservation district has yet become the kind of multiple-purpose unit which Wallace expected, because neither has ever had an exclusive soil conservation mission to perform. In recent years, the districts undoubtedly have been useful in promoting a wider view of the contributions which various conservation agencies can make to farmers. But district boards have negotiated relatively few agreements with state conservation departments to give any priority of service to farmers who are district cooperators. The old rivalries with the extension services and ASC committees have abated, but too few district supervisors have initiated closer relations with other agencies. The National Association of Soil and Water Conservation Districts (NACD) has rather consistently opposed the Agricultural Conservation Program without offering any constructive suggestions for changes which would not concentrate federal assistance on the farms of district cooperators on the basis of first-come-first-served, regardless of the needs of other farmers. Supporters of districts do not help public understanding of conservation needs and programs by creating the impression that district programs are materially different from, and superior to, all other soil conservation activities.

Since 1951, the Agricultural Conservation Program, administered for the Secretary by ASCS, has provided coordination of the agencies offering technical advice and assistance to farmers. Since ASCS has never developed its own professional staff, it has depended upon others for technical direction and advice. Within a framework set by the Secretary, ASCS negotiates within each state the responsibilities of the different agencies for providing technical assistance and recommending practices and specifications. On the whole, this arrangement seems to have resulted in a tolerable adjustment of the differences which arose among SCS, the Forest Service, and the state extension services over their competence and jurisdiction in the Department's conservation programs. Although the federal and state technical agencies still differ over recommendations for specific conservation practices, so that there is some duplication in their efforts, the situation is bearable and may be useful in bringing alternative measures to the attention of farmers. As farm population decreases, however, it will be increasingly hard to justify the present plurality of agencies. Of course, it is quite possible that the agencies will shift their attention to conservation activities, such as water supply and recreation, which appeal especially to urbanities. If there is still half or more of the conservation job to be done, ample work remains for all agencies without too much concern for a tidy organization chart.

The conflicts over reorganization which were only incompletely settled in 1953 are evidence that for some time still to come the Secretary of Agriculture is unlikely to have much leverage to reorganize some of the related agencies in the interest of either efficiency or economy. No Secretary has reorganized the conservation agencies to

achieve comprehensiveness and unity within a single organization because he is effectively isolated by successive layers of departmental
bureaucracy from developing an effective base of support among farmers. Each of the Department's bureaus and the "cooperating" colleges
of agriculture has built support with segments of the public having
fragmented and localized interests in agricultural policy. Each of these
agencies has nurtured ties with its own constituency groups which are
usually formal sponsors of its programs. Normally, their relations are
harmonious, but their organizational structure must accommodate not
only diversity, but also internal conflict on occasion. Their interest in
self-preservation, however, is overriding and it is best served in the
short run by the fragmentation of power which makes rationalization
of authority in the executive branch a virtual impossibility.

The role of state governments in agricultural conservation has been
cooperative, but relatively limited. Conservation departments and
agricultural colleges supply technical advice and some direct assistance. State soil conservation committees have had only limited functions since their inception, although their administrative assistance to
districts has increased during the past ten years. The amount and
quality of this work are very uneven, however, because most committees have respected the tradition that districts should be "independent," even in conducting their own elections. Since most state committees have not developed distinctive institutional identities and initiative in policy-making, they have functions that are useful to districts,
but minor insofar as other agencies are concerned.

Some decisions of state governments will continue to have significant
effects on national conservation policy. The states are responsible, of
course, for legislation authorizing soil conservation districts and other
special districts, and about three-fifths of the states provide districts
with funds. One of the questions which states will have to answer in
the future is whether they should increase, continue, or decrease the
number of governmental units authorized to cooperate in resources
administration. There is some reason for argument against increasing
the number because of declining rural population. There should be a
review of the policy of urging all state and county governments to make
appropriations for soil conservation districts and to create and finance
sub-watershed districts or conservancy districts with the power to tax,
or to assess and collect benefits. The capacity of local governments to
assume local financial responsibilities for small watershed projects
should be explored further. Indeed, the question whether local government sponsors ought to be required by federal law raises fundamental

issues which deserve reconsideration at all levels of government.

Confusion is created, when the Department of Agriculture encourages the reorganization of soil conservation districts to coincide with counties and at the same time encourages the establishment of new special districts organized as watersheds—especially since the idea of watershed districts failed in the past. If the states are to be asked to change the name and authority of soil conservation districts to some kind of general-purpose resource districts, what powers and purposes should such districts be given? What reason would there be for the formation of districts, especially if they are to coincide with county boundaries? Would the general-purpose districts coordinate locally the programs of several resource agencies, or function primarily with a single agency having broad powers? Would they effectively harmonize all interests in land and water resources, urban and rural, or chiefly rural?

Experience with special districts makes it doubtful that they are capable of coordinating the programs of several agencies. To do this, a unit of local government must have an independent base of power and offer some advantage which the agencies need and can obtain by no other means.[3] When special districts are created to assist an agency to carry out its programs, they are not, and should not be, independent, for independence breaks up power. Without power, national objectives cannot be secured. This is not to say that there is no place for units of government which permit continuous consultation between officials and affected interests. The objectives of zealous civil servants devoted to the missions of their agencies need to be tempered with the wishes of the affected public. In the American system of government, however, there is not one point of access, but many, to officialdom. Because of the federal division of powers, there is a tendency to believe that all local interests must have independent or self-governing bodies at the grass roots to make the power of federal agencies responsive and accountable to functional constituencies. Actually, the primary point of access often is Congress, if not the executive agencies directly.

New problems of coordination, cooperative relations, and program goals were added by the Watershed Protection and Flood Prevention Act of 1954—the Small Watershed Act. Soil conservation districts have helped to sponsor projects, but usually they do not have adequate authority and funds to act as exclusive sponsors. Apparently the

[3] Lyle E. Craine, "The Muskingum Watershed Conservancy District: A Study of Local Control," *Law and Contemporary Problems*, Vol. 22 (Summer 1957), pp. 378–404.

Department of Agriculture has not advocated standard amendments to the state laws creating soil conservation districts designed to make districts exclusive sponsors, and has preferred to enter into agreements with a variety of local governments. This policy has left in doubt the future role of soil conservation districts in the Small Watershed Program. The amendments to the original Small Watershed Act have greatly altered the Department's authority. A program which started as agricultural flood control on small tributaries or headwaters has been altered to provide a wide range of benefits, especially to urban communities. It is too early to tell, therefore, whether this program will contribute primarily to needed soil conservation, provide public works which may improve agricultural land use, or heighten opportunities for employment in areas of surplus agricultural labor.

Certainly one of the greatest weaknesses of the Small Watershed Program is that it really is not a program; it is the authority for as many individual projects as Congress will support with funds. The Department has not announced national goals for this program to insure that these small projects will be developed steadily in accordance with national criteria and priorities. Theoretically, the initiative for a project starts locally. The state soil conservation committees, which now do not represent the urban population, screen projects on the basis of criteria and priorities which have not yet been publicized. They then send them to the Department of Agriculture for review by one or more agencies, and finally to Congress, where designated committees pass upon all but very small projects. Although it is called the Small Watershed Program, this is really a collection of individual projects planned and executed piecemeal, in the same way that soil conservation practices are usually applied to farms under the Agricultural Conservation Program.[4]

When the Department of Agriculture first insisted that the states organize soil conservation districts as a condition for receiving technical and other kinds of assistance, it was expected that the districts would fit into a developing pattern of cooperative federalism characterized by functional specialization according to level. Some advocates of districts assumed that they would eventually use grants-in-aid in spite of their lack of independent taxing power. Instead, districts have become the constituents of agencies and are expected primarily to secure favorable legislation and funds within the states to supplement

[4] Congressional preference for projects over programs is not new: see Arthur A. Maass, "Congress and Water Resources," *American Political Science Review*, Vol. 44 (September 1950), pp. 576–93.

federal appropriations to the Soil Conservation Service. They have not evolved beyond this stage because they have lacked the strength necessary to overcome opposition from other agencies and their constituency groups and to obtain a minimum and approximately uniform level of supplementary financial assistance. About three-fifths of the states appropriate funds for districts, and in about half of the states some counties provide a degree of financial support. Probably not more than 600 districts provide conservation aides; possibly as many as 1,000 have part-time clerks; somewhat fewer provide equipment for farmers at favorable rates. The level and kinds of support vary widely, but the greatest amount is provided in the Great Plains, where the Department's erosion control programs were obviously necessary in the thirties and where water conservation is now popular. Even among the Plains states, however, the contrasts between such neighbors as Nebraska and South Dakota are striking and significant.

The amount of assistance which districts have provided in cooperating with the Department of Agriculture necessarily has been only limited and uneven. This is because the existence of districts threatened the influence of other agencies which had deep roots in both party and functional constituencies of their own at county and state levels. Presumably, the Department could have made the soil conservation districts its exclusive channels for giving all federal conservation assistance to farmers. If it had done so, the Department might also have been successful in requiring state governments and districts to meet conditions which would make the districts really effective instrumentalities for planning and executing the conservation programs. Instead, the Soil Conservation Service has been left, or permitted, to struggle to establish districts in every agricultural county. At the end of a quarter of a century, SCS has not quite achieved this goal. In turn, districts have been left weak and dependent. Their weakness stems from a very elementary fact which is not often mentioned in discussions of intergovernmental relations.

It is commonly assumed that once Congress has authorized a program calling for the cooperation of state and local governments with a federal agency and provided the funds to administer it, administrators have nothing left to do but push buttons marked "go." The fact is ignored that members of Congress rarely, if ever, have any real influence on their state legislatures and county governments. Yet it is the states and counties which must provide enabling legislation and funds if they are to participate cooperatively. The party system fragments political power so thoroughly that county and state organizations also fail to assist federal agencies by obtaining necessary legislation or appropriations. If the facts in this case study are typical, it is apparent that Congress leaves the federal agencies to fight whatever battles they must outside Washington to develop a clientele and constituency groups with access to the political processes in local and state governments. Where these ties to state and local supporters are weak or missing, it is more necessary than ever for federal agencies to develop strong contacts with members of Congress, especially on the appropriations committees. Such constituency groups as the soil conservation districts are prized, not because they provide direct accountability to the public at the grass roots, but because they supply leadership in support of the federal agencies. Contrary to some prevailing theories, the agencies do not always respond—or merely respond—to pressures from the grass roots; they create the pressures, as they must to achieve their program goals in the face of apathy or hostility.[5] In such circumstances, the leadership provided by constituency groups may be more valuable to an agency than any amount of financial support from state or local governments.

CONGRESS AND THE NEW DECENTRALIZATION

One hundred years before the Department of Agriculture initiated a new system of local governments to administer its action programs, Alexis de Tocqueville eulogized New England town government, observing that it "is not the administrative, but the political effects of decentralization which I so much admire in America."[6] If it were possible for him to return and again provide his keen insights, undoubtedly he would be intrigued by the ways in which Congress contributes to the new decentralization.

First of all, congressional influence on the key conservation policies of the Department of Agriculture has measurably increased during the past twenty years. From 1933 to about the middle of the forties, the initiative was largely with the executive branch. All the major decisions were made there, with support, of course, from a few key members of Congress. Erosion control was started and the complex questions of relations with the state agricultural colleges were worked out in the

---

[5] See Charles E. Gilbert, "The Framework of Administrative Responsibility," *Journal of Politics*, Vol. 21 (August 1959), pp. 373–407 (especially p. 403).

[6] Alexis de Tocqueville, *Democracy in America*, Phillips Bradley, ed. (New York: Knopf, 1945), Vol. I, p. 94.

executive agencies; the colleges and the American Farm Bureau Federation were powerless to stop these vital decisions. Because of the lack of conclusive evidence, it cannot be said whether a key member of Congress or someone in the Roosevelt Administration decided late in 1943 that there should be a postwar drive to organize soil conservation districts. Beginning in 1948, however, a bipartisan coalition of senators and representatives from the Southern and Central states blocked consolidation of the Production and Marketing Administration (PMA) with SCS; prevented "decentralization" of SCS to the state extension services; and later maneuvered small watershed legislation through Congress over the opposition of the Secretary of Agriculture. In this 1954 watershed measure, the members of Congress had the help of President Dwight D. Eisenhower, who supported the legislation, perhaps, because he did not fully understand the intentions of his Secretary of Agriculture, Ezra Taft Benson. When Secretary Benson attempted to reduce the practices subsidized by ACP, he had only limited success against bipartisan opposition in Congress. Members of the Senate are now insisting that each state receive its "fair" share of small watershed projects.

Every executive agency is constrained by its relations with Congress to make its program "national," just as the Soil Conservation Service has done since 1938. No service agency can survive without providing benefits which are valuable in the home districts of a maximum number of congressmen, unless it serves an interest so vital that it is well protected in Congress. A member of the House summed up the matter in these words: "If the law were changed so that this program were carried out in those areas where it is so vitally needed . . . we all know that whoever the Member of Congress is, if you are going to have a national program, they would never consent to a change in the law to make such a program applicable only to those areas where it is so badly needed, because, representing other districts . . . they would want to see that it was expanded on a national basis."[7] He was talking about soil conservation.

Critics of Congress are disposed to write off this attitude as evidence of the myopia of legislators whose sole concern is to please their home folks. The "pork barrel" mentality is dismissed as a fact of life, although scarcely a worthy one.

The tendency of Congress to reinforce the centrifugal forces in American politics is not new. In *The Federalist*, No. 45, James Madi-

[7] *Department of Agriculture Appropriation Bill for 1949*, House, 80 Cong. 2 sess. (1948), Pt. 2, p. 160.

son remarked that the tendency in all confederacies is to weaken the
central power.[8] His concern was that the Constitution might not provide for sufficient strength for the central government, not that it would
absorb the states. Congressmen, as members of political parties, represent a confederated form of organization. Executive agencies are unitary in structure and powers, and as Congress has developed techniques of surveillance over them, Madison's theory has been put to a
test. The result has been that both Congress and the agencies share
power which is neither wholly confederated nor quite unitary.

A key element in this relationship is the political neutrality of the
civil service. In the case of the Soil Conservation Service, this neutrality
has been reinforced by working through soil conservation districts
which have governing boards who are, for the most part, not openly
partisan. The National Association of Soil and Water Conservation
Districts has fostered this reputation as the best possible insurance that
changes of party in the executive agencies will not eliminate the
benefits they receive in the form of assistance from the Soil Conservation Service. Congressmen appreciate this attitude, with its reassurance
that the President will not often use SCS in a way to unsettle relations
between them and their constituents. The power of congressmen to
negotiate the distribution of this kind of public benefits in their
constituencies is an important element in their political power, and for
many is essential for winning re-election. It is in the interest of
congressmen to see that the distribution of benefits is not made by
either a quasi-independent bureau or a higher executive authority in a
way that will aid political rivals at home. National planning of resource
programs, especially those involving construction projects, has foun-
dered on the resistance of congressmen who distrust the political
motives of executive agencies. Such congressmen view executive independence to determine the location, size, and type of such projects as a
means of building presidential power at their expense.

The ability of SCS to maintain its reputation for political neutrality
has been aided in great measure by the generally nonpartisan system of
selecting district supervisors. Except in New York, Pennsylvania, and
Wisconsin, open partisanship in the selection of boards is normally
avoided, so that SCS does not ordinarily have to "clear" with local
courthouse rings before servicing farmers. This insulation of administration from overt partisanship denies to local party organizations
control over the distribution of this federal benefit. The overwhelming
majority of district boards also fail to control the allocation of SCS

[8] *The Federalist*, Edward E. Meade, ed. (New York: Random House, Modern
Library edition, 1948), p. 299.

benefits to farmers and, therefore, lack the means of building effective and independent political power in their own right. In this situation, SCS is sensitive to congressional criticism, but is largely independent of direction by political party leaders holding offices in local governments. At the risk of some oversimplification, it can be said that SCS shares its power with Congress, but not with the courthouses.

By functioning through districts, SCS has the opportunity to work with both leading farmers and leaders of farm organizations. The former are cultivated to become district cooperators and thereby to stamp SCS with the seal of legitimacy by respected farmers. The leaders of friendly farm organizations can provide the vital contacts with political officialdom on all levels—local, state, and national. In seeking such support, SCS is not breaking new ground, of course, since the techniques were learned long ago in the state extension services. The ability of SCS to forge a close relationship with many key members of Congress, however, has been essential for survival, since its rivals— especially the extension services—have a decentralized base of power rooted in county governments. For example, when the House Committee on Agriculture reported out the Poage Bill to amend Public Law 566 on small watersheds in 1956 (discussed in Chapter 7), the Executive Committee of NASCD was meeting in Washington. The bill needed a special order from the Rules Committee before it could be debated and voted upon. Various NASCD officers and members of the Board of Directors discussed the matter and then dispatched two spokesmen to secure the support of the chairman of the Rules Committee, Representative Howard W. Smith (Democrat) of Virginia. Both of these men were officers of their state association of districts and of the NACD, as well as being prominent farm leaders in Mr. Smith's congressional district. Soil conservation has never been part of a mass movement by farmers; agencies concerned with it, therefore, are particularly dependent upon earning the good will of men who maintain good relations with their representatives and senators.

It is not hard to understand why federal agencies accommodate their programs to the wishes of congressmen who insist upon sharing intimately in the decision-making processes. The two branches of government are mutually dependent; their relation is symbiotic. The result for resources planning of this relationship has been known for a longer time than most advocates of reform care to recall. The heart of the matter, as a small band of academic critics has maintained, is a party system in which there is an almost infinite number of independent centers of power. National party organs scarcely exist, much less do they link the President and Congress. So long as this situation prevails, members of Congress, representing a kaleidoscopic variety of interests whom they must satisfy to remain in office, will try not to yield to the executive branch the power to develop national resources programs which channel federal funds into some, but not all, districts and states. Congress has the power to block resources programs, while insisting upon resources projects, and to maintain an intimate role in determining the identity of the projects' beneficiaries and the scope of their benefits. This power and its use are rooted in the character of the political parties in the United States.[9]

For resources policy, these centrifugal forces could be mitigated to a considerable degree and national objectives could be better achieved with some structural changes. The President needs a resources planning staff adequate to serve national needs in a future which is certain to place increasing strains on natural resources. The nation's earlier experience with resources planning was limited, and the effort was not maintained as it should have been. Any future move to institute an adequate staff, however, will not pass unresisted, as Secretary Charles F. Brannan's experience with the Missouri Basin Program showed. But, until an effort is made, the nation cannot be prepared to use its resources wisely and for the benefit of its increasing population.

It is possible that agricultural conservation may become more amenable to unified direction as the power of long-standing farm organizations and congressional committees over farm policy gives way to new forces. To an increasing degree, future decisions will be made on higher executive levels as agricultural policy becomes a matter of more fully integrated party policy.[10] Much depends upon the future of legislative reapportionment and the development of national political parties. If the parties develop the cohesiveness and centralized discipline which some observers see budding, the conditions will certainly be improved for developing organs of national resources planning which will be more clearly accountable to the nation. Certainly, as in 1936, a new balance needs to be established, but for that to happen, there must be new direction and effort.

[9] "The American legislator is uniquely on his own, and he lives and dies politically through the display of talents more numerous and more demanding than party regularity. He must therefore make his own adjustment among the forces that play upon him, even if this means defiance of his party's leadership. . . . Our administrators, too, feel the pressures that arise out of this strange complex of weak parties and strong interests. They cannot resist these pressures if they are not protected by their masters in the legislature . . ." Clinton Rossiter, *Parties and Politics in America* (Ithaca: Cornell University Press, 1960), p. 22.

[10] See discussion in Dale Hathaway, *Agriculture: Government and Economic Policy in a Democratic Society* (New York: Macmillan, 1963), pp. 233–36.

APPENDIX A

Citations from State Codes
for Soil Conservation District Laws

Alabama: *Code of Alabama,* Title 2, 658 *et seq.,* The Michie Co., 1960.

Alaska: *Alaska Statutes,* Title 41, Chapter 10, 41.10.010 *et seq.,* The Michie Co., 1962.

Arizona: *Arizona Revised Statutes Annotated,* 45–2001 *et seq.,* West Publishing Co., 1956.

Arkansas: *Arkansas Statutes Annotated,* 1947, Volume 2, 9–909 *et seq.,* 1956 Replacement Volume, Bobbs-Merrill Co., 1956.

California: *California Code Annotated,* Volume 56, 9000–9715, 9850–9953, West Publishing Co., 1956.

Colorado: *Colorado Revised Statutes,* 1953, Volume 5, Chapter 128, Callaghan and Co., 1954.

Connecticut: *Connecticut General Statutes Annotated,* Title 22, Titles 25–103 *et seq.,* West Publishing Co., 1960.

Delaware: *Delaware Code Annotated,* Volume 3, Title 7, Chapter 39, 3901 *et seq.,* West Publishing Co., 1953.

Florida: *Florida Statutes Annotated,* Volume 17, Chapter 582, 582.01 *et seq.,* The Harrison Co., 1962.

Georgia: *Code of Georgia Annotated,* Volume 2A, Title 5, Chapter 5–18, 5–1801 *et seq.,* The Harrison Co., 1962.

Hawaii: *Revised Laws of Hawaii,* 1955, Volume I, Chapter 28, Book Publishing Co. 1963 Supplement.

Idaho: *Idaho Code, General Laws of Idaho Annotated,* 1947, Volume V, Title 22, Chapter 27, 22–2701 *et seq.,* Bobbs-Merrill Co., 1948. 1963 Supplement.

Illinois: *Illinois Annotated Statutes,* 1941, Chapter 5, 106 *et seq.,* Burdette Smith Co., 1955. 1964 Pocket Supplement.

Indiana: *Annotated Indiana Statutes,* 1964 Replacement Volume 5, Part I, Title 15, Title 18, 1801 *et seq.,* Bobbs-Merrill Co., 1964.

Iowa: *Iowa Code Annotated,* Volume 25, Chapter 467A, 467A.1 *et seq.,* West Publishing Co., 1949.

Kansas: *General Statutes of Kansas Annotated,* Chapter 2, Article 19, 2.1901 *et seq.,* Corrick Co., 1949. 1961 Supplement Volume.

Kentucky: *Kentucky Revised Statutes Annotated,* Title XXI, Chapter 262, 262.010 *et seq.,* Banks-Baldwin Co., 1963.

Louisiana: *Louisiana Revised Statutes,* 1950, Volume I, Chapter 9, 1201 *et seq.,* West Publishing Co., 1951.

Maine: *Revised Statutes of Maine,* Volume 1, Chapter 34, 1 *et seq.,* The Michie Co., 1954. 1963 Cumulative Supplement.

Maryland: *Annotated Code of Maryland,* Volume 6, Art. 66c. 84 *et seq.,* The Michie Co., 1957.

Massachusetts: *Massachusetts General Laws Annotated,* Volume 19, Title XIX, Chapter 128, 128B, 1 *et seq.,* West Publishing Co., 1958.

Michigan: *Michigan Statutes Annotated,* Volume 9, Title 13, Chapter 110a, 13.1781 *et seq.,* Callaghan and Co., 1958.

Minnesota: *Minnesota Statutes Annotated,* Volume 4, Part I, Chapter 40, 40.01 *et seq.,* West Publishing Co., 1963.

Mississippi: *Mississippi Code of 1942, Annotated 1956,* Volume 4, Title 19, Chapter 14, 4940, The Harrison Co., 1956. 1962 Pocket Supplement.

Missouri: *Vernon's Annotated Missouri Statutes,* Volume 14, Title XVII, Chapter 278, 278.060 *et seq.,* Vernon Law Book Co., 1963.

Montana: *Revised Codes of Montana, 1947 Annotation,* Replacement Volume 5, Title 76, 101 *et seq.,* The Allen Smith Co., 1964.

Nebraska: *Revised Statutes of Nebraska,* Chapter 2, Article 15, 2–1501 *et seq.,* Dennis and Co. 1962 Reissue.

Nevada: *Nevada Revised Statutes,* Volume 4, Title 49, Chapter 548, 548.010 *et seq.,* Legislative Counsel Bureau, 1960.

New Hampshire: *New Hampshire Revised Statutes Annotated,* Volume 4, Title 60, Chapter 430, 430:1 *et seq.,* Lawyers Co-operative, 1955.

New Jersey: *New Jersey Statutes Annotated,* Title 4, Chapter 24, 4:24–1 *et seq.,* West Publishing Co., 1959.

New Mexico: *New Mexico Statutes Annotated 1953,* Volume 7, Chapter 45, Article 5, 45–5–1 *et seq.,* The Allen Smith Co., 1954.

New York: *McKinney's Consolidated Laws of New York Annotated,* Book 52–B, 1 *et seq.,* The Edward Thompson Co., 1949.

North Carolina: *General Statutes of North Carolina,* Replacement Volume 3C, Chapter 139, 139–1 *et seq.,* The Michie Co., 1964.

North Dakota: *North Dakota Century Code Annotated,* Volume I, Title 4, Part V, 4–22–01 *et seq.,* The Allen Smith Co., 1959.

Ohio: *Page's Ohio Revised Code Annotated,* Title 15, Chapter 1515, 1515.01 *et seq.,* The W. H. Anderson Co., 1954.

Oklahoma: *Oklahoma Statutes,* Volume 1, Title 2, Chapter 20, 801 *et seq.,* West Publishing Co., 1964.

Oregon: *Oregon Revised Statutes,* Volume 4, Title 46, Chapter 568, 568.010 *et seq.,* Oregon Legislative Counsel Committee, 1963.

Pennsylvania: *Purdon's Pennsylvania Statutes Annotated,* Title 3, Chapter 13, 849 *et seq.,* West Publishing Co., 1961.

Rhode Island: *General Laws of Rhode Island,* Volume I, Title 2, Chapter 4, 2–4–1 *et seq.,* Bobbs-Merrill Co., 1957.

South Carolina: *Code of Laws of South Carolina*, Volume 12, Title 63, 63–51 *et seq.*, The Michie Co., 1962.

South Dakota: *South Dakota Code of 1939*, Volume I, Title 4, Part IV, Chapter 4.15, 4.1501 *et seq.*, State Publishing Co., 1939. 1960 Supplement, Mitchell Publishing Co.

Tennessee: *Tennessee Code Annotated*, Replacement Volume 8, Title 43, Chapter 15, 43–1501 *et seq.*, Bobbs-Merrill Co., 1964.

Texas: *Vernon's Civil Statutes of the State of Texas Annotated*, Volume I, Article 165a–4, 1 *et seq.*, West Publishing Co., 1959.

Utah: *Utah Code Annotated 1953*, Replacement Volume 7, Title 62, Chapter 1, 62–1–1 *et seq.*, The Allen Smith Co., 1961.

Vermont: *Vermont Statutes Annotated*, Volume 3, Title 10, Chapter 21, 401 *et seq.*, Equity Publishing Co., 1958.

Virginia: *Code of Virginia 1950*, Title 21, Chapter 1, 21–1 *et seq.*, The Michie Co., 1950.

Washington: *Revised Code of Washington*, Volume 10, Title 89, Chapter 89.08, 98.08.010 *et seq.*, West Publishing Co., 1962.

West Virginia: *West Virginia Code of 1961 Annotated*, Volume I, Chapter 19, Article 21A, 21B, 2193 (1) *et seq.*, The Michie Co., 1961.

Wisconsin: *West's Wisconsin Statutes Annotated*, Volume 15, Chapter 92, 92.01 *et seq.*, West Publishing Co., 1957.

Wyoming: *Wyoming Statutes 1957*, Volume 4, Title 11, Chapter 17, 11–234 *et seq.*, The Michie Co., 1959.

APPENDIX B

A Note on the Special Questionnaires for this Study

Two special questionnaires for this study were sent by Resources for the Future to a sample of soil conservation districts and to all state conservation committees.

The 22-page questionnaire designed to obtain information about operations of soil conservation districts was based on an examination of several written sources. These included the study by W. Robert Parks, *Soil Conservation Districts in Action* (Ames: The Iowa State College Press, 1952), operating instructions of the Soil Conservation Service, and materials prepared by the National Association of Soil and Water Conservation Districts.

Detailed questions were asked covering such major topics as agency responsibilities, program and planning activities, internal administration, equipment programs, the Agricultural Conservation Program, educational activities, and criticisms by district supervisors of existing programs and procedures.

This special questionnaire was sent to about 700 districts constituting approximately 25 per cent of the districts which were organized as of July 1, 1960. Included with the questionnaire was a covering letter from the President of the National Association of Soil and Water Conservation Districts. The questionnaire was mailed to districts during the summer of 1961; in the spring of 1962 a follow-up copy, also with a covering letter from NACD's President, was sent to all districts which had not responded to the first questionnaire.

The sample was drawn to insure a distribution by states and by the age of districts. The 277 districts which responded in time to have their answers tabulated constituted nearly 10 per cent of all districts in the United States at that time. Appendix Table 1 shows the distribution by states of the total number of soil conservation districts, the number in the sample, and the number which responded.

The special questionnaire which was sent to the fifty state soil conservation committees requested information about their organization, staffing, functions, financing, and operations. They were asked also to supply information about their supervisory relations with districts, district financing, and key district operations as a cross-check with the responses on the questionnaires returned by soil conservation districts. Five states (Arkansas, Delaware, Kentucky, Tennessee, and Utah) did not respond to the initial questionnaire or a follow-up request.

APPENDIX TABLE 1. SOIL CONSERVATION DISTRICTS WHICH RECEIVED AND RESPONDED TO SPECIAL QUESTIONNAIRE FOR THIS STUDY, BY REGIONS AND STATES, 1961–62

Location of districts	Number of districts fully organized July 1, 1960[a]	Number of districts receiving questionnaire	Number of districts responding to questionnaire	Percentage of districts responding
Northeast region:				
Connecticut	8	2	0	
Delaware	3	1	0	
Maine	15	4	2	
Maryland	24	6	3	
Massachusetts	15	4	2	
New Hampshire	10	3	0	
New Jersey	14	3	1	
New York	46	11	4	
Pennsylvania	48	13	3	
Rhode Island	3	0	0	
Vermont	13	4	1	
West Virginia	14	4	3	
12 Northeastern states	213	55	19	34.5

Appendix Table 1 (cont'd)

	Number of districts fully organized July 1, 1960[a]	Number of districts receiving question- naire	Number of districts responding to question- naire	Percentage of districts responding
Southeast region:				
Alabama	52	11	7	
Arkansas	75	19	13	
Florida	59	15	4	
Georgia	27	5	0	
Kentucky	121	31	17	
Louisiana	26	6	4	
Mississippi	74	18	1	
North Carolina	37	9	5	
South Carolina	45	12	8	
Tennessee	95	23	13	
Virginia	29	7	0	
11 Southeastern states	640	156	72	46.1
Central region:				
Illinois	98	25	14	
Indiana	80	20	8	
Iowa	100	25	14	
Michigan	76	19	8	
Minnesota	80	20	10	
Missouri	37	9	3	
Ohio	87	22	16	
Wisconsin	71	19	4	
8 Central states	629	159	77	48.4
Plains region:				
Kansas	105	26	12	
Nebraska	87	21	14	
North Dakota	75	19	9	
Oklahoma	87	22	15	
South Dakota	68	17	6	
Texas	176	45	12	
6 Plains states	598	150	68	45.3
Mountain and Pacific region:				
Alaska	8	2	0	
Arizona	47	12	0	
California	158	39	8	
Colorado	95	24	4	
Hawaii	16	4	0	
Idaho	51	13	1	
Montana	60	15	4	
Nevada	35	9	5	
New Mexico	57	14	1	
Oregon	57	14	5	
Utah	41	11	7	
Washington	77	19	5	
Wyoming	44	14	1	
13 Mountain and Pacific states	746	190	41	21.6
Totals in 50 states	2,826	710	277	

[a] Excludes districts in Puerto Rico and the Virgin Islands.

Sources of Figures

FIGURE 1. Based on responses to a special 1961–62 questionnaire for this study sent to soil conservation districts, which is explained in Appendix B.

FIGURE 2. Based on responses to a special 1961–62 questionnaire for this study sent to soil conservation districts, which is explained in Appendix B.

FIGURE 3. Based on responses to a special 1961–62 questionnaire for this study sent to soil conservation districts, which is explained in Appendix B.

FIGURE 4. Based on data supplied by the Agricultural Stabilization and Conservation Service. (The data used are the number of ASCS referrals to SCS for determination of need for ACP practices requested by farmers.)

FIGURE 5. Based on 1959–62 reports of Agricultural Stabilization and Conservation Service, *Agricultural Conservation Program, Summary by States.*

FIGURE 6. Based on information supplied to the author by the Agricultural Stabilization and Conservation Service, May 8, 1964.

FIGURE 7. Based on information supplied to the author by the Agricultural Stabilization and Conservation Service, May 8, 1964.

FIGURE 8. Based on data from the U.S. Department of Agriculture.

FIGURE 9. Based on data from the U.S. Department of Agriculture.

FIGURE 10. Reproduced from R. Burnell Held and Marion Clawson, *Soil Conservation in Perspective,* Figure 38.

FIGURE 11. Map from Soil Conservation Service.

FIGURE 12. Map from Soil Conservation Service.

FIGURE 13. Map from Soil Conservation Service.

FIGURE 14. Map from Soil Conservation Service.

FIGURE 15. Map from Soil Conservation Service.

Index